667.9
.M683P

000881636

DATE DUE

RETURN TO
KENT
TECHNICAL LIBRARY
THE BOEING COMPANY   MS 8K-38

Powder Coatings

Powder Coatings

Chemistry and Technology

Tosko Aleksandar Misev
DSM Resins BV, Zwolle, The Netherlands

JOHN WILEY & SONS
Chichester · New York · Brisbane · Toronto · Singapore

Copyright © 1991 by John Wiley & Sons Ltd.
Baffins Lane, Chichester
West Sussex PO19 IUD, England

All rights reserved.

No part of this book may be reproduced by any means,
or transmitted, or translated into a machine language
without the written permission of the publisher.

Other Wiley Editorial Offices

John Wiley & Sons, Inc., 605 Third Avenue,
New York, NY 10158-0012, USA

Jacaranda Wiley Ltd, G.P.O. Box 859, Brisbane,
Queensland 4001, Australia

John Wiley & Sons (Canada) Ltd, 22 Worcester Road,
Rexdale, Ontario M9W 1L1, Canada

John Wiley & Sons (SEA) Pte Ltd, 37 Jalan Pemimpin 05-04.
Block B, Union Industrial Building, Singapore 2057

Library of Congress Cataloging-in-Publication Data:

Misev, Tosko Aleksandar.
    Powder coatings : chemistry and technology / Tosko Aleksandar
Misev.
      p.   cm.
    Includes bibliographical references and index.

    ISBN 0 471 92821 6 (cloth)
    1. Plastic powders. 2. Plastic coating. I. Title.
TP1183.P68M57   1991   90-12700
668.4′9—dc20   CIP

British Library Cataloguing in Publication Data:

Misev, Tosko Aleksandar
  Powder coatings.
  1. Powder coatings
  I. Title
  667.9

ISBN 0 471 92821 6

Typesetting by Thomson Press (India) Ltd, New Delhi.

Printed and bound in Great Britain by Biddles Ltd, Guildford, Surrey

*To my parents,
Mila and Aleksandar Misev*

Contents

Preface ix

1 Introduction 1
    1.1   Historical Background 1
    1.2   Market Situation and Powder Economics 3
          References 8

2 Thermoplastic Powder Coatings 9
    2.1   Vinyl Powders 9
    2.2   Polyolefinic Based Powder Coatings 21
    2.3   Nylon Based Powder Coatings 28
    2.4   Polyester Powder Coatings 34
          References 39

3 Thermosetting Powder Coatings 42
    3.1   Curing Reactions used in Powder Coatings 44
          References 78
    3.2   Monitoring the Curing Process 82
          References 106
    3.3   Crosslinkers for Powder Coatings 107
          References 129
    3.4   Industrial Thermosetting Powder Coatings 131
          References 170

4 Parameters Influencing Powder Coating Properties 174
    4.1   Molecular Weight of the Binder 175
    4.2   Functionality of the Coating Composition 177
    4.3   Glass Transition Temperature 181
    4.4   Viscosity 191
    4.5   Resin/Crosslinker Ratio 196

	4.6	Catalyst Level	199
	4.7	Surface Tension	204
	4.8	Pigment Volume Concentration and Pigment Dispersion	211
	4.9	Particle Size	215
	4.10	Stoving Temperature Profile	217
		References	220
5	**Technology of Production of Powder Coatings**		**224**
	5.1	Premixing	226
	5.2	Hot Melt Compounding of Powder Coatings	238
	5.3	Fine Grinding	251
	5.4	Particle Size Classification	261
	5.5	Powder Collection and Dedusting	279
	5.6	Quality Control	284
		References	300
6	**Powder Coatings Application Techniques**		**304**
	6.1	Surface Preparation	305
	6.2	Application of Powder Coatings	324
	6.3	Design of the Spraying Booths	349
	6.4	Troubleshooting	354
		References	360
7	**Future Developments**		**362**
	7.1	The Market	362
	7.2	Binders	364
	7.3	Crosslinkers	365
	7.4	Production and Application	366
	7.5	Powder Coatings	367
	7.6	Conclusion	368

Subject Index **370**

Preface

More than three decades have passed since the appearance of the first powder coating. The period of reluctance in the very beginning was followed by a period of overenthusiasm; both are now past. Powder coatings are coming of age, being the most successful coating systems used in the last decade. However, the commercial success of powder coatings was not followed by extensive publishing activity, as might be expected. The two books of Miller and Harris (E. P. Miller, and D. D. Taft, *Fundamentals of Powder Coatings*, Society of Manufacturing Engineers, Dearborn, Michigan, 1974; S. T. Harris, *The Technology of Powder Coatings*, Portcullis Press, Redhill, 1976) treating the technology of powder coatings appeared in the 1970s when thermosetting powder coatings were making an impact on the market. They were followed by two monographs in the 1980s exclusively dealing with the application of powder coatings (J. E. Hughes, *Electrostatic Powder Coating*, Research Studies Press Ltd, Letchworth, Hertfordshire, 1984; E. Miller, *Users Guide to Powder Coatings*, Society of Manufacturing Engineers, Dearborn, Michigan, 1985).

Rapid developments in the 1980s have made it necessary to produce a book that summarizes the achievements of the last two decades, with emphasis on the chemistry and film formation of thermosetting powder coatings. Dr Swaraj Paul, the technical director of PPPolymer AB and the director of the course 'Alternatives for Conventional Coatings' organized by CEI/Elsevier, pointed out the need for such a book. I would like to use this occasion to express my gratitude to him for encouraging me to write this book, for his assistance in organizing the material in the presented form and for his recommendation to the publisher.

This book summarizes the enormous work done and published by many scientists and enthusiasts who believe in powder coatings. I am thankful to them and to the journal publishers, book publishers and equipment producers for granting permission to reproduce their material. I would also like to express my thanks to DSM Resins BV for the financial support they have given me. Thanks are also due to my colleagues in the Research and Application Laboratories of DSM Resins BV in Zwolle and DSM Research BV in Geleen for the pleasant and creative atmosphere without which the writing of this book would have been impossible. Finally, I would like to express my gratitude to my wife and children for being patient, having understanding and sacrificing part of family life during the period of preparation of the manuscript.

Zwolle,   T. A. Misev
The Netherlands   April, 1990

1
Introduction

1.1 Historical Background 1
1.2 Market Situation and Powder Economics 3
References 8

1.1 HISTORICAL BACKGROUND

The appearance of powder coatings is often associated with the ecological and energy related events of the late 1960s and early 1970s. The famous Rule 66 which was brought in by the Town Council of Los Angeles in 1966 was the first legislative act regulating the environmental aspects of the coatings. Later on similar regulations were introduced in most of the industrially developed countries.

Although the history of powder coatings has been strongly influenced by environmental aspects, first developments in the field began early in the 1950s when powdered polyethylene was successfully applied in a fluidized bed process on a preheated metal surface [1]. Very soon the fluidized bed technique for application of thermoplastic powders including PVC and nylon powder coatings was well established in the USA.

In the late 1950s the first thermosetting powder coatings appeared on the market, mainly as a result of the research work done by Shell Chemicals. The target was development of superior protective organic coatings for the company's own underground natural gas and oil pipelines. The first systems were relatively simple physical dry blends of epoxy resins, hardeners and pigments dispersed by ball milling techniques. Due to a considerable degree of heterogenicity, the application results were rather inconsistent.

The hot melt mixing methods of the present day for production of powder coatings were preceded by a technique that employed liquid epoxy resins and hardeners. The homogeneous liquid binder/crosslinker blend is prereacted until partially cured solid material is obtained, which undergoes fine grinding in the next step. A drawback of this technique was the lack of reproducibility and difficult control of the process [2]. Extrusion methods for production of powder coatings which are in current use were developed in the Shell Chemical Laboratories in England and The Netherlands in the period 1962–1964. In 1964 the first decorative epoxy/DICY powder coatings appeared on the European market. In the same period SAMES in France developed the first equipment for electrostatic spraying of organic powder coatings. This made a considerable contribution to the success of decorative thermosetting powder paint since for the first time coatings with an acceptably 'thin' layer thickness could be applied.

One serious drawback of epoxy/DICY thermosetting powder coatings is their sensitivity to attack by u.v. light. Exposed to sun they chalk and deteriorate rapidly. This, followed by poor yellowing resistance, restricted their use mainly to protective coatings or decorative coatings for interior use where resistance towards yellowing was not a prime target.

Attempts to overcome these problems led to polyester melamine systems which were introduced in 1970 by SCADO BV and UCB in The Netherlands and Belgium [3, 4]. Almost at the same time Huneke reported that the gloss retention and yellowing resistance of powder coatings based on blends of epoxy and polyester resins are considerably improved compared to pure epoxy powders [5].

The real breakthrough in the area was made in 1971 by SCADO BV in The Netherlands. It was discovered that coatings with exceptionally good decorative properties can be obtained by hot melt compounding of carboxyl functional polyester resins and epoxy compounds in the form of biphenol A resins (polyester/epoxy hybrids) or triglycidyl isocyanurate (TGIC) systems [6]. Shortly after, in 1972 a weathering resistant powder coating system produced by VP-Landshut in West Germany was used for protection of aluminium extrusions and claddings in outdoor architecture in Switzerland [7]. These two systems at the moment dominate the market in Europe.

At the same time in 1971, Bayer AG and BASF AG in Germany each offered thermosetting acrylic powder coatings on a commercial scale. Although later on attempts to commercialize the acrylic systems failed in Europe and the USA, they were widely accepted in Japan for outdoor use.

In the last decade polyurethane powder coatings established a solid position in the American market and in Japan with a marginal market share in Europe.

Recent developments in the area include a wider acceptance of PVDF powder coatings [8] mainly used for monumental architectural objects and ethylene chlorotrifluoroethylene (ECTFE) powders for protective purposes [9]. The development of new powder coating systems was quickly followed by improvements in production and application equipment. Although the melt extrusion method 'borrowed' in the 1960s from the plastic industry remained almost the only process used for powder production, contemporary plants are changing substantially and very often use a continuous concept of powder paint production.

Latest developments in application equipment led to the introduction of so-called frictional tribo guns. The frictional method of electrical charging of powder

INTRODUCTION

coatings has reduced considerably the Faraday cage effect, improving the penetration of powder coatings into recessed areas during spraying. These developments were followed by corresponding efforts to overcome the inefficient tribo chargeability of polyester powder coatings, which at this moment dominate the market [10–13].

Reclaiming powder during application is an advantage that leads to almost 100% utilization of the material. This creates, however, serious problems with respect to the colour change in the spraying booths. The producers of application equipment have made considerable efforts to speed up the colour change. This resulted in the development of systems that allow colour change within 5 minutes, which is quite close to the liquid systems [14].

If one excludes the initial period of development when powder coatings were considered rather as a technical curiosity, the real market history of powder coatings is a little longer than 20 years. Western Europe was the first market area to switch to powder coatings in the real sense of the word in the late 1960s. This process took place in the USA roughly 10 years later, mainly as a result of the air quality regulations which restricted the use of volatile organic compounds. According to Shanley, Europe is at the moment the biggest producer of powder coatings, with 53% of the world production, followed by the USA with 20% and Japan with 15%. More than 30% of Europe's industrial coatings lines use powder [15].

1.2 MARKET SITUATION AND POWDER ECONOMICS

The different powder coating systems that have been developed in the last 20 years employ most of the polymers used in the conventional solvent borne coatings. At the same time the continuing market growth considerably lowered production costs. Combined with sharp price increases of solvents after the 'oil shock' in the 1970s this resulted in powder coating prices comparable with those of the wet coatings. Therefore powder coatings today are recognized not only as environmentally friendly systems but also as materials that can compete successfully in price with the solvent and water borne coatings.

Next to environmental and price aspects, quality is an even more important element, especially in high-tech areas of use where powder coatings are commonly used. In this respect, powder coatings satisfy many of the most stringent requirements by the end users, failing only in cases where an extremely good flow of the coating is expected.

Therefore the answer to the question 'why powder?' contains the popular 'four E's' introduced by Bocchi [16]:

Ecology
Excellence of finish

Economy
Energy

Comparing the relative importance of these factors, it can be said that, at the moment, the high quality and economic aspects perhaps contribute more to the acceptance of powder coatings by industrial users than the regulatory compliances alone.

As presented in Table 1.1, the market share of powder coatings compared to other types of coating materials is not large.

However, in a relatively mature field like the paint industry, powder coatings are among the few still enjoying a considerable growth in volume per year. According to available marketing data, the worldwide market for powder coatings was 74 000 tons in 1980 and 142 000 in 1985, with a prediction of 250 000 tons for 1990 [17, 18]. The annual growth rate in the period 1980–1985 was 11% in Europe, 12% in the USA, 25% in the Far East and 14% for the rest of the world. The overall growth rate for this five year period was 14%. The figures for the period 1985–1990 are 9.5% for Europe, 15% for the USA, 13.6% for the Far East and 12% worldwide [19]. They contradict somewhat the published data by Bodnar and Harris which claim an annual growth rate of 15.4% for Europe and 23% for the USA for 1988 [20, 21]. Bodnar projects for Europe a 10–12% average annual growth rate in consumption of powder coatings for the period 1990–1996, with a total production at the end of this period of 230 000 tons [20].

Europe, where the powder coatings were invented, is still the biggest market. Table 1.2, worked out from the data published by Bocchi [19] and Jotischky

Table 1.1 Market share of powder coatings [17]

	1985	1990
Europe	3.7	5.3
USA	1.4	2.5
Worldwide	1.4	1.8

Table 1.2 The world market for powder coatings by region for 1988

	Tonnes	%
Europe	112 000	50.9
North America	46 000	20.9
Far East	34 000	15.5
Rest of world	28 000	12.7
Total	220 000	100.0

INTRODUCTION

[22], gives an indication of the production and market shares of powder coatings in different geographical areas.

The types of powder coatings commonly used in different geographical locations differ considerably. For example, the three major market areas in the world, Europe, the USA and Japan, have developed completely different systems for exterior use, due to historical reasons and raw material availability. In Europe systems based upon carboxyl functional polyester resins/TGIC are the most popular, in the USA polyurethanes with aliphatic polyisocyanates as crosslinkers dominate, while in Japan powder coatings with glycidyl functional acrylic resins as binders and dibasic acids as crosslinkers are market leaders. Big differences can also be noticed between the coatings for indoor applications. The so-called hybrids based on carboxylated polyesters/epoxy resins are the most popular in Europe. In the USA pure epoxies crosslinked with dicyane diamide are the main representatives competing with the polyurethanes.

Table 1.3 presents average figures for the market shares of five types of powder coatings worked out from different published data [18, 19, 23–25].

The metal furniture industry was one of the first areas to be invaded by powder coatings, and it is steel that is very important. Hybrids are the main types used for this purpose, while for outdoor furniture (garden furniture, for example) polyester/TGIC or polyurethanes are the common systems used.

The industry of domestic appliances is an important consumer of powder coatings. Appliance applications in total currently represent 30% of the total market for powder coatings in the USA [26]. In Europe hybrids are the main types used since a mild orange peel effect is acceptable. In the USA a smooth finish is preferred and therefore polyurethanes are much more common. Typical products of this industry that are coated with powder coatings are food freezers and refrigerators, water heaters, toasters, washing machine lids, domestic cookers, electric and gas heaters, etc.

General engineering is one of the largest markets for powder coatings. Since the role of the coating is mainly functional the low cost hybrids and epoxy systems are preferred.

The automotive market is one of the most demanding concerning the protective and decorative characteristics of the coating. The lack of good flow is

Table 1.3 Market shares (in %) of different types of powder coatings

	Europe	USA	Japan
Epoxide	30	38	36
Epoxy/polyester	47	14	26
Polyester/TGIC	18	15	11
Polyurethane	5	33	15
Acrylic	—	—	12

an inborn weakness of powder systems and the orange peel effect is one of the major concerns. Other problems encountered when powder paints are used as automotive body coatings are inconsistent transfer efficiency, colour change time, colour contamination and difficulties to reach areas around the doors and under the hood and trunk areas. Together with relatively high curing temperatures these are the main reasons why powder coatings are not successful competitors to the wet paints for car body finishing. However, in 1977 General Motors began to use epoxy powder primer surfacer over the cathodic electroprimed truck cab and body. Two other car manufacturers, Honda Motors and Fiat, have developed a system consisting of powder primer surfacer covered by a conventional wet paint for the top coat. Powder clearcoat is used at General Motors as a top coat applied over an electrocoat base coating for metal parts such as door handles, window channels, division channels and lift-off support panels [27]. Hood and Blount have reported trials at Volkswagen with powder coating applied as a clear top coat over pigmented solvent based material in a 'dry-on-wet' concept [28]. There is no evidence so far for using a powder car body top coat, although GM has installed a complete powder top coat line with facilities for eight colours [29]. Although the potentials in the automotive field are enormous the present situation shows no major breakthrough in this area. However, for body trim items like bumpers, window frames, wipers, wing mirrors and for the wheel trims, powder coatings are widely used. The biggest success has been in coating the underbody components of the car, where epoxy and hybrid systems play the most important role. An overview of the main automotive applications of powder coatings has been given by Gill [30].

The market growth of exterior powder coatings has continued to expand due to the wide acceptance of polyester/TGIC systems in Europe and polyurethanes in the USA. The experience has shown that on a hundred buildings located in Europe where aluminium elements were coated with polyester/TGIC powder coating, the residual gloss after 5–7 years is kept at 65–85% and after 10 years at 50–65%. The coated thickness reduction was only 2–3 μm after 10 years [7].

Another big market for powder coatings is in corrosion resistant types for pipes and reinforcement bars (rebars). This is the area where pure epoxies dominate. In Europe the rebar market has been largely ignored, but its importance from a technical point of view has been recognized and this is an area where considerable growth can be expected.

It is very difficult to estimate the participation of different application segments in the consumption of powder coatings because of differences in the statistical classification in different areas. The figures in Table 1.4 represent a rough estimation which can serve only as an indication [19, 31].

When powder coatings economics is compared with the other VOC (volatile organic compound) compliance coating options, one has to take into account all of the expenses connected with the coating process. On-line operating advantages reported for powder coatings compared with other environmentally friendly

INTRODUCTION

Table 1.4 Percentage share of different application segments in total consumption of powder coatings

	Average	Europe	USA
General engineering	25–35	38	42
Corrosion resistant	10–20	14	—
Domestic appliances	10–20	10	30
Automotive	10–15	8	24
Exterior	10–20	13	4
Metal furniture	10–15	12	—
Pipes	—	5	—

systems are an application transfer efficiency or utilization of the material from 95% up to 99%; 30–50% less energy consumption combined with a reduction of labour expenses of 30–40%; reduction of waste material accounts for almost 90% and there are approximately four times fewer rejects due to surface defects [32]. Figures with respect to the economy of the powder coatings process given by Bocchi [32] are summarized in Table 1.5. Comparative data for four different systems—powder, high solids, water borne and conventional coatings—refer to a model coating line where formed sheet steel parts are coated automatically and touched up manually. The production rate is 1 million ft^2 coated surface per month, based on five days per week in one shift.

Other costs that are not included here but which favour powder coatings are the fire insurance costs, minimum operator training and minimum supervision for a powder line.

Recent developments in application equipment that make the clean-out of the line easier and the change of the colour faster and the trend in developing powders with lower curing temperatures and higher reactivity, which allows higher line speeds thus saving energy, make the economics concerning powder coatings even more attractive. The 'four E's'—excellence of finish, ecology,

Table 1.5 On-line cost comparison of four different coating systems

	Powder coating	High solids coating	Water borne coating	Conventional coating
Capital costs ($)	120 000	110 000	110 000	150 000
Annual material costs ($)	327 600	292 800	427 200	418 800
Total annual operating costs ($)	455 600	529 500	677 364	678 958
Applied costs per 100 ft^2 ($)	3.80	4.41	5.64	5.66

economy and energy—are attributes the powder coatings future can certainly rely upon.

REFERENCES

1. Crowley, J. D., Teague, G. S., Curtis, L. G., Foulk, R. G., and Ball, F. M., *Journal of Paint Technology*, **44** (571), 56, 1972.
2. Harris, S. T., *The Technology of Powder Coatings*, Portcullis Press Ltd, Redhill, England, 1976.
3. Scado BV, US. Pat. 3 624 232, 1970.
4. UCB SA, Ger. Pat. 2 352 467, 1972.
5. Huneke, von H., *Metalloberflache*, **24** (9), 315, 1970.
6. Unilever NV, Ger. Pat. 2 163 962, 1971.
7. Bodnar, E., *Product Finishing*, August 1988, p. 22.
8. Meiyden, van der B., in *Thermoset Powder Coatings*, Ed. John Ward, FMJ International Publications Ltd, Redhill, 1989, p. 54
9. Maguire, G., *Product Finishing*, October 1988, p. 66.
10. Misev, A. T., Binda, P. H. G., and Hardeman, G., to Stamicarbon BV, NL Pat. 8 800 640, 1988.
11. Misev, A. T., and Binda, P. H. G., to Stamicarbon BV, NL Pat. 8 802 913, 1988.
12. Binda, P. H. G., and Misev, A. T., to Stamicarbon BV, NL Pat. 8 802 748, 1988.
13. Hoechst AG, Eur. Pat. 315 082, 1987; Hoechst AG, Eur. Pat. 315 083, 1987; Hoechst AG, Eur. Pat. 315 084, 1987.
14. Floyd, R., *Polymers Paint Colour Journal*, **179** (4236), 270, 1989.
15. *Chemical Business*, October 1989, p. 15.
16. Bocchi, G. J., *Modern Paints and Coatings*, **76**, 44, 1986.
17. Adapted from Dijkman, H., *Powder Coatings Bulletin*, **9** (4), 1, August 1986.
18. Bocchi, G. J., *The Powder Coating 88 Conference*, 1–3 November 1988, Cincinnati.
19. Bocchi, G. J., *Double Liaison*, No. 399–400, January–February 1989, pp. X–XVII.
20. Bodnar, E., *European Coatings Journal*, **2** 152, 1989.
21. Harris, S., *Powder Coatings Bulletin*, **11** (10), 7, 1989.
22. Jotischky, H., 'Powder coatings in the marketplace', in *Thermoset Powder Coatings*, Ed. John Ward, FMJ International Publications Ltd, Redhill, 1989, p. 134.
23. Harris, S. T., EMS-Chemie AG Customer Seminar, June 1987, published in *Powder Coating Bulletin*, **10**, 4, 1987.
24. Harris, S. T., *Powder Coatings Bulletin*, **8** (4), 1, 1985.
25. Franiau, R., in *Thermoset Powder Coatings*, Ed. John Ward, FMJ International Publications Ltd, Redhill, 1989, p. 62.
26. Bocchi, G., *American Paint and Coatings Journal Convention Daily*, 20 October 1988, p. 14.
27. Adams, J., *Metal Finishing*, **7**, 14, 1989.
28. Hood, J. D., and Blount, W. W., *Powder Coatings Buyer's Guide*, **5** (2), 28, 1983.
29. Bureau, M. W., *Powder Coatings Bulletin*, **8** (6), 1, 1986.
30. Gill, D. E., *Metal Finishing*, **86** (8), 41, August 1988.
31. Adapted from Harris, S., *Powder Coatings Bulletin*, **8**, 4, 1985.
32. Bocchi, G. J., *Product Finishing*, April 1987, cited in *Powder Coatings Bulletin*, **10** (2), 3, 1987.

2
Thermoplastic Powder Coatings

2.1 Vinyl Powders	9
2.1.1 PVC powder coatings	10
2.1.2 PVDF powder coatings	18
2.2 Polyolefinic Based Powder Coatings	21
2.3 Nylon Based Powder Coatings	28
2.4 Polyester Powder Coatings	34
References	39

The first powder coatings produced were based on thermoplastic polymers, which melt at the application temperature and solidify upon cooling. Several factors such as relatively simple methods of manufacturing and application, no involvement in complicated curing mechanisms, raw materials that in many cases belong to commodity polymers, acceptable properties for many different applications, etc., contributed to the popularity of these coatings in the market very soon after their appearance in the beginning of the 1950s. At the same time, however, weaknesses such as high temperature of fusion, low pigmentation level, poor solvent resistance and bad adhesion on metal surfaces necessitating the use of a primer can be listed. These problems inherent to the thermoplastic powder coatings were successfully overcome later on by the thermosetting powders which very quickly took the largest part of the market.

Despite the disadvantages, thermoplastic powder coatings can offer some distinguished properties. Some of them possess excellent solvent resistance (polyolefins), outstanding weathering resistance (polyvinylidene fluoride), exceptional wear resistance (polyamides), a relatively good price/performance ratio (polyvinyl chloride) or high aesthetic appearance (polyesters). These properties, combined with the simplicity of the system, created a considerable market share for thermoplastic powder coatings.

2.1 VINYL POWDERS

Two binders are used for the manufacture of the so-called vinyl powder coatings: polyvinyl chloride (PVC) and polyvinylidene fluoride (PVDF). On the basis of their polymer nature both powder coatings can be included in the same group, although they differ considerably in their performance. While PVC powder coatings are predominantly intended for indoor application because of their limited outdoor durability, PVDF powder coatings are among the best coating systems with respect to their weathering resistance.

Vinyl polymers belong to a group of resins having a vinyl radical as the basic structural unit. Polyvinyl chloride and copolymers of vinyl chloride are the most significant members of this group, being among the first thermoplastics to be applied by powder techniques.

2.1.1　PVC POWDER COATINGS

Polyvinyl chloride powder coatings were introduced on the market at the time when thermosetting powder coatings were in very early stages of development. PVC based coatings offered many advantages over the other thermoplastic materials available as binders for coating production. These coatings have very good resistance to many solvents, which is a rather poor characteristic of the thermoplasts, combined with resistance towards water and acids. They have excellent impact resistance, salt spray resistance, food staining resistance and good dielectric strength for electrical applications.

Polyvinyl chloride $(-CH_2-CHCl-)_n$ is one of the cheapest polymers produced by the industry on a large scale. Its basic properties include chemical and corrosion resistance, good physical strength and good electrical insulation. Polyvinyl chloride (PVC) is by nature a brittle polymer, but the flexibility of the material can be easily adjusted by using an appropriate amount of a suitable plasticizer.

Polymerization of vinyl chloride into PVC homopolymer or its copolymerization with different comonomers is carried out by a free radical mechanism. Most PVC resins are produced by emulsion or suspension polymerization of vinyl chloride in an aqueous system containing an emulsifying agent or suspension stabilizer. However, bulk and solution polymerization processes are also carried out on an industrial scale.

Emulsion polymerization of vinyl chloride can be performed in both a batch and continuous way. The reaction temperature is maintained between 40 and 50°C by cooling the reactor in order to remove the heat developing during the polymerization. The reaction medium is deionized water containing enough proper surfactant to obtain a stable emulsion. The initiators used are peroxides soluble in water, such as hydrogen peroxide or different persulphates. Since the monomer itself is a gas at room temperature, the polymerization is performed under pressure in autoclaves. The pressure in the reactor falls as the polymerization proceeds. After reaching conversion of ca. 90% the content of the reactor is discharged and the unpolymerized vinyl chloride is recovered.

Suspension polymerization of vinyl chloride is an important process in the commercial manufacture of PVC. In principle it is a batch process, although attempts have been made to develop a continuous technique for suspension polymerization of PVC. The polymerization is carried out by first charging to the reactor the required amount of deionized water and adjusting the pH depending on the suspending agent used and then the dispersing agent and the initiator. The monomer is charged after sealing the reactor and evacuating the oxygen. The polymerization reaction is carried out under pressure at 40–60°C, controlling the temperature by appropriate cooling of the reaction mass. Although a great amount of research has been done in this area, it is interesting to note that the compositions of the reaction mass do not differ substantially from those used in the very early days of development of suspension PVC. The differences are mainly limited to the choice of the suspending agent, which is in most cases polyvinyl alcohol obtained by saponification of polyvinyl acetate, gelatin, methyl cellulose and copolymers of vinyl acetate with maleic anhydride. The initiator is water insoluble peroxide such as lauryl peroxide or azobisisobutyronitrile.

The type of suspending agent plays a very important role in obtaining primary particles with high porosity. Gelatin normally produces glassy spherical particles which have poor plasticizer absorption characteristics, whereas polyvinyl alcohol gives particles of a porous nature which readily absorb plasticizers to give dry powder blends. Thus, a patent of Air Products and Chemicals, Inc. [1] discloses a process for suspension polymerization of vinyl chloride giving a polymer specially suitable for production of powder coatings. The suspension system according to the patent consists essentially of two different polyvinyl alcohols, namely 0.03–0.05% polyvinyl alcohol which is 65–75 mol% hydrolysed and at least 0.15% polyvinyl alcohol with 30–40 mol% hydrolysed. The use of this seemingly excessive amount of secondary suspending agent causes two effects necessary for the critical powder coating application. Firstly, it reduces the size of the primary particles within the polyvinyl chloride grain, thus raising the surface/volume ratio and allowing plasticization of the primary particles by the plasticizer. Secondly, very high porosity is gained, allowing complete and uniform plasticization of the resin grain in its entirety.

Solution polymerization of vinyl chloride is almost exclusively used for manufacturing copolymers containing vinyl acetate. The comonomers are dissolved in a suitable solvent such as cyclohexane or n-butane, and the polymerization is carried out at 40–60°C catalysed by a free radical initiator which is soluble in the reaction mass. The copolymer begins to precipitate after a certain molecular mass is reached, which depends on the comonomers' ratio, polymerization temperature and type of solvent. The last step of the process includes filtering the final product and then washing to remove the residual diluent and any traces of organic peroxide which would have a detrimental effect on the heat stability.

The bulk polymerization of vinyl chloride has been developed by Pechiney-St Gobain, and plays an important role in the commercial production of PVC.

Although bulk polymerization is associated with a homogeneous system, this is not the case with the bulk polymerization of vinyl chloride; namely at a very early stage of the reaction the polymer formed precipitates from the reaction medium in a form of insoluble material dispersed in the monomer. The system is therefore heterogeneous for a significant portion of the whole conversion. Since the polymer precipitates from the monomer and there are no solvents present in the system, the concentration of the monomer available for polymerization remains constant with time. Therefore, at a constant temperature the average molecular weight of the polymer obtained is apparently independent of the degree of the monomer conversion.

Bulk polymerization of vinyl chloride is performed in variants of the original Pechiney two-stage process. Vinyl chloride containing initiator is charged in a reactor and the polymerization is carried out at 50–60°C. Already at 1% conversion the polymer begins to precipitate. After reaching conversion of about 10% the entire content of the reactor is discharged in an autoclave where a further quantity of vinyl chloride is added. When the conversion is at about 20% the material is in the form of a wet powder; at 40% conversion the residual monomer is completely absorbed by the polymer and the mass takes on the form of a dry powder. The design of the reactor permits further polymerization which usually proceeds up to 90% conversion. The residual monomer is then removed under vacuum and the polymer is discharged and sieved. Polymer particles with sizes larger than desired undergo additional grinding and sieving.

Polyvinyl chloride is an amorphous polymer containing very small amounts of imperfect crystallites. The average molecular weights for the most commercial PVC resins are in the range 50 000–120 000. Better physical properties are obtained by PVC having a high molecular weight, but, on the other hand, for applications in powder coatings, PVC resins with low molecular weight are more suitable because of a lower melt viscosity and lower softening temperatures. Although different glass transition temperatures ranging from 68 to 105°C are reported, depending on the method of polymerization and especially the temperature at which the polymerization is carried out, the normal commercial grades of PVC have T_g of between 80 and 85°C.

Figure 2.1 represents the differential thermal analysis curve of polyvinyl chloride. Five major regions of different thermal behaviour can be noticed above room temperature. Matlack and Metzger [2, 3] assign the glass transition temperature to the change of the slope at point A. The endothermic effect between 165 and 210°C is a result of the melting of the crystallites present in the polymer. At 250°C they define the point C as a region of oxidative attack, while at 300°C (point D) the dehydrohalogenation takes place with evolving gaseous hydrogen chloride. Finally, at about 450°C (point E) the endotherm is attributed to the occurrence of depolymerization.

Polyvinyl chloride is soluble in a wide range of organic solvents. The best solvents for PVC are tetrahydrofuran and cyclohexanone, but it is also easily

THERMOPLASTIC POWDER COATINGS

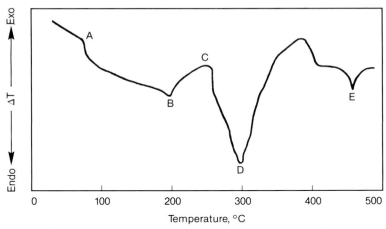

Figure 2.1 Differential thermal analysis curve of unplasticized PVC [3] (Reproduced by permission of Federation of Societies for Coatings Technology)

solubilized by other ketones, halogenated hydrocarbons and nitrated aromatic solvents. However, from the standpoint of a powder coating producer it is important that PVC polymers are not soluble in alcohols and aliphatic hydrocarbons, and powder coatings having PVC as a binder possess considerable solvent resistance with respect to the most common solvents used in the industry.

The performance characteristics of the vinyl powder coatings have been compared by Christensen and are summarized in Table 2.1 [4]. Comparisons of this type are always possible. In this case, the outdoor durability of the vinyl polymers should be reconsidered, keeping in mind new developments, particularly those of polyester thermosetting coatings. The table, however, gives a good indication of the characteristics of the vinyl powders with respect to the other thermoplastic materials.

Table 2.1 Performance characteristics of various powder coating materials [4]

	Vinyl	Polyester	Epoxy	Nylon	Polyethylene
Exterior durability	E	VG	F	G	F
Salt spray resistance	E	G	VG	VG	G
Impact resistance	E	G	G	E	F
Flexibility	E	F	F	VG	E
Food staining	E	E	E	VG	E
Dielectric strength	VG	E	E	G	E

E = excellent, VG = very good, G = good, F = fair.

The main problems of PVC powder coatings are related to the stability of the binder during processing, and also its brittle nature. These weak points were successfully overcome due to the enormous amount of work that was done on stabilization and plasticization of polyvinyl chloride. The result was development of a number of heat stabilizers, plasticizers and lubricants permitting problem-free hot compounding of the vinyl resins.

Because it is a rigid material not having sufficient flexibility polyvinyl chloride as such cannot be used for powder coating purposes. In fact, only in a few applications can PVC be used without the addition of plasticizers. A wide range of high boiling materials has been developed for plasticization of PVC, such as esters of phthalic acid, phosphoric acid and aliphatic diacids. Liquid polymers or oligomers with relatively low glass transition temperatures are also used for the same purpose.

Phthalate esters are the cheapest plasticizers for PVC, proving to be very satisfactory over all properties. Di-2-ethylhexyl phthalate and diisooctyl phthalate are the most widely used, offering a good price/performance ratio. Phthalic esters based on linear C7–C11 alcohols with a lower migration rate of the plasticizer are also used for special applications. They are also characterized with better oxidation stability and low temperature flexibility.

Phosphate esters are used to obtain PVC products with self-extinguishing properties. Trixylyl phosphate is the most widely used material among the great variety of phosphate esters. The same self-extinguishing effect is obtained by the use of chlorinated paraffins. They in general cannot completely replace the so-called primary plasticizers of phthalate or phosphate type, but they are quite competitive for price reasons. The common chlorinated paraffins used for plasticizing PVC have 42–56% chlorine content and a chain length of C15–C25.

Linear polyesters produced by esterification of dibasic acids with dibasic alcohols are plasticizers that provide good solvent resistance. They are specially suitable for PVC coatings in contact with hydrocarbon fuels or other organic solvents. However, polyvinyl chloride plasticized with polyesters exhibits relatively poor low temperature flexibility.

The addition of the plasticizer to PVC improves the flexibility and the impact resistance of the polymer, but at the same time reduces the tensile strength, modulus and hardness of the plasticized material. The desired balance between hardness and flexibility can be achieved by careful choice of the type and amount of plasticizer. Although many efforts have been made to predict the behaviour of the plasticized system by systematic approaches to this problem, the desired set of properties is usually achieved by making a series of formulations.

The curve representing modulus–temperature behaviour of externally plasticized PVC with dioctyl phthalate (DOP) exhibits a typical drop of the modulus at the glass transition temperature of 80–90°C with the rubbery plateau region afterwards in the non-plasticized PVC (Figure 2.2) [3]. A further increase in the temperature leads to melting of the crystallites, resulting in an additional drop of

THERMOPLASTIC POWDER COATINGS

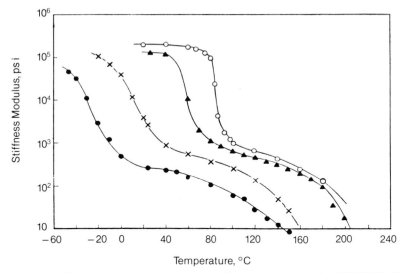

Figure 2.2 Modulus–temperature behaviour of externally plasticized PVC [3] (Reproduced by permission of Federation of Societies for Coatings Technology)

the modulus. The addition of DOP shifts the rubbery plateau to lower temperatures, but at the same time decreases the modulus, in this way affecting the hardness of the film.

The type and amount of added plasticizer must be determined in a way that will result not only in a good compromise between flexibility, impact resistance and hardness but also in a way that will not affect the powder storage stability.

Polyvinyl chloride that exhibits good mechanical properties has a molecular weight above 40 000. In pigmented systems, due to such a high molecular weight there is a considerably more pronounced increase of the melt viscosity compared to the thermosetting binders. The melt viscosity of PVC pigmented with TiO_2 shows an asymptotic increase at concentrations of the pigment between 15 and 25% (by weight). This is almost four times lower than the concentrations of the alkyd resins [3].

High melt viscosity of the pigmented PVC powder coatings is one of the reasons for their application in thick films in order to obtain good flow. Very often the low price of PVC powder coatings is compensated by the thick film application, losing in this way the advantage of having cheap and good quality protection. Smith and Koleske have suggested that for film thicknesses below 50 μm, the melt viscosity during cure should be in the range between 500 and 5000 poises [3]. This corresponds to a weight averaged molecular mass of between 20 000 and 40 000 (Mn = 10 000–20 000), which is two to four times lower in value than the commonly used molecular masses.

PVC powder coating with a proper formulation should always contain a suitable stabilizer. An enormous amount of work has been done to elucidate the mechanism of thermal degradation of polyvinyl chloride. It is beyond the scope of this book to go into detail about the degradation phenomena of PVC. The overall reaction of decomposition involving hydrogen chloride will be presented simply as an illustration to help explain the role of the stabilizers:

$$-CH_2-CHCl- \longrightarrow -CH=CH- + HCl$$

The common property of all materials used for stabilization of polyvinyl chloride is the ability to react with hydrogen chloride. It seems rather strange that the stabilizer reacts with the product of degradation, i.e. it is active after the degradation has happened. Therefore it can be assumed that the stabilizer decreases the rate of degradation, lowering the concentration of hydrogen chloride that may act as a catalyst for the degradation process. This can be accepted for some simple stabilizers such as the metal stearates. The experiments with labelled stabilizers of a metal soap nature or dialkyl tin dicarboxylates showed, however, that the soap or carboxylate moities are retained to a greater extent than the metal or the alkyl groups. This indicates that the stabilizer is involved in some coordinate complex with the polymer chain and that its action is not limited to a simple scavenging action of the evolved hydrogen chloride only.

A wide variety of materials has been developed for thermal stabilization of PVC. Such materials include mixed salts of calcium and zinc with rather inferior stabilizing properties, barium–cadmium soaps, lead sulphates, tin mercaptides, dibutyltin derivatives, di(n-octyl)tin maleate, amino crotonic esters, epoxy compounds, etc.

A typical vinyl powder coating formulation containing a plasticizer and stabilizer is given in Table 2.2 [5].

Table 2.2 Vinyl powder coating formulation [5] (Reproduced by permission of Communication Channels, Inc.)

Ingredient	Parts by weight
Bakelite vinyl resin QYNJ[a]	100.3
Bakelite vinyl resin QYJV[a]	9.3
Flexol plasticizer EP-8[a]	6.0
Flexol plasticizer 10–10[a]	36.0
Mark C[b]	2.0
Thermolite 31, PVC stabilizer[c]	6.0
TiO_2—rutile	20.0

[a] Union Carbide Corp.
[b] Argus Chemical Corp.
[c] M&T Chemicals, Inc.

THERMOPLASTIC POWDER COATINGS

Additives known as lubricants are often used in powder coating formulation. They influence both the behaviour during processing and the properties of the finished product. The primary requirements of the lubricants are to improve the flow properties of the fused compound and to help the movement of the material through the extruder during the production of the powder coating by lubricating the interface between the PVC and the metal surface.

Two grades of lubricants are usually used for this purpose. The internal types which are compatible with PVC reduce the melt viscosity of the compound, improve the flow properties and reduce the friction in the processing equipment. The external lubricants which are not compatible with PVC act in this case only as a lubricating layer on the surface of the PVC where it comes into contact with the metal surface of the processing equipment.

Representatives of internal lubricants are long-chain fatty acids, calcium stearate, alkylated fatty acids and long-chain alkyl amines. External lubricants include fatty acid esters, high molecular weight alcohols, synthetic waxes, low molecular weight polyethylene and lead stearate.

PVC powder coatings applied by the fluidized bed method have been the most widely used materials for coating wire goods in thick layers. In this application they offer a good compromise between price and performance. They are characterized by uniformity of the coating, no sagging and good edge coverage. The usual thickness of the PVC powder coatings applied by the fluidized bed method is between 200 and 400 μm. To achieve good uniformity of the coating in thinner layers, the electrostatic spray technique is required.

Typical products coated with PVC powders are appliances and dishwasher racks. In the latter case, PVC coated materials can withstand the normal testing conditions of total vapour and water immersion at 80–100°C without blistering. Detergent resistance and resistance to food stains is excellent. This is combined with the important characteristic of allowing the water to run off from the coated products. PVC powder coatings provide another important characteristic to dish racks by acting as sound dampers, thus preventing the development of noise during filling or discharging of the racks. Chemical and detergent resistance make them suitable for protection of dishwasher tubes.

The non-toxic nature of PVC powder coatings allows them to be used for refrigerator racks, freezers and shelves. The stain resistance and low odour of PVC powder coatings are appreciated characteristics in coatings for shelves made of wire, expanded metal or sheets. For the same reasons they are used as metal furniture coatings.

Good dielectric properties of PVC powder coatings are among the prerequisites they fulfil for application in the electric and electronic industries. The soft touch feeling makes them suitable for covering various types of handles, automotive seat belt anchors, bus bars, tools, etc.

Outdoor durability is not a strong point of PVC powder coatings. The main reason is the sensitivity of the binder to the ultraviolet spectrum below 350 nm.

The development of effective u.v. stabilizers has opened up new possibilities for outdoor use of PVC powders. A large market for exterior use of PVC powders developed in the USA includes the electrical transformer whose manufacturers give a 20 year guarantee for their products coated with PVC powder [6]. The successful development of thermosetting powder coatings specially designed for outdoor application is certainly a big obstacle for PVC powders to compete in the market for outdoor coatings.

2.1.2 PVDF POWDER COATINGS

The best representative of vinyl polymers exhibiting outdoor durability is polyvinylidene fluoride (PVDF). Powder coatings based on PVDF have the highest resistance to degradation on outdoor exposure, very good abrasion resistance, excellent chemical resistance, very low surface friction, ice release properties and very low fluid absorption.

PVDF is produced by free radical polymerization of vinylidene fluoride, using peroxides as initiators, or coordination catalysts of the Ziegler–Natta type. Different patents describe various methods for polymerization of vinylidene fluoride, including emulsion, suspension or solution polymerization [7–12]. The structure in the repeating unit of polyvinylidene fluoride is $-CH_2-CF_2-$. Hydrogen and fluorine atoms are spatially symmetrical, which optimizes crossbonding forces between polymer molecules. Polyvinylidene fluoride is a crystalline polymer with a melting point between 158 and 197°C. It exists in two different crystalline forms: the so-called alpha form has a *trans-gauche–trans-gauche* conformation, while the other beta form possesses a planar zigzag conformation [13]. The polymorphism of PVDF is the reason why the melting point, which is in a considerably broad temperature range, is not well defined. The relatively high melting point allows PVDF to be used permanently in a relatively broad temperature range from -40 up to 150°C, which corresponds with its glass transition temperature and the lower border of the melting range.

Polyvinyl fluoride is characterized by a high mechanical and impact strength and very good abrasion resistance combined with excellent flexibility and hardness. It is resistant to attacks by most corrosive chemicals such as acids, alkalies, strong oxidizing agents, etc. It is also not soluble in the common solvents used in the coating industry. Some highly polar solvents can only soften the surface of polyvinylidene films temporarily. The only chemicals that can damage the surface of PVDF films are fumic sulphuric acid and strong primary amines. PVDF complies with the requirements of the Food and Drug Administration for materials that can be used in the food processing industry and allowed to come in direct contact with food.

PVDF powder coatings have been recognized as materials with exceptional properties, such as low friction and abrasion, oil and water repellency, extremely good outdoor durability, excellent flexibility, corrosion and chalking resistance, and resistance to chemicals and aggressive industrial atmospheres rich in sulphur dioxide. They are very easy to maintain because of extremely low dirt pick-up properties.

These unique properties are due to the small bond polarization of the fluorine–carbon bond, which is responsible for the low surface energy of the PVDF used as a binder in these coatings. The very high bond energy of the carbon–fluorine bond (477 kJ/mol) provides exceptional weathering resistance [14, 15]. PVDF can be used as a sole binder in the manufacture of powder coatings, especially when high weathering resistance is required, although this is not a common case in practice. The main reasons include the high melt viscosity resulting in the formation of pinholes during thin film applications, rather poor adhesion to metal substrates and the relatively high price.

In order to improve the melt flow properties, adhesion and aesthetic appearance of the powder coatings, acrylic resins are usually blended with PVDF. It is common for the binder for PVDF powder coatings to contain 30% acrylic resins. Higher amounts of acrylic resin decrease the weathering resistance of the coatings, although it still remains superior to other manufactured organic coating materials known so far. The results of a study of the weathering resistance of PVDF coatings containing 25–40% acrylic resin baked at 300°C are presented in Figure 2.3 [14]. Similar results with respect to accelerated weathering in QUV apparatus are obtained by PVDF powder coatings containing 30% acrylic resin, baked for 10 minutes at 230°C (Figure 2.4) [16, 17].

PVDF powder coatings have low gloss in the range of 30 ± 5% (at 60°C). This is certainly a restriction for a whole range of applications of PVDF powder coatings for decorative purposes.

The process of production of PVDF powder coatings does not differ from that of the other powder coatings. It consists of extrusion of the resins and pigment

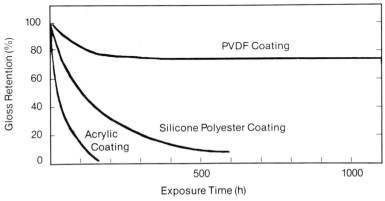

Figure 2.3 Gloss retention of PVDF coating on exposure in a Dew Cycle Weather-O-Meter [14] (Reproduced by permission of Elsevier Sequoia SA)

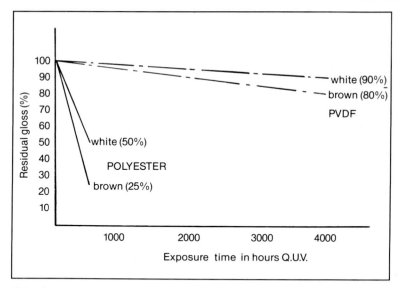

Figure 2.4 Accelerated weathering (QUV) of PVDF white and brown coatings compared with polyester powder coatings [17] (Reproduced by permission of Technologie Communikatie Markt BV)

premix in a single or twin screw extruder followed by granulation and drying of the granulate. The next step is cryogenic grinding and sieving in order to obtain powder particles below 50 μm [17].

The very low surface energy of PVDF which provides low dirt pick-up properties to the coating at the same time causes difficulties concerning adhesion to the substrate. This, a general weak point of thermoplastic powder coatings, is even more emphasized in the case of PVDF powders. As already mentioned, the blends with acrylic resins improve adhesion, but even in this case direct application of PVDF powder coatings to a metal substrate is not recommended. For good adhesion, PVDF powder coatings are applied over an epoxy stoving primer [16, 17]. Systems based on polyurethane primer with a PVDF top coating are also reported [18].

The problem of bad adhesion can be overcome in the way described in the US patent from 1974 [19] where a two-coat system based on polyvinylidene fluoride is used. The first primer coating is produced by physical mixing of polyvinylidene fluoride of a particle size from a 60–200 mesh and silica sand of a particle size from a 150–325 mesh. The adhesion of the coating was tested by exposure to steam at 100°C. The time before the occurrence of blisters, taken as the relevant parameter for estimation of the adhesion properties of the system, ranged from 7 hours in the case of the primer containing no silica sand to more than 480 hours for the one having a PVDF/silica ratio of 100:40 or more. Similar results were obtained with a primer in which graphite of the same particle size was used instead of silica sand [19].

Although for the coil coatings based on PVDF as a binder a 20 year guarantee is normally given [14, 18, 20], for PVDF powder coatings containing 30% acrylic resins a 10 year guarantee for a maximum 50% loss of the initial gloss has been granted [16].

PVDF powder coatings are mainly used as architectural coatings for monumental-type structures. Architectural panels for roofing and wall claddings and aluminium extrusion window frames are the main surfaces on which PVDF powder coatings have been applied.

2.2 POLYOLEFINIC BASED POWDER COATINGS

The discovery of the new coordination catalysts for polymerization of olefins in the first half of the 1950s resulted in the development of a new class of polyolefins. This invention has not only made a big commercial impact but has also contributed very much to the development of polymer science.

By the use of coordination catalysts, or the so-called Ziegler–Natta catalysts after the names of the inventors, polypropylene and low pressure polyethylene are produced on an industrial scale, being two of the most important thermoplastic polymers of this century. They are the main representatives of the polyolefins used in the manufacture of powder coatings. Being non-polar, high molecular weight crystalline polymers, with carbon–carbon bonds in the polymer backbone, they possess a unique balance of toughness and chemical and solvent resistance. It is obvious that such materials could be of great interest when used as binders for protective coatings. However, their insolubility in the solvents commonly used by the paint industry restricted their application to the powder coatings only. In fact, the first powder coating which was used in a fluidized bed application at the beginning of the 1950s was based on powdered polyethylene [21].

The empirical formula of polyethylene $-(CH_2-CH_2)_n-$ is rather simple and does not indicate the many implicit structural possibilities that can exist, offering products with different properties and uses. The molecular weight of polyethylene and molecular weight distribution, the amount and length of side chains that depend on the manufacturing conditions, together with different degrees of crystallinity are parameters that influence to a great extent the properties of the polymer. The crystallinity of polyethylene, in particular, directly reflects the degree of packing of polymer chains and thus the density of the material and is an important parameter distinguishing several classes of commercial products in the market. According to the ASTM classification, commercial grades of polyethylenes are grouped into four types based upon densities regardless of the molecular weight or molecular weight distribution:

Type	Density (g/cm^3)
I	0.910–0.925
II	0.926–0.940
III	0.941–0.959
IV	0.960 and higher

When the density or crystallinity of polyethylene is reduced, then the impact resistance, cold flow, tackiness (blocking), clarity and permeability increase. On the other hand, a reduction in density produces a decrease in stiffness, tensile strength, hardness, abrasion resistance, brittleness temperature, softening point, fusion temperature, chemical resistance and shrinkage.

Both low and high density grades of polyethylene are successfully used for the manufacture of powder coatings. The high density grade is suitable for industrial applications while the low density grade is used for domestic applications [22]. The first process with a commercial potential of polymerizing ethylene under high pressure and temperature was discovered in the laboratories of Imperial Chemical Industries Ltd in 1933. Polyethylene produced in this way (temperatures of 100–300°C and pressures above 1000 atm) has a high molecular weight, broad molecular weight distribution, a high degree of branching, moderate crystallinity and low density.

By the discovery of Ziegler in 1953 polyethylene was produced in an astonishingly simple process at atmospheric pressure and almost room temperature by using alkylaluminium compounds combined with titanium tetrachloride as catalyst. This was an outstanding event not only from the technological point of view but also because it offered polymer that was highly oriented with a high degree of crystallinity, almost without side chains, and properties that differed very much from those obtained by the high pressure process.

Another method of production of polyethylene was developed in the USA about the time of Ziegler's discovery, and employed medium pressure and temperature during polymerization in the presence of metal oxides as catalysts.

Table 2.3 Typical properties of polyethylene obtained by the three basic processes [23] (Reproduced by permission of John Wiley & Sons, Inc.)

	High pressure process	Medium pressure process	Ziegler process
Crystallinity	65%	95%	85%
Relative rigidity	1	4	3
Softening temperature	104°C	127°C	124°C
Tensile strength (lb/in^2 gauge)	2000	5500	3500
Elongation	500	20	100
Relative impact strength	10	3	4
Density (g/cm^3)	0.92	0.96	0.95

THERMOPLASTIC POWDER COATINGS

Table 2.4 Important differences between different polyethylene types with respect to their application in powder coating [23] (Reproduced by permission of John Wiley & Sons, Inc.)

	Low density	Medium density	High density
Resistance to acids	Resistant	Very resistant	Very resistant
Oxidizing acids	Attacked	Slowly attacked	Slowly attacked
Resistance to alkalies	Resistant	Very resistant	Very resistant
Resistance to organic solvents	Resistant below 60°C	Resistant below 60°C	Resistant below 80°C
Clarity	Translucent	Translucent	Opaque
Crystalline melting point (°C)	108–126	126–135	126–136
Heat resistance (continuous use, °C)	82–100	104–121	121
Density (g/cm^3)	0.910–0.925	0.926–0.940	0.941–0.965
Elongation (%)	90–800	50–600	15–100

These processing conditions lie between high pressure polymerization and the process of Ziegler.

Table 2.3 presents the basic differences in the properties of polyethylene polymers obtained by these three different methods [23]. Other important differences from the standpoint of the powder coating chemist are presented in Table 2.4 [23].

The other representative of polyolefins that is used as a binder for powder coatings, polypropylene, can be produced by a wide variety of catalysts, including cationic, anionic and free radicals. The technique for polymerizing polyethylene under high pressure has been unsuccessful for polypropylene. The explanation is that polypropylene radicals tend to chain transfer to C_3H_6, giving the unreactive allyl radical. Therefore, the free radical polymerization of propylene has almost no commercial importance.

The polymerization process catalysed by cationic catalysts of the Friedel–Crafts type results in amorphous polypropylene of low molecular weight. There is a typical drawback of the polypropylene obtained in this way which restricts the application of the polymer in the powder coating field; namely the polymer chains obtained by cationic polymerization are without structural order. As a consequence, crystalline polymers cannot be made by this route.

The production of polypropylene with properties useful for most modern applications became possible with the discovery of the Ziegler–Natta catalysts.

The polymerization of propylene by the anionic coordination mechanism produces a stereoregular crystalline polymer with a melting point high enough from a practical standpoint.

Three types of polypropylene can be distinguished with respect to the stereoregularity of the chains. Atactic polypropylene, without a regular structure and obtained by cationic polymerization, is an amorphous, soft and tacky material with little or no physical strength. On the other hand, isotactic and syndiotactic polypropylenes are obtained by anionic polymerization. These structures are schematically presented below:

$$\begin{array}{c} \text{C} \quad\quad \text{C} \quad\quad \text{C} \quad\quad \text{C} \\ | \quad\quad | \quad\quad | \quad\quad | \\ -\text{C}-\text{C}-\text{C}-\text{C}-\text{C}-\text{C}-\text{C}-\text{C}- \end{array}$$

Isotactic

$$\begin{array}{c} \text{C} \quad\quad\quad \text{C} \quad\quad\quad \text{C} \\ | \quad\quad\quad | \quad\quad\quad | \\ -\text{C}-\text{C}-\text{C}-\text{C}-\text{C}-\text{C}-\text{C}-\text{C}-\text{C}-\text{C}- \\ \quad\quad | \quad\quad\quad\quad | \\ \quad\quad \text{C} \quad\quad\quad\quad \text{C} \end{array}$$

Syndiotactic

The ordered structures of the syndiotactic and isotactic polypropylene allow a great degree of crystallinity, resulting in a polymer with high physical strength, solvent resistance and chemical resistance.

The anionic polymerization of propylene mainly gives the isotactic type of polymer. This is one of the lightest commercial plastics with a density of only 0.9 g/cm^3. The melting point is in a range from 165 to 170°C for the commercial grades, while the melting point of 100% isotactic polypropylene is 183°C.

Syndiotactic polypropylene is not produced on a commercial scale. Its melting point is about 20°C lower than that of the isotactic type and it is more soluble in ether and hydrocarbon solvents.

The same catalysts used for production of high density polyethylene at atmospheric pressure are suitable for polymerization of polypropylene. The catalytic system usually includes titanium chloride, an aluminium compound, preferably trialkylaluminium or aluminium chloride, and optionally a Lewis base or other modifier.

Polypropylene has a number of desirable properties that make it a versatile material for production of powder coatings. It possesses good surface hardness and is scratch and abrasion resistant. It is essentially unaffected by most chemicals and possesses excellent solvent resistance. It is somewhat more brittle than polyethylene at normal temperature, which is caused by the relatively high glass transition temperature ranging between 25 and 35°C depending on the degree of crystallinity. The brittleness of the commercial grades and impact resistance are markedly improved by copolymerization with other olefins.

Considerable amounts of polypropylene are sold as copolymer containing 2–5% of ethylene. The resulting polymer has increased flexibility, toughness, impact resistance and clarity, and a somewhat lower melting point.

The fact that the fragility of polyethylene and polypropylene is increased by cooling with liquid nitrogen or by soaking with alcohol before grinding is used as the basis for several techniques to make fine powders. Also some polymerization processes yield polyethylene directly as a fine powder. High pressure polyethylene, however, is produced as a solid resin and must be ground to produce powder.

Because of the carbon–carbon bonds in the molecular chains, polyethylene and polypropylene powder coatings exhibit very good chemical resistance. Due to the unpolar paraffinic nature of the binder they have a reputation for coatings with excellent solvent resistance. Therefore, a very important application field for polyethylene and polypropylene powder coatings is for the protection of chemical containers, pipes and pipelines for keeping and transporting different chemicals and solvents.

Being inert materials, polyolefins have poor adhesion to metal or other substrates. Therefore, before polyethylene and polypropylene powder coatings can be used successfully the surface to be coated must first be primed or adhesion promoters must be added to the powder coating to improve adhesion. A family of adhesion promoting polymers, called Hercoprime, has been developed which, when blended with polyolefins, and specially with polypropylene, give a one-coat system with good adhesion [6, 24]. These polymers are produced in the form of small particles whose size is of a range typical for powder coatings. Their density is similar to that of polypropylene and they can be used by simple tumble blending with polypropylene powder coatings. For pigmented systems, a 15% addition of these adhesion promoting resins provides good adhesion for most of the substrates. Clear propylene coatings need 5–10% of primer promoter to exhibit satisfactory adhesion properties [25]. Different modifications of polyolefins have been tried in order to improve the adhesive properties. In some cases this is a simple blend of polyethylene or polypropylene with polymers containing acid groups as adhesion promoters. A Dainippon patent [26] describes a modified polyolefin composition prepared from 60–97 parts of polyolefin and 3–40 parts of copolymer based on 20–60% acid monomers, 40–80% styrene and 0–40% alkylmethacrylates. It is claimed that the powder coating possesses excellent adhesiveness, coatability and colour stability. In examples of another patent referring to improved adhesion of polyolefin powder coatings [27], 0.05–40 parts of a polypropylene modified product containing 15–30% by weight grafted maleic anhydride and 99.5–60 parts of polyolefins are used as a binder composition for polyolefin powders. Excellent adhesion of the coating is combined with excellent mechanical properties.

A powder composition with improved adhesion properties is obtained by physical mixing of polyethylene powder with powder of copolymer comprising

60–95 wt% of ethylene and 5–40% of glycidyl methacrylate [28]. The coating is applied in a fluidized bed on a preheated surface at 300°C followed by postcuring at 200°C for 20 minutes. It is claimed that the adhesion properties of this coating are much improved in comparison with pure polyethylene powder.

Polyethylene powder coatings are even more problematic with respect to adhesion properties than polypropylene powders. Priming of the surface is usually necessary prior to application of the polyethylene powder coating, since the adhesion promoting resins do not perform as well as in the case of polypropylene. A patent assigned to Hoechst describes a two-step method for heavy duty protection of pipelines for the transport of oil [29, 30]. In the first step, a bisphenol A epoxy resin with amino hardener is sprayed over preheated metal pipe at 270°C in a 100 μm thick layer. The still soft, highly viscous layer is coated with a powder coating consisting of 75% polyethylene and 25% copolymer of vinyl acetate–ethylene–acrylic acid–alkyl acrylate in a weight ratio of 3:76:1:20. The total thickness is very high (above 2 mm), providing extremely good mechanical protection. Pipes can be transported without damage over rocky ground, they are heat stable and they resist desert conditions.

Three-layer polyolefin pipe coatings have gained widespread international acceptance. The first layer is thermosetting epoxy powder or solventless liquid coating with a thickness between 60 and 100 μm. The intermediate polymer layer with a thickness between 250 and 400 μm is a polyethylene powder coating and is applied over the epoxy primer, followed by the third layer as a top coat which is applied by extrusion with a coating thickness between 1.5 and 3.0 mm. These systems are well established in Europe, the Middle East and Asia, offering excellent corrosion resistance and interfacial properties of epoxy coatings, combined with excellent chemical, solvent and mechanical resistance of the thick polyethylene coatings. The Hajira–Bijaipur–Jagdispoor 1700 km long pipeline system is an example of one of the major operations to be coated with three-layer polyethylene coatings. [31].

A one-step application procedure refers to a coating system that combines the excellent anticorrosive properties of zinc dust, adhesion of the epoxy powder coating and good solvent, chemical and weather resistance of the polyethylene powder coating [32]. An epoxy powder coating comprising bisphenol A epoxy resin as the main binder, dicyanediamid as the hardener, barium sulphate and carbon black as the filler and pigment, catalytic amounts of amine accelerator and flow additive with a particle size of 75–150 μm is dry blended with clear polyethylene powder and zinc dust in the ratio of 70:30:5. Powder coating prepared in this way has been applied on steel panels using an electrostatic spray gun in a one-step operation. After curing for 3 minutes at 150°C and raising the temperature at a linear rate to 215°C for 10 minutes, a flat finish is obtained with a clear polyethylene top coating and a zinc-rich bottom layer covered with an epoxy intermediate thick layer. The main proposed application is to coat the interior of underground oil or gas pipes. The zinc layer produces a galvanized

THERMOPLASTIC POWDER COATINGS 27

finish on the interior of the pipe, while the epoxy part overlaying the zinc serves to protect the zinc from abrasion as well as providing an integral coating with high corrosion resistance. The polyethylene top coating is an electrically non-conducting layer, and thus prevents or minimizes electrolytic corrosion, providing at the same time excellent solvent resistance. The same coating technique is proposed for applications where the polyethylene top coating contributes in improving exterior durability, for example in coating automobile wheel rims.

Powder coatings based on polyolefin polymers have a low aesthetic appearance, and therefore are used mainly for protective rather than deocrative purposes. For pigmentation, dry blending techniques with high shear mixers can be successfully used. This is not of course a suitable technique for highly decorative coatings, but in general, for protective polyolefin powder coatings, it seems to be a method that gives a good compromise between processing costs and coating performances.

In the case of polypropylene powder coatings, high gloss can be obtained by water-quenching immediately after fusion. Because of the rapid cooling, the spherulite size of the crystalline polypropylene is smaller, resulting in a film with less surface roughness and imperfections combined with higher flexibility. Coatings with better aesthetic appearance can be obtained by using higher molecular weight polyolefins, since the crystallization rate considerably decreases due to the reduced polymer chain mobility in the melt.

Physical ageing of high molecular weight polypropylene has in principle a less emphasized effect on decreasing impact strength. High molecular weight polypropylene powder coatings retain their initial impact strength over a six-month period after quenching, while lower molecular weight grades exhibit slight decreases with time.

Polyolefinic powder coatings are resistant to a variety of hostile environments. Solvent resistance at room temperature is excellent against all common solvents used in industry. However, at elevated temperatures, non-polar solvents, such as hydrocarbons and aromatics, will swell or in extreme cases dissolve the coating. Resistance against water, salt solutions, detergents, acids and bases is excellent. Only very strong oxidizing agents, such as nitric acid, will attack the coatings at room temperature.

Although polyolefinic powder coatings are not recommended for outdoor use where decorative purposes are requested, stabilized versions of polypropylene can be used for the manufacture of powder coatings with reasonable gloss retention when exposed to weathering conditions. The main use of polyethylene and polypropylene powders, however, remains in the field of functional coatings with excellent protective properties. Applications for these functional coatings include appliance parts, such as dishwasher racks, refrigerator racks, washer tubes, metal containers, drums, pipes, electrical applications and industrial equipment components [33]. Polyolefinic powder coatings comply with FDA regulations and can be used for covering surfaces which come in direct contact with food products.

2.3 NYLON BASED POWDER COATINGS

The early work on polyamides resulting in production of the first industrial polymers known under the name of nylon was done by W. H. Carothers in the laboratories of Du Pont. A period of 10 years of intensive work starting from 1929, in which the basic patents disclosing the process of polycondensation of diacids and diamines or self-condensation of amino acids were issued [34–37], was necessary to build up the base for commercialization of polyamides. In the same period a process of preparation of polyamides from lactams was protected by I. G. Farbenindustrie in Germany [38]. Although the experimental work on polyamides began with amino acids capable of self-condensation to form linear polymers, the first polymer produced on an industrial scale was the so-called nylon-6,6, prepared from adipic acid and hexamethylene diamine; production started in 1939 in Du Pont's plant for manufacturing nylon products [39]. The success of nylon textile yarns initiated enormous research work on the development of new types of polyamides and the extension of application outside the textile industry. This resulted in the development of a very large family of polyamides which were used as moulding powders, extruded films, surface coatings, adhesives, printing inks, etc. As usually happens, the commercial name nylon became a synonym for the whole group of polymers, and it is used interchangeably with the name polyamide, a practice widely followed in this book.

There are three general routes used to produce polyamides. The first commercial polyamide, the so-called nylon-6,6, was produced by condensation of adipic acid and hexamethylene diamine. This is done in a two-step procedure. In the first step a salt between the diacid and diamine is formed, usually in an aqueous medium. In the second step of the process the salt is dehydrated by heating. The elimination of water as a by-product results in the formation of an amide group connecting the monomers in a repeating unit:

$$HOOC(CH_2)_4COOH + H_2N(CH_2)_6NH_2 \longrightarrow HOOC(CH_2)_4\text{-}$$
$$[COO^-NH_3^+(CH_2)_6NH_3^{+\,-}OOC(CH_2)_4COO^{-\,+}NH_3]_n(CH_2)_6NH_2 \longrightarrow$$
$$HO[OC(CH_2)_4CONH(CH_2)_6NH]_{n+1}H$$

The formation of the salt and its purification is an essential step in the process providing a stoichiometric ratio between the reaction partners, necessary to avoid terminating the polymer chain by having an excess of either reagent present. The polymerization reaction is carried out at relatively high temperatures between 260 and 290°C, without a catalyst or in the presence of acid compounds which exhibit the catalytic effect.

There is no principal difference between the process of polycondensation of diamines and diacids and the method of producing polyamides starting from amino acids:

$$mH_2N(CH_2)_nCOOH \longrightarrow H[HN(CH_2)_nCO]_mOH + (m-1)H_2O$$

In both cases it is important that the combination of diamines and diacids or the amino acid itself is not capable of forming a five- or six-membered ring as a result of the condensation reaction. Since these rings are thermodynamically preferred structures, the condensation will lead to the formation of cyclic materials rather than polymers. Therefore, amino acids suitable for polymerization into polyamides should contain more than six carbon atoms.

A very important commercial route for preparing polyamides is the thermal polymerization of lactams. The ring opening may be affected by initiation with water leading to hydrolysis of the lactam to the corresponding amino acid, or by direct addition of amino acid as the polymerization promoter:

$$\begin{array}{c} CO \\ (CH_2)_n NH + H_2O \longrightarrow H_2N(CH_2)_nCOOH \end{array}$$

$$\begin{array}{c} CO \\ m(CH_2)_n-NH + H_2N(CH_2)_nCOOH \longrightarrow H[HN(CH_2)_nCO]_{m+1}-OH \end{array}$$

A worldwide accepted nomenclature for polyamides was established very soon after their introduction in the market. When the polymer is obtained by condensation of diamines and diacids, the word polyamide (or generic name nylon) is followed by two numbers. The first represents the number of carbon atoms in the diamine and the second the number of carbon atoms in the diacid. For example, nylon-6,6 is the name of the polyamide produced by condensation of hexamethylene diamine and adipic acid. When the product is obtained by self-condensation of amino acids or corresponding lactams, then the number which follows the name polyamide (or nylon) is equal to the number of carbon atoms in the amino acid in question. For example, nylon-6 is the name for the polymer obtained by condensation of 6-aminohexanoic acid or its lactam.

The considerable number of different raw materials available for synthesis of polyamides makes it possible to produce polymers which cover a wide variety of structures having different physical properties. Because of the structure regularity, most of the commercial types of polyamides are crystalline materials with relatively well-defined melting points. Compared to the crystalline aliphatic polyesters, melting temperatures of polyamides are much higher. This is caused by the high polarity of the amide groups and the existence of interchain hydrogen bonds. As expected, polyamides melt at increasingly higher temperatures as the

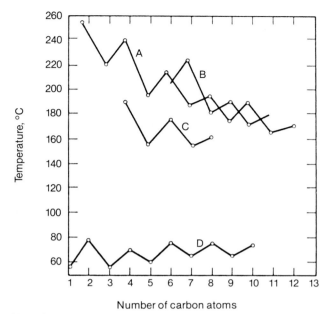

Figure 2.5 Effect of frequency of polar linkages on melt temperature for (A) polyamides, (B) poly(ω-amino acids), (C) polyurethanes, (D) polyesters [23, p. 522] (Reproduced by permission of John Wiley & Sons, Inc.)

concentration of the amide linkages increases. As in the case of crystalline polyesters, polyamides with an even number of CH_2 groups between the amide polar linkages have higher melting points compared to those with an odd number of CH_2 groups. Figure 2.5 gives a comparison between the melting temperatures of different polymers with different numbers of carbon atoms in the repeating unit.

Normal baking temperatures for powder coatings are below 200°C. Therefore those polyamides having low temperatures of melting are given preference over the others because of energy saving measures. Nylon-11 has a relatively low melting point of 185°C. Together with nylon-12 which has a melting point of 178°C they are practically the only polyamides that have found wide applications as binders for powder coatings. Despite their wide availability and comparatively low price, nylon-6, nylon-6,6 and nylon-6,10 with melting points of 215, 250 and 210°C respectively, have not been significantly accepted by powder coatings producers.

Nylon-11 is a condensation product of 11-aminoundecanoic acid. It was introduced to the market in 1955 by the company Aquitaine-Organico in France under the name Rilsan. Nylon-12 was pioneered during the early 1960s by Chemische Werke Huls AG in Germany and can be defined as polylauryllactam.

THERMOPLASTIC POWDER COATINGS

The monomer for synthesis of nylon-11 is prepared from castor oil by its hydrolysis to ricinoleic acid followed by oxidation to undecylenic acid, hydrobromation to 11-bromoundecanoic acid and its amination to 11-aminoundecanoic acid:

$$CH_3(CH_2)_5(OH)-CH_2CH=CH-(CH_2)_7-COOH \xrightarrow{oxidation}$$

$$CH_3-CH=CH-(CH_2)_7-COOH + CH_3(CH_2)_5-CHO \xrightarrow{HBr}$$

$$Br-CH_2(CH_2)_9-COOH \longrightarrow H_2N-(CH_2)_{10}-COOH$$

Melt polycondensation of 11-aminoundecanoic acid is carried out at 215°C under nitrogen in order to suppress the occurrence of side reactions.

The production of nylon-12 is based on self-condensation of laurillactam according to the reaction scheme already given.

Both nylon-11 and nylon-12 are practically insoluble in the common organic solvents. However, they are soluble or easily attacked by phenols, formic acid, mineral acids and simiral compounds, even at room temperature. At higher temperatures they are soluble in mixtures of alcohols and halogenated hydrocarbons, nitro alcohols and calcium chloride–methanol mixtures.

When exposed to light, nylon-11 and nylon-12 like all other polyamides have a tendency to lose tensile strength. This is especially the case in the presence of air. The deterioration of mechanical properties is usually accompanied by discoloration of the polymer. Light, oxidative and thermal degradation of polyamides have been the subject of many investigations. The most probable mechanism involves the free radical chain reaction which is peroxide catalysed. It is believed that oxidation sites exist at the methylene group adjacent to the carbonyl or the amide group [39, p. 41]. The peroxide catalysed chain degradative mechanism is supported by the fact that small amounts of antioxidants drastically reduce thermal degradation of polyamides. Different types of stabilizers have been developed in order to increase the service life of the polyamides. Copper and manganese salts and organic antioxidants such as N,N-di-β-naphthyl-p-phenylenediamine have been reported as effective stabilizers for polyamides [40].

Nylon-11 and nylon-12 are unaffected by water at room temperature, or even by boiling water. Stability against the action of alkalies is also very good. However, the stability in acidic media is not a strong point of polyamides in general.

The first nylon powder coatings were introduced in Europe in 1956 using nylon-11 as a binder [6, p. 182]. Several unique properties of nylon based powder coatings give them an advantage over other powder coatings. They are characterized by very good toughness combined with superior impact resistance, even at low temperatures. Their low coefficient of friction and unusual wearing properties make them excellent interfacial coatings to reduce metal-to-metal wear or noise. Good electrical insulation and good thermal insulation are other distinguishing features of nylon powders. Additionally, most of these properties are retained over a wide range of working temperatures. Nylon powder coatings have exceptionally good resistance against most of the solvents commonly used in practice. This is to be expected and emanates from the presence of the amide groups characterized by the formation of strong hydrogen bonds. The resistance to organic acids, inorganic salts and alkalies in low concentration is also good.

The possibility of stabilization of nylon-11 makes this material attractive not only for indoor functional coatings but also for outdoor use, where the weather resistance of the coating is combined with good chemical and humidity resistance, high impact and abrasion resistance and high wearing properties. Although weathering data for nylon powder coatings are not widely available, there are indications that this could also be a strong point in connection with this system. Blackmore [41] has reported excellent results with coatings based on nylon-11, having a service life time of more than 10 years. Similar results are obtained with external cladding panels coated with Rilsan (nylon-11) based powders [42].

Problems which could be encountered with nylon powder coatings derive from their easy contamination by other materials such as PVC, epoxies, polyethylene, etc. Even the epoxy booth-sprayed cyclons ruined nylon powder coatings reclaimed through them, as reported by Blackmore [41]. However, modern nylon coatings have suitable anticrater additives incorporated into them and do not suffer from these problems [43].

Like most thermoplastic powder coatings, the use of a primer is recommended in order to improve adhesion to the substrate. In general, suitable primers for nylon powder coatings are solvent based coatings used in conventional systems. In some cases where the end user has not given high demands with respect to adhesion, a single coating procedure applying the nylon powder coating to an unprimed surface can be used.

A possible improvement of adhesion properties of nylon powder coatings is suggested by W. E. Wertz [44] by melt blending minor amounts of a reactive epoxy resin and polyvinyl acetal into the coating composition while omitting an epoxy curing agent from the formulation. It could be expected that during the melting of the coating the epoxy compound would not react with the hydrogen belonging to the amide group of the polyamide. Surprisingly it does not affect the mechanical properties, but it does provide an improvement in the adhesion. Epoxy resins with epoxy equivalent weights of between 200 and 2000 are recommended for this purpose, in an amount from 2 to 15 parts by weight per hundred parts of nylon, with inclusion of 1–10 parts by weight of polyvinyl acetal. The formulation given in Table 2.5 [44] illustrates an example of such a powder coating based on nylon-11 as the main binder.

The fluidized bed method is the most commonly used application technique for nylon powder coatings. In this way, thick coatings of the order of 200–700 μm can be applied in a one-step coating procedure. Depending on the velocity of the conveyor chain on which the articles are hung and the mass of the article to be

Table 2.5 Nylon powder coating with improved adhesion [44]

Material	Parts by weight
Nylon-11[a]	100
Epoxy resin[b]	7
Polyvinyl butyral[c]	3
Calcium carbonate	20
TiO_2	8

[a] Rilsan BMNO—Organico.
[b] Epon 1001—Shell.
[c] Mowital B_3OH—Hoechst.

coated, the fusion temperatures of nylon-11 powder coatings are between 200 and 230°C [43]. The temperature most recommended is 220°C [6, p. 204]. For thin film application, electrostatic spraying can be successfully used with either positive or negative charge. However, it has been noticed that the positive charge deposits greater quantities of powder in a given time. Normal spraying equipment for the other powder coatings, operating at between 30 and 70 kV, is also suitable for nylon powders. The film thickness obtained by electrostatic spraying is usually between 100 and 150 μm.

An advantage of nylon powder coatings concerning the application properties is the high line speed, that can not be achieved with thermosetting powders. As thermoplastic materials, nylon powder coatings need only melting and fusion into continuous film without additional crosslinking. Moreover, the immediate cooling of the coated surface by immersing in water or by spraying with cold water without damaging the surface (because of low dimensional changes with temperature) also reduces the residence time of the article on the conveyor chain. The quenched cooling in water gives a high gloss finish, while the slower air cooling produces matt finishes.

It has been found that infrared ovens are not suitable for fusion of nylon powder coatings. They can be used in cases where the coated articles are flat sheets with a uniform thin film thickness. However, the high thermal insulation coefficient of the polymer and the sharp melting point are two factors that prevent the successful use of infrared curing ovens when three-dimensional parts have to be covered with a thick nylon layer. In such cases conventional recirculating air ovens of the tunnel or box design with an air temperature between 200 and 230°C are the best choice [43].

Due to their specific properties, nylon powder coatings have found broad application fields. Their resistance to salt water and inertness to fungus, bacteria and barnacles makes them very suitable coatings for articles that will be immersed or in touch with sea water during their service lives.

Nylon powder coatings are non-toxic and odour and taste free. This, combined with the fact that they are not attacked by fungus and do not facilitate bacterial growth, makes them very attractive candidates for coating machine parts and pipings used in the food processing industry, or any other place where food comes in direct contact with the coated surface. Powder coatings based on nylon-11 have approval in all industrial countries for use with potable water and foodstuffs [43].

The excellent electrical insulation properties fulfil the most stringent requirements for application on electrical parts.

Another strong point for the use of nylon powder coatings is their excellent impact resistance, which is retained over a broad temperature range from -38 to $150°C$. Stability in atmosphere containing oxygen permits continuous use in air up to $80°C$, while in the absence of air they can be continuously used up to $150°C$.

The low coefficient of friction, excellent abrasion resistance and low dirt pick-up promote their use for coating car wheels, motorcycle frames, architectural items, luggage trolleys, metal furniture, safety devices, sports equipment, farm machinery, etc.

Valve stems and seats, pump housings, degreasing baskets and trays, pump housing for domestic washing machines and large pipes for pipelines are items that are also coated with nylon power coatings, having the advantage of extremely good solvent resistance and resistance to weak alkalies and detergents.

An interesting application of nylon powder coatings is in the coverage of handles of all types. Not only are abrasion and scratch resistance important elements in encouraging the use of nylon powders for this purpose, but their low thermal conductivity gives the handle a warm feeling, which makes these materials very interesting for coating tool handles, door handles, steering wheels, etc.

2.4 POLYESTER POWDER COATINGS

Power coatings based on thermoplastic polyesters as binders are prepared from linear high molecular weight polymers produced by polycondensation of dibasic acid and diols.

There are two main methods by which high molecular weight polyesters can be produced. A direct esterification of diacids with diols results in the formation of polyester and water as a by-product:

$$n\,HOOC-R_1-COOH + (n+1)HO-R_2-OH \longleftrightarrow$$
$$HO-R_2-(O-CO-R_1-CO-O-R_2)_n-OH + 2nH_2O$$

The other method is transesterification of diesters of diacids and diols under vacuum at high temperature, leading to a formation of polyester and corresponding diol as a by-product:

THERMOPLASTIC POWDER COATINGS

$$n\text{HO}-\text{R}_2-\text{O}-\text{CO}-\text{R}_1-\text{CO}-\text{O}-\text{R}_2-\text{OH} \longleftrightarrow$$
$$\text{HO}-\text{R}_2-(\text{O}-\text{CO}-\text{R}_1-\text{CO}-\text{O}-\text{R}_2)_n-\text{OH} + n\text{HO}-\text{R}_2-\text{OH}$$

Direct esterification is a less suitable production method because it is very difficult to keep the correct mole ratio between the reactants in order to reach the desired molecular mass. On the other hand, the process of esterification is reversible; in order to obtain a high degree of polycondensation, the by-product (water in this case) has to be permanently removed from the reactor so that the equilibrium state is avoided. Schulz [45] has derived an equation giving the dependence between the degree of polymerization P, the equilibrium constant K and the mole concentration of the reaction water C_w in the reactor:

$$P = \left(\frac{K}{C_w}\right)^{1/2}$$

In order to reach a high degree of polymerization the concentration of water should be as low as possible. This can be obtained by means of a vacuum. However, together with the reaction water a certain amount of glycol is removed too. This makes it even more difficult to keep a well-defined mole ratio between the diacids and diols. This problem can be successfully solved by the so-called azeotropic cooking technique. Here recycling of an inert solvent is used to help the removal of water. Since the resins for making powder coatings must be solvent free, an additional step in removing the solvent at the end of the process is necessary. It is very difficult to perform this step in the same reactor under vacuum. This is due to the high molecular weight of the resin and its high melt viscosity. The resulting heavy foaming during the stripping off of the solvent under vacuum can not be prevented. Therefore only a spray drying technique can be used for this purpose, which makes the process expensive.

Transesterification under high vacuum is a much more convenient way of producing high molecular weight polyesters. During the process, the resulting diol is permanently stripped off from the reactor by means of a vacuum, thus approaching the correct ratio for obtaining the desired molecular weight. The molecular weight of the polyester is dependent only on the efficiency of the glycol removal. Using the equation of Schulz in the case of polyethyleneglycol terephthalate at transesterification at 280°C when the equilibrium constant has a value of 4.9, Griehl and Förster [46] have calculated that to obtain a molecular mass of 15 000 the concentration of ethylene glycol should be lower than 0.000 75 mol/litre. In order to keep the concentration of the glycols at this low level, a vacuum in the range of 0.5–1 mmHg has to be used at relatively high transesterification temperatures. Technically such problems have been successfully solved so that at present this is the process most commonly used for production of high molecular weight polyester resins on an industrial scale.

The film forming process of thermoplastic polyester coatings does not involve any additional increase in molecular weight. Therefore, to obtain a coating

Table 2.6 Melting points (in °C) of polyester resins made from different pairs of glycols and diacids [48–55]

Glycol	Diacid								
	1	2	3	4	5	6	7	8	9
Ethylene glycol	159	−22	102	−19	47	25	70	256	330
Trimethylene glycol	66	−25	43	35	36	41	49	217	246
Tetramethylene glycol	103	−24	113	36	58	38	64	222	255
Pentamethylene glycol	49	−26	32	22	37	39	53	134	160
Hexamethylene glycol	70	−48	52	28	55	52	65	148	195

1 = oxalic acid
4 = glytamic acid
7 = sebacic acid
2 = malonic acid
5 = adipic acid
8 = terephthalic acid
3 = succinic acid
6 = pimelic acid
9 = p,p'-diphenyl dicarboxylic acid

having good mechanical properties it is necessary to start with a polyester resin with a relatively high molecular weight. Molecular weights higher than 15 000 are typical for such an application.

Another important characteristic of polyester resins for thermoplastic powder coatings is their ability to crystallize. A certain degree of crystallinity is necessary to assure a high enough melting point, which at the last instance determines the stoving temperature range of the coating during its application. For example, the melting point of the crystalline polyethlene terephtalate is 256°C. However, polyethylene isophthalate with a much lower degree of crystallinity has a melting point of 108°C, while the fully amorphous polyethylene *ortho*-phthalate melts at a temperature of only 63°C [47]. By the correct choice of diacids and diols and their combinations, polyesters with melting points varying over a rather wide range can be synthesized. Table 2.6, compiled on the basis of a series of published results by Korshak and coworkers, gives an indication of the melting points of different polyester resins based on aliphatic and aromatic diacids. It is quite remarkable that polyester resins obtained from diacids with an odd number of carbon atoms have lower melting points, not only than their neighbours with an even number of carbon atoms but also compared to all polyesters that are based on diacids with an even number of carbon atoms. The same tendency although not so emphasized can be noticed with the glycols. It can also be seen that aromatic diacids with a carboxyl group in the para position impart very high melting points to polyester resins. It is interesting that mixed polyesters based on aliphatic and aromatic diacids or different glycols have melting points which are not only in between the melting points of the pure polyesters but for certain combinations the melting points are even lower than the lowest melting point of the pure polyester, showing a minimum for a certain fixed ratio [55]. This phenomenon is often used to produce polyester resins with a rather high degree of crystallinity and a relatively low melting point.

Thermoplastic polyester resins for powder coatings differ in the raw materials used for their preparation and their ratio in the polyester formulation. Known processes include a mixture of polyesters made out of neopentyl glycol and terephthalic acid, copolyesters of cyclohexyl dimethanol, terephthalic isophthalic acid in combination with acetylcellulose, where high gloss of the coating is obtained, and copolyesters of ethylene glycol and terephthalic acid containing 8–20% of isophthalic acid, which exhibit good adhesion to metal substrate and good bending processability [56]. The main objection that could be addressed to these polyester resins is the rather inferior impact resistance and high fusion temperature in the neighbourhood of 250°C; a fluidized bed application would require a preheated surface up to 350°C [57]. An interesting approach involves copolyester derived from a dicarboxylic acid composition comprising 40–60 mol % of terephthalic acid and at least 70 mol % of 1,4-butane diol with a melting point in the range between 100 and 150°C determined by differential thermal analysis [58]. In the two-step production process, the blend of dimethyl terephthalate, dimethyl isophthalate and, if necessary, dimethyl adipate is transesterified in the first step with 1,4-butane diol using zinc acetate as the ester exchange catalyst at a temperature of 150–220°C. In the second step, with addition of antimony trioxide as the polymerization catalyst, the reaction mixture is heated to 270°C and subjected to glycol elimination under a vacuum of 0.5 mmHg until a specified molecular mass is obtained.

Thermoplastic polyester powder coatings are based in most cases on high molecular weight crystalline or semicrystalline linear polyester resins, alone or in combination with other binders, in order to achieve desirable properties of the fused films. Problems which have to be overcome when thermoplastic polyester resins are used as binders are poor adhesion to the substrate, crack formation upon rapid cooling or bending and poor lustre is spite of the fact that the polyester resins offer reasonable weathering resistance. Addition of cellulose esters such as cellulose acetate butyrate or propionate improves the gloss of the coating to a considerable extent.

Thermoplastic polyester powder coatings should have a certain degree of crystallization when cooled down to room temperature after the fusion of powder particles has been completed. The melting and crystallization of thermoplastic polyester resins is a reversible process. The crystalline regions act as crosslinking points eliminating the necessity of chemical crosslinking and the thermal curing step. The degree of crystallization is important to obtain surface hardness and a reasonably high service temperature of the coating. On the other hand, too high a crystallinity reduces adhesion to the substrate due to an increase in the cohesion energy during crystallization.

Good adhesion to the substrate can be obtained when the diacids used to synthesize the polyester resin are a combination of isophthalic acid and terephthalic acid in a certain optimal ratio that adjusts the crystallinity of the system. It is claimed that polyethylene isophthalate copolyesters containing 8–20 mol % of an isophthalic acid component with intrinsic viscosity of 0.7–1.0 give coatings with good adhesion and excellent surface hardness and corrosion and weathering resistance [56]. However, it seems that although this recipe can reduce some disadvantages, it still cannot provide good overall properties, particularly impact resistance, thermal shock resistance and a sufficiently low fusion temperature. The improvement of adhesion and impact resistance by incorporation of isophthalic acid is limited by the very slow crystallization rate and low melting point of the polyester. The latter decreases by increasing the isophthalic content to values that are not acceptable for practical reasons. On the other hand, a large amount of terephthalic acid would increase the rate of crystallization enormously, lowering the impact and shock resistance of the formed film.

An interesting approach in solving this problem is described by Toray Industries, Inc. [58] where a ratio between the isophthalic and terephthalic acids is kept quite high (40:60) in order to improve adhesion and the impact resistance of the coating. The reduced rate of crystallization is compensated for by using nucleating agents. In this way enough crystallinity can be obtained during the film forming process after the fusion of the powder particles. Useful nucleating agents are said to be talc, aluminium stearate, barium stearate, kaolin or clay. A typical powder coating composition according to the examples of the patent consists of polyester resin, titanium dioxide, 0.1–1.5 (by weight based on the resin content) nucleating agent and flow additives.

Thermoplastic polyester powder coatings have found an application in the processing of food cans where they are used for covering the welding seams. The enormously high speeds at which cans are manufactured allow only a few seconds to melt down the powder. Thermosetting coatings cannot complete the curing process within such a short time. High molecular weight polyesters with a molecular weight of 25 000 and a fusion temperature of 175°C have been developed for this purpose [59]. Thermoplastic polyester powder coatings are characterized by very high decorative properties giving a porcelain-like finish. The surface of the coating has a high gloss and excellent flow. Adhesion to metal substrates, in contrast to other thermoplastic powder coatings, is excellent, so previously priming the substrate with an adhesion promoter is not necessary. The abrasion resistance, scratch resistance and hardness of the coatings are excellent, combined at the same time with good flexibility and impact resistance. They possess excellent electrical properties, considerable chemical resistance, long-term durability and an unlimited shelf-life [25]. They do not undergo chemical curing during film formation and therefore are very suitable for high-speed running lines. The melting temperature of the powder coating can be easily adjusted by the proper choice of polyester resin, which makes them suitable for a variety of stoving schedules.

Several weak points, however, are inherent in the thermoplastic powder coatings. Due to the high molecular mass, acceptance of the pigments and fillers and dispersion efficiency during processing are very low. The low pigmentation level has as a consequence low hiding power of the coating. From a practical point of view this is not always a problem, since the powder coatings are in principle applied in somewhat thicker layers, but it has of course a negative influence on the final price.

Low grinding efficiency during the fine grinding step after extrusion is another typical problem related to thermoplastic polyester powder coatings. The grinding of the coating is hampered by the high modulus of the polyester resins, thus decreasing the capacity of the plant. Often, to keep plant capacity at a certain level cryogenic grinding has to be performed; this prevents the powder particles sticking as a result of heat development during the grinding operation.

Polyester powder coating have the lowest solvent resistance among thermoplastic coatings. This is easy to understand if the structure of the binder is kept in mind. Fortunately, since these coatings are predominantly used for decorative purposes, this weakness is not very important from a practical point of view.

Relatively low hydrolytic stability of the ester group restricts the application of these coatings mainly to indoor use. In the new types of polyesters, glycols with bulky substituents in the α position with respect to the hydroxyl group are used. In these cases the ester groups formed during the process of polyesterification are considerably sterically hindered. Although the hydrolytic stability of these polyesters is considerably improved, outdoor use should be recommended only after careful weathering tests.

Because of their excellent aesthetic properties, thermoplastic polyester powder coatings are used for metal furniture, guardrails, inside architectural elements and panels, shelves, aluminium extrusions, tools and machine components. Good dielectric properties make them suitable for applications including transformers, housing for electrical equipment and instruments. Although the weathering resistance of polyester resins is satisfactory, thermoplastic powder coatings still cannot compete with the thermosetting counterparts with respect to exterior durability. They are used mainly for inside applications or on articles which are not permanently exposed to weathering conditions.

REFERENCES

1. Air Products and Chemicals, Inc., Eur. Pat. 0 264 861, 1987.
2. Matlack, J. D., and Metzger, A. P., *Polymer Letters*, **4**, 875, 1966.
3. Smith, O. W., and Koleske, J. W., *Journal of Paint Technology*, **45** (582), 60, 1973.
4. Christensen, J. K., *Polyester Powder Coatings*, Paper FC72-941, SME Powder Coating Conference, Cincinnati, Ohio, March 1972.
5. Hagan, J. W., *Paint and Varnish Production*, **4**, 45, 1972.
6. Miller, P. E., and Taft, D. D., *Powder Coating*, Society of Manufacturing Engineers, Dearborn, Michigan, 1974, p. 165.
7. Pennwalt Corp., US Pat. 3 193 539, 1965.

8. Pennwalt Corp., US Pat. 3 245 971, 1966.
9. Kureha Chem. Comp., Fr. Pat. 1 419 741, 1965.
10. Daikin Kogyo Co. Ltd, Br. Pat. 1 178 227, 1967.
11. E.I.du Pont de Nemours & Co., Inc., Ger. Pat. 1 806 426, 1969.
12. Kali-Chemie AG, Br. Pat. 1 057 088, 1967.
13. Miyazawa, T., and Ideguchi, Y., *Journal of Polymer Science*, **B3**, 541, 1965.
14. Munekata, S., *Progress in Organic Coatings*, **16**, 113, 1988.
15. Sheppard, W. A., *Organic Fluorine Chemistry*, Benjamin, New York, 1969, p. 1.
16. 'Sigma PVDF-powder coating', Brochure of Sigma Coatings BV, July 1987.
17. Sietses, W., *Aluminium*, No. 3, p. 16, May 1988.
18. Naden, V., *Industrial Finishing*, **3**, 20, 1988.
19. Kureha Kagaku Kabushiki Kaisha, US Pat. 3, 824 115, 1974.
20. Moorman, R., *Industrial Finishing*, **3**, 20, 1988.
21. Crowley, J. D., Teague, G. S., Curtis, L. G., Foulk, R. G., and Ball, F. M., *Journal of Paint Technology*, **44** (571), 56, 1972.
22. Anon., *Finishing Industries*, June 1977, p. 20.
23. Adapted from *Encyclopedia of Polymer Science and Technology*, Vol. 6, Ed. N.M. Bikales, John Wiley & Sons, New York, 1967, p. 280.
24. Ball, F. M., and Curtis, G. L., *Journal of Powder Coatings*, **2** (1), 1, 1979.
25. Paul, S., in *Surface Coatings—Science and Technology*, John Wiley & Sons, Chichester, 1985, p. 671.
26. Dainippon Ink Chem., Jap. Pat. 57 202 335, 1981.
27. Dainippon Ink Chem., Jap. Pat. 58 049 736, 1981.
28. Sumitomo Chemical, Can. Pat. 1 142 675, 1980.
29. Hoechst AG, Eur. Pat. 0 057 823, 1981.
30. Hoechst AG, Eur. Pat. 0 100 992, 1982.
31. Connely, G., and Gaillard, G., *Proceedings of UK Corrosion 87*, Pt 26, no. 37, October 1987, p. 377.
32. Beatrice Foods Company, US Pat. 3 860 557, 1975.
33. Jezl, J. L., in *Encyclopedia of Polymer Science and Technology*, Vol. 11, Ed. N.M. Bikales, John Wiley & Sons, 1969, p. 611.
34. Du Pont, US Pat. 2 071 250, 1936 (patent appl. from 1931).
35. Du Pont, US Pat. 2 071 253, 1937.
36. Du Pont, US Pat. 2 130 523, 1938.
37. Du Pont, US Pat 2 130 948, 1938.
38. I.G. Farbenindustrie, Ger. Pat. 748 253, 1943.
39. Floyd, D. E., in *Polyamide Resins*, Reinhold Publishing Corp., New York, 1966, p. 2.
40. Todareva, L. G., Mikhailov, N. V., Potemkin, Z. I., and Kovaleva, M. V., *Visokomolekul. Soedin.*, **2**, 1728, 1960.
41. Blackmore, C. E., *Product Finishing*, August 1988, p. 37.
42. Westmore, E. R., *Finishing Industries*, February 1979, p. 53
43. Anon., *Finishing*, **13** (6), 39, 1989.
44. The Polymer Corporation, US Pat. 4 248 977, 1981.
45. Schulz, G. V., *Z. Phys. Chem.*, **A 182**, 127, 1938.
46. Griehl, W., and Förster, F., *Faserforschung und Textiltechnik*, **7** (10), 463, 1956.
47. Mleziva, J., Hanzlik, V., Kincl, J., and Miklas, Z., *Poliesteri*, Tehnika, Sofia, 1969, p. 90.
48. Korshak, V. V., Vinogradova, S. V., and Vasova, E. S., *Dokl. Akad. Nauk. SSSR*, **94**, 61, 1954.
49. Koshak, V. V., and Vinogradova, S. V., *Dokl. Akad. Nauk. SSSR*, **89**, 1017, 1953.
50. Korshak, V. V., Vinogradova, S. V., and Frunze, T. M., *Zurnal Obsh. Him.*, **27**, 1600, 1957.

51. Korshak, V. V., and Vinogradova, S. V., *Zurnal Obsh. Him.*, **26**, 539, 544, 1956.
52. Korshak, V. V., and Vinogradova, S. V., *Izv. Akad. Nauk. SSSR, Otd. Him. Nauk.*, **1953**, 121; **1954**, 1097.
53. Korshak, V. V., *Izv. Akad. Nauk SSSR, Otd. Him. Nauk.*, **1950**, 51.
54. Korshak, V. V., Vinogradova, S. V., and Beljakov, V. M., *Izv. Akad. Nauk. SSSR, Otd. Him. Nauk.*, **1957**, 730, 737.
55. Korshak, V. V., and Vinogradova, S. V., *Heterocepnie Polyerfiri, Izv. Akad. Nauk. SSSR*, Moskva, 1958, p. 275.
56. Jap. Pat. 097 023, 1974, cited in Eur. Pat. 0 213 887, 1986.
57. Watase, H., Jap. Pat. 59 041 367, 1982.
58. Toray Industries, Inc., Eur. Pat. 0 213 887, 1986.
59. Schmitthenner, M., *Proceedings of 14th International Conference on Organic Coatings Science and Technology*, Athens, 1988, p. 293.

3

Thermosetting Powder Coatings

3.1 Curing Reactions used in Powder Coatings 44
    3.1.1 Acid/Epoxy Curing Reaction 45
    3.1.2 Acid Anhydride/Epoxy Curing Reaction. 48
    3.1.3 Epoxy/Amino Curing Reactions 51
    3.1.4 Polyphenols/Epoxy Curing Reaction. 53
    3.1.5 Polyetherification 54
    3.1.6 Isocyanate/Hydroxyl Curing Reaction 56
    3.1.7 Curing with Amino Resins 68
References 78
3.2 Monitoring the Curing Process 82
    3.2.1 Differential Scanning Calorimetry (d.s.c.) 83
    3.2.2 Thermogravimetry Analysis (t.g.a.) 92
    3.2.3 Thermal Mechanical Analysis (t.m.a.) 97
References 106
3.3 Crosslinkers for Powder Coatings 107
    3.3.1 Crosslinkers of the Epoxy Type 108
        3.3.1.1 Triglycidyl Isocyanurate (TGIC) 108
    3.3.2 Polyisocyanates 110
        3.3.2.1 Caprolactam Blocked IPDI Derivatives 111
        3.3.2.2 Uretidione of IPDI. 113
        3.3.2.3 TDI Derivatives 114
        3.3.2.4 TMXDI and other Polyisocyanates 115
    3.3.3 Polyamines 117
        3.3.3.1 Dicyandiamide and its Derivatives 117
        3.3.3.2 Modified Aromatic and Aliphatic Polyamines 118
    3.3.4 Polyphenols 120
    3.3.5 Acid Anhydrides. 123
    3.3.6 Amino Resins 124
References 129
3.4 Industrial Thermosetting Powder Coatings 131
    3.4.1 Epoxy Powder Coatings 131
    3.4.2 Polyester Powder Coatings 144
        3.4.2.1 Interior Polyester Powder Coatings 152
        3.4.2.2 Exterior Polyester Powder Coatings 154

THERMOSETTING POWDER COATINGS

    3.4.3   Acrylic Powder Coatings. 162
    3.4.4   Unsaturated Polyester Powder Coatings 167
References. 170

Thermoplastic powder coatings do not undergo curing during stoving of the coating. The film formation is a result of melting and sintering of the powder particles using surface tension as the main driving force for this process. However, for good mechanical properties, the polymer must have a high molecular weight. Since there is no additional increase in the molecular weight of the thermoplastic powder coating, the initial molecular weight of the binder must be high enough to ensure good flexibility, hardness, impact and scratch resistance of the formed film.

In most cases the molecular weight of the polymers exhibiting good mechanical properties is in the range between 20 000 and 200 000 [1]. This is the range of molecular weights of the thermoplasts, which are in use for production of powder coatings. Certain exceptions are the linear polyesters which in some cases possess good impact resistance and tensile strength at molecular weights higher than 10 000 [2]. The high molecular weight of a polymer binder worsens the processing characteristic of the coating; namely high molecular weight polymers have very poor wetting properties. Therefore the amount of pigment and filler that can be successfully incorporated in the coating based on a thermoplastic binder in principle is much lower than in the case of thermosets. This has an influence on the covering power of the coating, on the aesthetic value (especially in the case of high gloss coatings) and also on the price of the coating formulation.

In general, powder coatings based on thermoplastic binders have very poor adhesion on metal surfaces. Only coatings which use thermoplastic high molecular polyester resins as binders can be directly applied onto a metal surface without the use of a primer. Using a primer as an adhesion promoter adds to the expense in the coating process and is usually avoided unless special corrosion protection is requested.

The problems inherent to the thermoplasts are successfully overcome by the thermosetting polymers. During stoving of the powder coatings based on thermosetting polymers a crosslinking reaction takes place. Therefore polymers with much lower molecular weights and consequently lower melt viscosities can be used. As a result larger amounts of pigments and fillers can be successfully dispersed and incorporated in the coating. The pigment/resin ratio is limited by the flow properties of the coating during the film formation rather than by the efficiency of the dispersion process.

The crosslinked nature of the cured film makes these coatings exceptionally resistant to solvents. This also improves chemical resistance in general, since the crosslinked film is not very permissive for the attacking agent.

Finally, there are many possibilities to introduce polar groups in the resin backbone, or polar groups are formed due to the curing reaction (like hydroxyl groups in the curing of carboxyl polyesters and epoxy curing agents). These considerably improve the adhesion properties. In general thermosetting powder coatings do not require primers as intercoats.

Next to the binder thermosetting powders contain crosslinkers involving different curing chemistries in the coating composition. This makes the process of film formation more complex. A broader knowledge and more sophisticated experimental techniques are necessary to understand, follow and elucidate the phenomena of film formation. Therefore this chapter includes sections describing the crosslinking chemistry, the monitoring of the curing process and the types of crosslinkers before referring to the thermosetting powder coatings themselves.

3.1 CURING REACTIONS USED IN POWDER COATINGS

The crosslinking of powder coatings enables a three-dimensional network to be formed. In this way the low molecular weight binders are transformed during film forming into crosslinked polymers exhibiting desirable physicochemical and mechanical properties.

Unlike many solvent based coatings, which are soft and tacky after evaporation of the solvent and which are converted into harder non-tacky materials upon subsequent crosslinking, powder coatings after fusion into continuous film are hard and non-tacky materials even without curing. However, the flexibility, impact resistance and solvent resistance of the uncured thermosetting powder coatings are far beyond the standard levels typical for industrial coatings. Here the main purpose of the crosslinking is to increase the toughness of the coatings and to make the cured film resistant to the attack of common industrial solvents.

Network formation depends on the average functionality of the system. When the average functionality of the system binder/crosslinker is two or only slightly over two, full conversion of the functional groups will lead to a high molecular linear or branched polymer but not to a network formation. Films obtained by such a system that depend on the molecular weight after curing may have good mechanical properties, but certainly will not exhibit good solvent resistance. On the other hand, too high a functionality and excessive crosslinking can lead to the formation of a brittle film exhibiting unsatisfactory postforming properties and low impact resistance.

Curing of powder coatings is a rather complex problem from the theoretical point of view. For complete understanding of the theories describing the curing process mathematical knowledge, which is not very typical for a resin chemist, is often required. The easiest treatment of the problem to understand is that given by Carothers who was among the first scientists to deal with so-called non-linear polymerization [3]. Although his theory suffers from serious errors, especially in choosing the criteria for infinite network formation, it is widely used as a basis for calculating formulations of many different types of polycondensation resins.

However, the theory of Carothers is not suitable for a satisfactory treatment of the crosslinking phenomena. Flory [4] and Stockmayer [5] have derived the basic relations between the extent of the reaction and the resulting structure of the polymers undergoing crosslinking or the basic equations describing the non-linear polymerization processes. In the 1960s Gordon developed a new treatment of the phenomenon using the theory of stochastic branching processes with cascade substitution to derive equations describing the process of crosslinking [6, 7]. In the 1970s Macosko and Miller, starting with elementary probability and utilizing the recursive nature of network polymers, developed simpler equations to predict the number average and mass average molecular weight of the polymers in the pre- and postgel stadium [8, 9]. The theory of Gordon was recently used to predict the curing behaviour of powder coatings [10].

The theoretical treatment of network formation is beyond the scope of this book. The interested reader should consult the enormous literature dealing with crosslinking phenomena and network formation which is partly cited in the references at the end of this section [11–30].

Various types of curing reactions are employed in thermosetting powder coatings. The crosslinking reactions typical of thermosetting liquid coatings in principle can be used for hardening powder coatings. However, certain differences emanating from the nature of powder coatings exclude many of the crosslinking reactions typically used in conventional liquid systems. This section will describe the mechanism of crosslinking reactions used in commercial powder coatings today.

3.1.1 ACID/EPOXY CURING REACTION

The reaction between epoxy and acid functional groups is the most important curing reaction used in practice for crosslinking thermosetting powder coatings. The so-called hybrid powder coatings in which carboxyl functional polyesters are cured with bisphenol A epoxy resins and polyester powder coatings where triglycidyl isocyanurate is used as a curing agent are the best representatives of this group of products.

The reaction mechanism between the acid and epoxy groups has been a matter for wide-ranging investigation [31–47]. Schechter and coworkers [31–33] anticipate four possible reactions in the curing of epoxy resins with polybasic acids:

1. Ring opening addition of epoxy group to the acid group leading to the formation of the corresponding hydroxyl ester:

$$RCOOH + CH_2\underset{O}{-}CHR_1 \longrightarrow RCOOCH_2\underset{OH}{CHR_1} \qquad (3.1)$$

2. Esterification between the acid and the hydroxyl group formed in the previous

reaction:

$$\text{RCOOH} + \text{RCOOCH}_2\underset{\underset{\text{OH}}{|}}{\text{CHR}}_1 \longrightarrow \text{RCOOCH}_2\underset{\underset{\text{OCOR}}{|}}{\text{CHR}}_1 + \text{H}_2\text{O} \quad (3.2)$$

3. Ring opening addition of the epoxy group to the hydroxyl group formed in reaction (3.1); resulting in the corresponding ether alcohol:

$$\text{RCOOCH}_2\underset{\underset{\text{OH}}{|}}{\text{CHR}}_1 + \text{CH}_2\overset{\diagdown\!\diagup}{\underset{\text{O}}{-}}\text{CHR}_1 \longrightarrow \text{RCOOCH}_2\underset{\underset{\text{OCH}_2\text{CHR}_1-\text{OH}}{|}}{\text{CHR}}_1 \quad (3.3)$$

4. Hydrolysis of the epoxy ring by water:

$$\text{H}_2\text{O} + \text{CH}_2\overset{\diagdown\!\diagup}{\underset{\text{O}}{-}}\text{CHR}_1 \longrightarrow \text{HO}-\text{CH}_2-\underset{\underset{\text{OH}}{|}}{\text{CHR}}_1 \quad (3.4)$$

The extent of the above reactions depends on the catalyst used in the system. In the absence of catalyst the first three reactions proceed at the same rate to a comparable extent. For a non-catalysed curing process, Shechter, Wynstra and Kurkjy [32] report that when the curing takes place in a stoichiometric ratio between the reactive groups, the ratio between the reaction products according to reactions (3.1), (3.2) and (3.3) is between 2:1:1 and 1:2:2 respectively.

Reaction (3.3) occurs only when the curing is catalysed by strong acid catalysts. This is almost never the case in powder coating formulations where the catalyst employed is of a basic nature. Therefore, the etherification reaction does not play an important role in practice.

Powder coatings where the carboxyl/epoxy curing reaction is used for crosslinking in most cases employ basic catalysts. In this case, when the acid/epoxy stoichiometry is 1:1, the reaction would selectively give hydroxyl esters (reaction 3.1), excluding the formation of polyethers (reaction 3.3) [48]. However, in an excess of epoxy groups and after the consumption of acid groups, the etherification reaction proceeds according to the scheme 3.3 [32]. Two fundamentally different reaction mechanisms are anticipated for the base catalysed acid/epoxy reaction [45, 48]. When tertiary amines are used as catalysts, it is believed that decomposition of the acid–catalyst salt generates active carboxylate anion species according to the following reaction:

$$\text{RCOOH} + \text{B} \rightleftharpoons [\text{RCOOH}/\text{B salt}] \overset{K_{eq}}{\rightleftharpoons} \text{RCOO}^- + \text{BH}^+ \quad (3.5)$$

In the next step of the reaction the carboxylate anion attacks the epoxy ring, resulting in the formation of an alkoxide anion and the corresponding ester:

$$\text{RCOO}^- + \text{CH}_2\overset{\diagdown\!\diagup}{\underset{\text{O}}{-}}\text{CH}-\text{R}_1 \overset{k_6}{\longrightarrow} \text{RCOOCH}_2-\text{CH}-\text{R}_1 \atop \underset{\text{O}^-}{|} \quad (3.6)$$

THERMOSETTING POWDER COATINGS

The alkoxide anion formed in reaction (3.6) reacts to generate new carboxylate anions or free base catalyst according to the following reaction schemes:

$$\text{RCOOCH}_2-\underset{\underset{O^-}{|}}{\text{CH}}-R_1 + \text{RCOOH} \xrightarrow{k_7} \text{RCOOCH}_2-\underset{\underset{OH}{|}}{\text{CH}}-R_1 + \text{RCOO}^- \quad (3.7)$$

$$\text{RCOOCH}_2-\underset{\underset{O^-}{|}}{\text{CH}}-R_1 + BH^+ \xrightarrow{k_8} \text{RCOOCH}_2-\underset{\underset{OH}{|}}{\text{CH}}-R_1 + B \quad (3.8)$$

It is supposed that both reactions (3.7) and (3.8) are much faster than reaction (3.6) which will be responsible for the overall kinetics of the system. Accordingly, the rate expression for this mechanism will be

$$S_2 = k_2 [\text{RCOO}^-][\text{epoxy}] = k_2 K_{eq} [\text{RCOOH}][\text{epoxy}][B]/[BH^+]$$

The reaction rate is obviously controlled by the concentration of the catalyst, as well as by the rate of decomposition of the acid–catalyst salt or the value of the equilibrium constant K_{eq}. In other words, the reaction rate will also depend on the type of base used as a catalyst.

A different reaction mechanism is supposed to be valid for the quaternary ammonium salts which are also widely used as catalysts. The first reaction to be considered is the dissociation of the salt:

$$AB \xrightleftharpoons{K_{eq}} A^+ + B^- \quad (3.9)$$

The next steps involve a direct attack of the base on the epoxy, generating alkoxide anion:

$$B^- + \text{CH}_2-\underset{\underset{O}{\diagdown\diagup}}{\text{CH}}-R_1 \xrightarrow{k_{10}} B\text{CH}_2-\underset{\underset{O^-}{|}}{\text{CH}}-R_1 \quad (3.10)$$

carboxylate anion:

$$B\text{CH}_2-\underset{\underset{O^-}{|}}{\text{CH}}-R_1 + \text{RCOOH} \xrightarrow{k_{11}} B\text{CH}_2-\underset{\underset{OH}{|}}{\text{CH}}-R_1 + \text{RCOO}^- \quad (3.11)$$

and again alkoxide and carboxylate anions in the next sequences:

$$\text{RCOO}^- + \text{CH}_2-\underset{\underset{O}{\diagdown\diagup}}{\text{CH}}-R_1 \xrightarrow{k_{12}} \text{RCOOCH}_2-\underset{\underset{O^-}{|}}{\text{CH}}-R_1 \quad (3.12)$$

$$\text{RCOOCH}_2-\underset{\underset{O^-}{|}}{\text{CH}}-R_1 + \text{RCOOH} \xrightarrow{k_{13}} \text{RCOOCH}_2-\underset{\underset{OH}{|}}{\text{CH}}-R_1 + \text{RCOO}^- \quad (3.13)$$

Reactions (3.11) and (3.13) are considered as fast, having no influence on the overall rate of curing. The dissociation step also plays no role because the result of dissociation, the anion B (reaction 3.9), reacts irreversibly with the epoxy, thus shifting the equilibrium completely to the right. The important reactions for the kinetics of curing are (3.10) and (3.12), involving the following expressions for the reaction rate:

$$S_{10} = k_{10}[B^-][\text{epoxy}]$$
$$S_{12} = k_{12}[RCOO^-][\text{epoxy}]$$

Regardless of the values of the constants k_{10} and k_{12}, the concentration of the carboxylate anion will be constant and equal to the original concentration of B, because it regenerates in the sequences of the reactions presented above. Therefore, the reaction rate will depend in practice on the concentration of the catalyst and the epoxy groups, but not on the type of catalyst. The type of catalyst will determine the ratio between the rate constants k_{10} and k_{12}. If $k_{10} < k_{12}$, a delay should be expected before the reaction runs at its proper rate. The experimental results obtained by Henig, Jath and Mohler [49] and van der Linde and Belder [45] support the above described mechanisms.

3.1.2 ACID ANHYDRIDE/EPOXY CURING REACTION

Anhydrides of polybasic acids have lost their importance as curing agents for epoxy powder coatings mainly for toxicological reasons. However, this system offers several advantages compared to curing with polybasic acid, such as higher crosslinking density and relative insensitivity with respect to the deviations of the anhydride/epoxy ratio from the stoichiometrical ratio.

Different articles deal with the mechanism and kinetics of the polyaddition reaction of epoxy resins with acid anhydrides. In the case of an uncatalysed system, the following reactions of curing have been proposed by several authors [50–55]:

$$R_3\text{---OH} + O{=}C\text{---}R_2\text{---}C{=}O \longrightarrow R_3OCO\text{---}R_2\text{---}COOH \quad (3.14)$$
$$\diagdown O \diagup$$

$$R_3OCO\text{---}R_2\text{---}COOH + CH_2\text{---}CH\text{---}R_1 \longrightarrow R_3OCOR_2COOCH_2\underset{\underset{OH}{|}}{C}HR_1 \quad (3.15)$$

$$R_3OH + CH_2\text{---}CHR_1 \longrightarrow R_3OCH_2\underset{\underset{OH}{|}}{C}HR_1 \quad (3.16)$$

Reaction (3.14) shows that for initiation of the curing a hydroxyl containing compound must be added to the system. In the case of higher homologues of bisphenol A epoxy resins, the hydroxyl groups belonging to the polymer backbone can initiate the reaction themselves. The reaction can be initiated by water as well as by carboxyl containing compounds, since their reaction with the epoxy groups leads to the formation of the hydroxyl group. The other two reactions, (3.15) and (3.16), are in fact the same, as in the case of curing of epoxides with polyacids (reactions 3.1 and 3.3).

A different curing mechanism has been proposed by Shechter and Wynstra [31] and Newey [56] in the case of base catalysed curing. They have found no evidence of the hydroxyl/epoxy reaction leading to the formation of ether linkages. However, Kaplan [57] in a study of the effect of water on the epoxy acid anhydride reactions concluded that dependent on the temperature, type of anhydride and epoxy, and the presence and type of solvent, an etherification reaction (reaction 3.3) proceeds to a certain extent.

The recent work of Fedtke and Domaratius [58] supports the findings of Fischer [59] that the etherification reaction does not occur essentially when the epoxy acid anhydride system is catalysed by tertiary amine. Studying the system of phenyl glycidyl ether, phthalic anhydride and N,N-dimethyl aniline as catalyst, they have proposed an initiation step that involves a reaction of the tertiary amine with the epoxy group, resulting in the formation of zwitterion containing quaternary nitrogen atom and an alkoxide anion. This reaction mechanism had been proposed somewhat earlier by Matejka et al. [60]:

$$R_1OCH_2-CH\underset{O}{\overset{}{-}}CH_2 + NR_3 \longrightarrow R_1OCH_2-CH(O^-)-CH_2NHR_3^+ \qquad (3.17)$$

This zwitterion can react either with the epoxide or with acid anhydride. Fedtke and Domaratius propose that the reaction with anhydride is much faster, thus practically excluding the formation of ether linkages due to the possible reaction of the zwitterion with epoxy groups:

$$R_1OCH_2-CH(O^-)-CH_2NHR_3^+ + R_2(C=O)_2O \longrightarrow R_1OCH_2-CH(OCO-R_2-COO^-)-CH_2NHR_3^+ \qquad (3.18)$$

They accept the suggestion of Tanaka and Kakiuchi [61] and Luston and Manasek [46] that the product of reaction (3.18) undergoes isomerization to an unsaturated compound:

$$R_1OCH_2-CH(OCO-R_2-COO^-)-CH_2NHR_3^+ \longrightarrow R_1OCH_2-C(OCO-R_2-COO^-NHR_3^+)=CH_2 \qquad (3.19)$$

The propagation steps are as follows:

$$R_1OCH_2-\underset{\underset{OCO-R_2-OCO^-NHR_3^+}{|}}{C}=CH_2 \quad + R_1OCH_2-CH-CH_2 \longrightarrow$$
$$\phantom{R_1OCH_2-\underset{\underset{OCO-R_2-OCO^-NHR_3^+}{|}}{C}=CH_2 \quad + R_1OCH_2-CH-CH_2} \diagdown O \diagup$$

$$R_1OCH_2-\underset{\underset{OCO-R_2-OCO-CH_2-\underset{\underset{O^-NHR_3^+}{|}}{CH}-CH_2OR_1}{|}}{C}=CH_2 \quad + R_2\underset{C=O}{\overset{C=O}{\diagup\diagdown}}O \longrightarrow$$

$$R_1OCH_2-\underset{\underset{OCO-R_2-OCO-CH_2-\underset{\underset{OCO-R_2-COO^-NHR_3^+}{|}}{CH}-CH_2OR_1}{|}}{C}=CH_2 \quad \text{etc}$$
$$\tag{3.20}$$

The termination step occurs by reversible decomposition of the active centres:

$$-O-CH_2-\underset{\underset{CH_2OR_1}{|}}{CH}-O^-NHR_3^+ \rightleftharpoons NR_3 + -O-CH_2-\underset{\underset{CH_2OR_1}{|}}{CH}-OH \tag{3.21}$$

$$-OCO-R_2-COO^-NHR_3^+ \rightleftharpoons NR_3 + -OCO-R_2-COOH \tag{3.22}$$

In the presence of proton donating compounds, the initiation step is somewhat different. The active centre of catalysis is formed by interaction of the proton donor with amine, which does not exclude the initiation according to reactions (3.17) to (3.19):

$$R_4H + NR_3 \longrightarrow R_4^-NHR_3^+ \tag{3.23}$$

$$R_4^-NHR_3^+ + CH_2-CH-CH_2OR_1 \longrightarrow R_4CH_2-\underset{\underset{CH_2OR_1}{|}}{CH}-O^-NHR_3^+$$
$$ \diagdown O \diagup \tag{3.24}$$

$$R_4^-NHR_3^+ + R_2\underset{C=O}{\overset{C=O}{\diagup\diagdown}}O \longrightarrow R_4-CO-R_2-COO^-NHR_3^+ \tag{3.25}$$

The propagation reaction remains the same, as in the case of base catalysed curing without the presence of a proton donating compound.

The ratio between the two initiation steps (3.17 to 3.19) and 3.23 to 3.25) depends on the acidity of the proton donor. The greater the interaction between the tertiary amine and the proton donating compound, the lower the initiation by the tertiary amine only.

The effect of the proton donating compounds should be considered, because in technical epoxy resins a proton donor is always present, either in the form of a secondary hydroxyl group belonging to the higher homologues of bisphenol A epoxy or in the form of a carboxyl group because of partial hydrolysis of the acid anhydride by humidity.

Fedtke and Domaratius have shown that the rate constant for the epoxy/acid anhydride reaction in the presence of benzoic acid as the proton donor is twice as high as in the presence of phenol for a system consisting of phenyl glycidyl ether, phthalic anhydride and N,N-dimethyl benzyl amine. The explanation is the higher acidity of benzoic acid compared to phenol. This is rather contradictory to the reactivity ratio suggested by Tanaka and Mika [48, p. 189], which decreases in the following order: phenol > acid > alcohol. On the other hand, acid anhydrides become less reactive in the order: maleic > phthalic > tetrahydrophthalic > hexahydrophthalic > methyltetrahydrophthalic [62]. The proposed mechanism involving isomerization of epoxy groups to a derivative of allyl alcohol is questionable, since according to some papers [63, 64] this has not been experimentally confirmed.

Other proposed mechanisms not involving the formation of allylic derivatives are those of Tanaka and Kakiuchi [54, 55], Fischer [65] and Feltzin [66]. The differences between these mechanisms are rather minor and can easily be incorporated into the mechanism proposed by Fedtke and Domaratius, avoiding reaction (3.19) which suggests the formation of an allyl derivative.

3.1.3 EPOXY/AMINO CURING REACTIONS

Epoxy compounds react readily with primary and secondary amines, giving secondary or tertiary amines as reaction products respectively:

$$R_1NH_2 + CH_2\overset{\diagdown}{\underset{O}{\diagup}}CHR \xrightarrow{k_1} R_1NH-CH_2-\underset{OH}{\overset{|}{CH}}-R \quad (3.26)$$

$$R_1R_2NH + CH_2\overset{\diagdown}{\underset{O}{\diagup}}CHR \xrightarrow{k_2} R_1R_2N-CH_2-\underset{OH}{\overset{|}{CH}}-R \quad (3.27)$$

It can be expected that the resulting tertiary amine can act as a catalyst promoting the reaction between the hydroxyl groups present in the epoxy resin or obtained as a result of the curing process, with the epoxy rings leading to the formation of ether linkages:

$$R_1R_2NCH_2-\underset{OH}{\overset{|}{CH}}-R + CH_2\overset{\diagdown}{\underset{O}{\diagup}}CHR \longrightarrow R_1R_2N-CH_2-\underset{O-CH_2-\underset{OH}{\overset{|}{CHR}}}{\overset{|}{CH}}-R \quad (3.28)$$

Curing between polyamines and polyepoxides is one of the most extensively studied reactions. In the very early work of Shechter, Wynstra and Kurkjy [32], Kakurai and Noguchi [67], Dannenberg [68] and the recent work of Gross and coworkers [69–71] it was confirmed that the rate of etherification is rather low at room temperature. However, at increased temperatures, as in the case of curing powder coatings and especially when the epoxy groups are in excess, the epoxy/hydroxy curing reaction should be taken into consideration. It has been shown that at higher temperatures, depending on the type of amine, even at a stoichiometric ratio between the amine and epoxy groups a considerable amount of the epoxy groups is used up in reaction with the hydroxyl groups available from the epoxy resin (in the case of higher homologues of bisphenol A epoxy resins) or is borne as the reaction product of curing [72]. The etherification reaction is also catalysed by tertiary amines, which can be used as accelerators for the curing process. On the other hand, the catalytic effect of the intramolecular tertiary amino group in the reaction products (β-amino alcohols) is less pronounced due to their rather high steric hindrance. It is suggested [69] that reactions with epoxies via additions of the hydroxyl groups at epoxy/amine reaction products only take place after completing the amino additions.

Gross, Kollek and Brockmann [69] have studied the curing of epoxy resins with amines at the model system of 1,2-epoxy-3-(4-methyl-phenoxy)-propane with n-propyl amine as the source of primary amino groups and diethyl triamine as the compound containing primary and secondary amino groups. They have concluded that there is a considerable difference between the reactivity of the primary and secondary amino groups. They have also found the same behaviour to be valid for the aromatic amines, with the remark that they react ten times more slowly than their aliphatic counterparts.

Dusek states that the ratio between the rate constants in reactions (3.26) and (3.27), k_2/k_1, varies between 0.5 and 0.15 depending on the structure of the amines. This value is even lower for aromatic amines than for aliphatic ones [73, 74].

Tanaka and Mika [48, p. 179] relate the difference in the reactivity between the primary and secondary amines to the steric factors, and support this finding by the fact that the reactivity of various amines towards phenyl glycidyl ether diminishes in the order: n-butyl amine > diethyl amine > di-n-propyl amine > di-n-butyl amine > diisoamyl amine. The same is valid for the aromatic amines, which have the following order of reactivity: methyl aniline > ethyl aniline > n-propyl aniline > n-butyl aniline.

Gross and coworkers add another element to the steric factors. According to them, the reactivity of the amines towards epoxy groups is also ruled by their nucleophilicity, since the increased steric hindrance of the secondary amino group also includes a decreased nucleophilic character.

The presence of alcohols or acids has a considerable accelerating effect on the curing reaction. The catalytic effect of the various alcohols is proportional to their acidity or their pK_a values. The added compounds have decreasing catalytic effect in the order: acids > phenols > water > alcohols > aromatic hydrocarbons > ethers. Schechter, Wynstra and Kurkjy [32] explain this phenomenon by the formation of a hydrogen bond in the transition state which accelerates the opening of the epoxy ring:

$$R_1R_2NH + CH_2\text{---}CHR + HO\text{---}X \longrightarrow R_1R_2NH\text{---}CH_2\text{---}CHR \longrightarrow$$
$$\underset{O}{\phantom{CH_2\text{---}CHR}} \qquad\qquad\qquad O\cdots HO\text{---}X$$

$$R_1R_2NH\text{---}\underset{\underset{OH^+OX^-}{|}}{CH_2}\text{---}CHR \longrightarrow R_1R_2N\underset{\underset{OH}{|}}{CH_2}\text{---}CHR + HO\text{---}X \qquad (3.29)$$

A similar mechanism is proposed by Smith [75].

Although the hydroxyl groups accelerate the curing reaction, the reactivity of the secondary amino group, which is the product of the reaction between the primary amino group and the epoxy ring (β-amino alcohol), is lower than the reactivity of the secondary amino groups belonging to amino hardeners containing no hydroxyl groups. Gross, Kollek and Brockmann [69] explain this by the formation of the intramolecular hydrogen bridge bond with the nitrogen atom of the secondary amino group. The nucleophilicity is thereby reduced, as is the affinity to epoxy groups. In the first approximation the following order of reactivity results:

$$-NH_2 > =NH > \underset{\underset{CH_2\text{---}CH\text{---}OH}{|}}{-NH^*} \gg \underset{\underset{CH_2\text{---}CH\text{---}OH^*}{|}}{-NH} > -OH$$

where (∗) indicates the reactive site.

It can be assumed that in the case of curing difunctional epoxies with primary diamines, the reaction products at the first stage of the reaction are primarily of a linear character. As the reaction proceeds, the linear high molecular material is crosslinked by the reaction of the formed secondary amino groups with the epoxide.

3.1.4 POLYPHENOLS/EPOXY CURING REACTION

The reaction between the epoxide and aliphatic hydroxyl groups does not proceed at a rate that is suitable for normal curing conditions used in practice with powder coatings, unless it is catalysed with strong acids such as BF_3. Because of the strong corrosive effect of the catalyst during the service life of the coating, these systems are not used in practice to formulate powder coatings.

However, epoxy groups readily react with phenolic hydroxyl groups when catalysed with base catalysts. This is a widely used reaction for preparing higher homologues of bisphenol A epoxy resins, starting from low molecular weight bisphenol A/epichlorohydrin precondensates.

Curing of the epoxy powder coatings with phenol or bisphenol A type novolacs is catalysed with tertiary amines or quaternary ammonium salts. The curing mechanism is almost the same as in the case of the carboxy/epoxy curing reactions. The reaction rate is proportional to the concentration of the catalyst and to the concentration of the epoxy and phenolic groups, involving the initiation step (reaction 3.30), the propagation step (reaction 3.31) and termination (reaction 3.32):

$$R_1OH + B \rightleftharpoons R_1O^- + BH^+ \qquad (3.30)$$

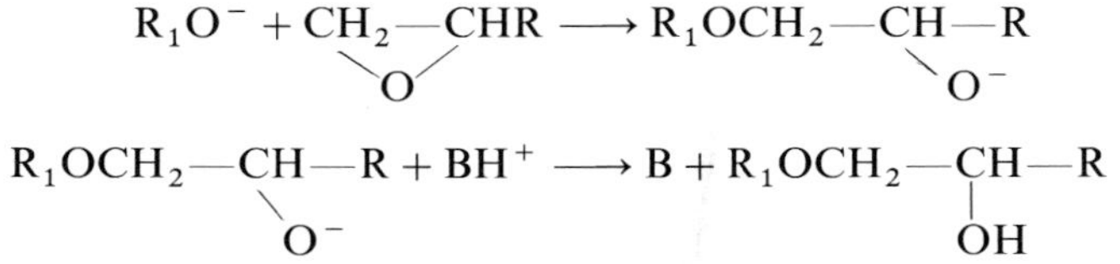

$$R_1O^- + CH_2\underset{\diagdown O \diagup}{-}CHR \longrightarrow R_1OCH_2-\underset{\diagdown O^-}{CH-R} \qquad (3.31)$$

$$R_1OCH_2-\underset{\diagdown O^-}{CH-R} + BH^+ \longrightarrow B + R_1OCH_2-\underset{OH}{CH-R} \qquad (3.32)$$

The base catalysed curing reaction between resins containing hydroxyl groups of phenolic type and epoxies proceeds at a proper rate in the temperature range between 140 and 200°C, which is a typical temperature range for curing powder coatings.

3.1.5 POLYETHERIFICATION

Epoxy resins can undergo catalysed polymerization with a variety of Lewis bases and acids as well as numerous salts and complex catalysts. Despite the fact that the Lewis bases and acids initiate anionic and cationic polymerizations respectively, the resulting network structure is of the same polyether nature. The cationic polymerization of epoxy resins is not used for curing powder coatings and will not be discussed in this book.

The anionic step polyetherification is a curing reaction used in epoxy powder coatings for obtaining a network structure by base catalysed self-crosslinking of epoxy resins. The catalysts used in practice are different imidazoles and tertiary amines. Polyetherification can be regarded as a stepwise process starting with an initiation (reaction 3.33), propagation (reaction 3.34) and termination (reaction 3.35):

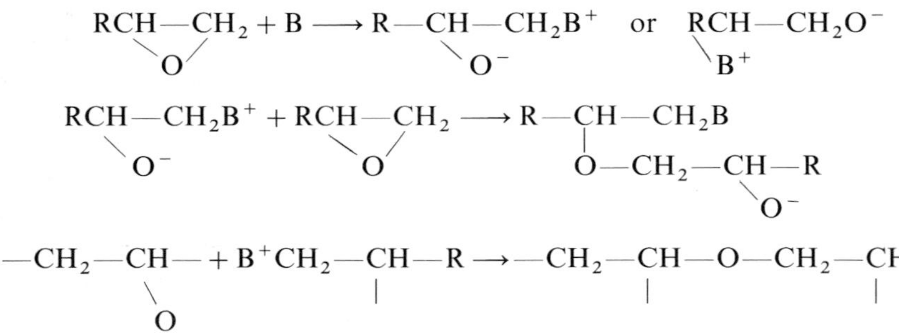

$$RCH\underset{\diagdown O \diagup}{-}CH_2 + B \longrightarrow R-CH\underset{\diagdown O^-}{-}CH_2B^+ \quad \text{or} \quad RCH\underset{\diagdown B^+}{-}CH_2O^- \qquad (3.33)$$

$$RCH\underset{\diagdown O^-}{-}CH_2B^+ + RCH\underset{\diagdown O \diagup}{-}CH_2 \longrightarrow R-\underset{O-CH_2-\underset{\diagdown O^-}{CH-R}}{\overset{|}{CH}}-CH_2B \qquad (3.34)$$

$$-CH_2-\underset{\diagdown O}{CH}- + B^+CH_2-\underset{|}{CH}-R \longrightarrow -CH_2-\underset{|}{CH}-O-CH_2-\underset{|}{CHR} + B \qquad (3.35)$$

Many excellent articles and monographs dealing with the anionic polymerization of epoxy resins explain the mechanism of the reaction [31, 76–90]. The above reaction scheme can be considered as a simplification of the actual

THERMOSETTING POWDER COATINGS

mechanism, which is rather complex and, despite all of the work done in this field, not yet completely understood. Next to the previously proposed scheme, it is very probable that the following reaction proposed by Berger and Lohse leading to the formation of unsaturated alcohols takes place [86]:

$$-CH-CH_2- + B^+CH_2-CH-R \longrightarrow -CH-CH_2 + B + R-C=CH_2$$
$$\ |\ |\ |$$
$$O^-OHOHOH$$
(3.36)

This is very similar to the termination mechanism proposed much earlier by Narracott [91] as an explanation for the accumulation of unsaturated compounds during polyetherification catalysed by tertiary amines:

$$R_3N(CH_2CHRO)_n-CH_2CHRO^- \longrightarrow$$
$$R_3N + CH_2=CRO(CH_2CHRO)_{n-1}CH_2CHROH \quad (3.37)$$

The statistical and kinetic theoretical approaches to polyetherification [88–90] which have been recently developed do not consider the complications that accompany the process, such as chain transfer, reinitiation and other side reactions. Dusek [73], comparing both approaches, has found that the kinetic treatment is correct within the assumption of the proposed mechanism, while the statistical method leads to serious deviations from experimental results. However, the statistical method does approach the results obtained by the kinetic treatment when larger fragments are taken as the building blocks of the polyether structure. Reasonably good results obtained by the kinetic method confirm the overall validity of the proposed model, within the scope of the assumptions involved.

Bressers and Goumans [92] have studied the curing mechanism of epoxides with imidazoles, using d.s.c. l.c. and n.m.r. techniques in a model system of phenyl glycidyl ether/1-benzyl-2-methyl imidazole. They give analytical support to the following curing mechanism:

$$C_6H_5-O-CH_2-CH-CH_2 \ + \ \underset{Me}{N\!\!\diagup\!\!\diagdown\!\!N}-CH_2-C_6H_5$$
(3.38)

(a) $C_6H_5-O-CH_2-CH-CH_2-N^+\!\!\diagup\!\!\diagdown\!N-CH_2-C_6H_5$ with O^- → Homopolymer

(b) $\underset{Me}{N\!\!\diagup\!\!\diagdown\!\!N^+}$ bearing $CH_2-C_6H_5$ and $CH_2-CH(-O^-)-CH_2-O-C_6H_5$ → Homopolymer

However, they could not give a full explanation of the multipeaked thermograms, which were also observed by other investigators [83], since the exact elucidation of the initiating species for polyetherification has been hampered by their instability.

3.1.6 ISOCYANATE/HYDROXYL CURING REACTION

The main feature of the isocyanates is their reactivity towards nucleophiles at room and elevated temperatures. The following reactions leading to the formation of urethanes (3.39), substituted ureas (3.40 and 3.41), amides (3.42), allophanates (3.43), biurets (3.44), acylureas (3.45), urediones (3.46), isocyanurates (3.47), thiourethanes (3.48) and oxazolidones (3.49) are typical for isocyanates:

$$RNCO + R_1OH \longrightarrow RNH-\underset{\underset{O}{\|}}{C}-O-R_1 \tag{3.39}$$

$$RNCO + R_1NH_2 \longrightarrow RNH-\underset{\underset{O}{\|}}{C}-NHR_1 \tag{3.40}$$

$$RNCO + H_2O \longrightarrow RNH-COOH \longrightarrow RNH_2 + CO_2$$
$$\text{(carbamic acid)}$$
$$\xrightarrow{RNCO} R-NH-\underset{\underset{O}{\|}}{C}-NH-R \tag{3.41}$$

$$RNCO + R_1COOH \longrightarrow RNH-\underset{\underset{O}{\|}}{C}-R_1 \tag{3.42}$$

$$RNCO + -NH-\underset{\underset{O}{\|}}{C}-O- \longrightarrow -\underset{\underset{\underset{NHR}{|}}{\underset{C=O}{|}}}{N}-\underset{\underset{O}{\|}}{C}-O- \tag{3.43}$$
$$\text{(urethane)}$$

$$RNCO + -NH-\underset{\underset{O}{\|}}{C}-NH- \longrightarrow -\underset{\underset{\underset{NH-R}{|}}{\underset{C=O}{|}}}{N}-\underset{\underset{O}{\|}}{C}-NH- \tag{3.44}$$
$$\text{(urea)}$$

THERMOSETTING POWDER COATINGS

$$\text{RNCO} + -\text{NH}-\underset{\text{(amide)}}{\overset{\overset{\displaystyle O}{\|}}{\text{C}}}- \longrightarrow \begin{array}{c} -\text{N}-\overset{\overset{\displaystyle O}{\|}}{\text{C}}- \\ | \\ \text{C}=\text{O} \\ | \\ \text{NH}-\text{R} \end{array} \quad (3.45)$$

$$2\,\text{RNCO} \longrightarrow \text{R}-\text{N}\overset{\overset{\displaystyle O}{\underset{\displaystyle \|}{\text{C}}}}{\underset{\underset{\displaystyle \|}{\underset{\displaystyle O}{\text{C}}}}{}}\text{N}-\text{R} \quad (3.46)$$

$$3\,\text{RNCO} \longrightarrow \begin{array}{c} \text{R} \\ | \\ \text{N} \\ O=\text{C}\diagup\ \ \diagdown\text{C}=O \\ \text{R}-\text{N}\ \ \ \ \text{N}-\text{R} \\ \diagdown\text{C}\diagup \\ \| \\ O \end{array} \quad (3.47)$$

$$\text{RNCO} + \text{R}_1\text{SH} \longrightarrow \text{RNH}-\underset{\underset{\displaystyle O}{\|}}{\text{C}}-\text{S}-\text{R}_1 \quad (3.48)$$

$$\text{RNCO} + \text{R}_1\text{CH}\overset{\displaystyle -}{\underset{\displaystyle O}{\diagdown\diagup}}\text{CH}_2 \longrightarrow \text{R}-\text{N}\overset{\text{CH}_2-\text{CH}-\text{R}_1}{\underset{\underset{\displaystyle \|}{\underset{\displaystyle O}{\text{C}}}}{\diagdown\ \ \ \ \ \ \diagup O}} \quad (3.49)$$

These reactions illustrate the richness of the isocyanate chemistry and indicate the large potential for using isocyanates as curing agents for coatings. Various types of coating systems utilize many of the reactions mentioned above for obtaining a crosslinked network at room and elevated temperatures. In the case of powder coatings, only reaction (3.39) between isocyanate and hydroxyl groups is used in practice in the so-called polyurethane powder coatings.

Reactions of isocyanates with hydroxyl compounds have been studied extensively by many researchers. Different analytical techniques and models have been used in attempts to determine the kinetic parameters of the curing process or to elucidate the reaction mechanism. The simplest is the titration method using di-n-butyl amine [93, 94]. Spectroscopic methods such as i.r., u.v. and n.m.r. have also been used for the same purposes [95–103]. The h.p.l.c. method combined with the conventional titration method have been used to determine the competing reactions in the system of phenyl isocyanate/n-butanol in the presence of different catalysts [104, 105]. A different approach is the modelling of the structure/property relations during network formation and a comparison of the experimental results with the theoretical predictions based on assumed reaction mechanism [106–110].

Reactions between isocyanates and alcohols proceed remarkably easily at ordinary temperatures with evolution of heat. The reaction rates differ, however, with the type of isocyanate and the coreactant. Aromatic isocyanates react with alcohols faster than aliphatic isocyanates. The curing reaction between the adducts of toluene diisocyanate with different polyols and hydroxyl containing polymers proceeds with very high speed, even at room temperature. In the same conditions, when aliphatic isocyanates are used as the crosslinker the curing proceeds only with the help of a suitable catalyst. Assuming a pseudo first-order reaction, Squiller and Rosthauser have compared the reaction rate constants of hexamethylene diisocyanate (HMDI)/n-butanol and toluene diisocyanate (TDI)/n-butanol reactions. In an uncatalysed system at 30°C having a stoichiometrical ratio between the reactants, they have confirmed the well-known fact of different reactivities of the two isocyanate groups in TDI, finding for the rate constants values of 28×10^4 and $7 \times 10^4 \, s^{-1}$, or an average value of $17 \times 10^4 \, s^{-1}$. This is almost a hundred times higher than the rate constant for HMDI, for which they found a value of $0.14 \times 10^4 \, s^{-1}$ [111].

The kinetics of the curing between aromatic isocyanates and alcohols has been extensively studied in a model system of phenyl isocyanate and n-butanol by many investigators. The reports agree with a second-order reaction in both catalysed and uncatalysed systems, dependent on both phenyl isocyanate and n-butanol [112–121]. Different catalysts are used to speed up the reaction between the isocyanates and alcohols. However, the same catalysts produce different activities with aromatic and aliphatic isocyanates. With aromatic isocyanates the formation of urethane linkage can be promoted by organometallic compounds or metal salts of organic acids. Thin compounds are widely used for this purpose, dibutyltin dilaurate and tin octoate being the predominant representatives among them. Tertiary amines are also used, although they are less effective than the metal compounds. Moreover, many of them catalyse the urethane formation as well as the formation of allophanates and trimerization, leading to cyclization into isocyanurate rings [105, 122]. For many years it has been known that the catalytic activity of tertiary amines increases with the basicity of the amine and decreases with the steric shielding of the amine nitrogen [123–125]. Less data are published with respect to the activity of the organometallic compounds as catalysts in aliphatic isocyanates/alcohol reactions. Based on kinetic studies, Squiller and Rosthauser [111] have established the following relative order of catalytic activity for the reaction of hexamethylene diisocyanate in excess of n-butanol, assuming the reaction to be of pseudo first order:

DMTDC > DBTDLM > DBTDL > DBTM > stannous octoate

This comparison is done on the basis of the relative rate constants at 30°C, when the catalysts are used in the same concentrations on a weight basis. The abbreviations have the following meanings: DMTDC = dimethyltin dichloride; DBTDLM = dibutyltin dilauryl mercaptide; DBTDL = dibutyltin dilaurate and DBTM = dibutyltin maleate.

Britain and Gemeinhardt have screened the effectiveness of various compounds showing catalytic effects by determining the gelation time of a system consisting of toluene diisocyanate (80:20 ratio of a 2,4/2,6 isomer mixture) and polypropylene triol in a stoichiometric ratio [126]. The gelation time was interpreted as a measure of the speed of the isocyanate/hydroxyl reaction. The same weight of compounds to be tested for catalytic activity was used in all tests. Table 3.1 presents part of the results obtained by screening hundreds of compounds.

The relative reactivities of the two different isocyanate groups of isophorone diisocyanate have been compared in several papers. It is assumed that the

Table 3.1 Catalytic activity of different compounds in a system of TDI/polyoxypropylene triol at 70°C, expressed as a gelation time in minutes (Adapted from Ref. 126)

Compound tested	Gel time	Compound tested	Gel time
Blank	>240	Titanium tetrachloride	50
Bismuth nitrate	1	Dibutyltitanium dichloride	13
Lead benzoate	8	Tetrabutyl titanate	8
Lead oleate	4	Ferric chloride	6
Triethylene diamine		Ferric acetylacetonate	16
(DABCO)	4	Antimony trichloride	13
Sodium propionate	32	Antimony pentachloride	90
Potassium acetate	10	Cobalt benzoate	32
Lithium acetate	60	Cobalt octoate	12
Tetrabutyltin	>240	Trioctyl aluminium	32
Tributyltin chloride	200	Aluminium oleate	70
Dibutyltin dichloride	109	Zinc naphthenate (14.6% Zn)	60
Butyltin trichloride	4	Zinc octoate (22% Zn)	65
Stannic chloride	3	1-Methyl-4-(dimethylamino-	
Stannous octoate	4	ethyl)-piperazine	90
Stannous oleate	8	N-Ethylethylenimine	32
Stannous tartrate	>240	N,N',N',N-Tetramethylene	
Dibutyltin di(2-ethyl		diamine	60
hexoate)	4	Triethylamine	120
Dibutyltin dilaurate	8	N-Ethylmorpholine	180
Dibutyltin diisooctyl		2-Methylpyrazine	>240
maleate	4	Dimethylaniline	>240
Dibutyltin oxide	3	Nicotine	240

primary isocyanate group is more reactive than the secondary isocyanate group. For reactions under mild conditions a reactivity ratio between the two groups of 10:1 has been noticed [127]. Lorenz, Decker and Rose have found that the selectivity of the reaction decreases with increasing temperature, and have reported that the relative reactivities of the two NCO groups at 40–110°C in excess of butanol are in the range of 4.5–5.1 to 1 [128]. Ono, Jones and Pappas report that the relative reactivities of the primary and secondary isocyanate group depend on the catalyst used [103]. They have found that when 1,4-diazabicyclo[2.2.2]octane (DABCO) is used as a catalyst, the primary NCO group has a higher reactivity, while in the case of dibutyltin dilaurate (DBTDL) they have identified the secondary NCO group as the more reactive one.

The high reactivities of isocyanates prevent their use in a powder coating composition. The relatively high reactivity towards water will result in homopolymerization of the polyisocyanates through formation of polyureas, and the reaction with the main binder will lead to crosslinking during the production of the powder coatings or during the storage before application.

The reactivity of the isocyanate group can be drastically reduced by blocking it with compounds that give a stable reaction product at room temperature and undergo deblocking at elevated temperatures, re-forming the parent isocyanate. Another possibility is to use the so-called internally blocked isocyanates, which are cyclic condensation products of polyisocyanates undergoing displacement reactions at elevated temperatures with alcohols to form polyurethanes.

A variety of blocking agents have been used in practice. The preparation of the blocked material and deblocking follows the scheme now presented in a simplified way where HBl is a blocking agent:

$$\text{RNCO} + \text{HBl} \longrightarrow \text{RNHCOBl} \xrightarrow{\text{heat}} \text{RNCO} + \text{HBl} \qquad (3.50)$$

However, the reaction of the blocked isocyanate with a compound containing an active hydrogen atom, such as an alcohol for example, could conceivably involve either a two-step process in which the urethane formation is preceded by unimolecular thermal dissociation of the blocked isocyanate to free isocyanate (reaction 3.51) or a direct bimolecular displacement reaction (reaction 3.52):

$$\underset{\underset{\text{Bl}}{|}}{\text{RN}-\text{C}=\text{O}} \longleftrightarrow \text{R}-\text{N}=\text{C}=\text{O} + \text{HBl} \xrightarrow{\text{R'OH}} \text{R}-\underset{}{\overset{\text{H}}{\text{N}}}-\underset{\underset{\text{O}}{\|}}{\text{C}}-\text{OR'} \qquad (3.51)$$

$$\underset{\underset{\text{Bl}}{|}}{\text{RN}-\text{C}=\text{O}} + \text{R'OH} \rightleftharpoons [\text{RN}-\underset{\underset{\text{Bl}}{|}}{\overset{\overset{\text{OH}}{|}}{\text{C}}}-\text{OR'}] \rightleftharpoons \text{R}-\underset{\underset{\text{O}}{\|}}{\text{NC}}-\text{OR'} + \text{HBl} \qquad (3.52)$$

The observation that various ester exchange catalysts facilitate curing by isocyanate adducts can be considered to support the theory of having an interchange mechanism [129]. Evidence of a displacement reaction mechanism has been provided by Gaylord and Sroog [130] in the case of a reaction between unsubstituted urethane and alcohol. At the same time, comparable reactions involving N-monosubstituted urethanes proceeded through primary dissociation to the free isocyanate. Since blocked isocyanates of the urethane classes are N-substituted, they would be expected to react through an initial dissociation step [131]. Anagnostou and Jaul [132] have studied deblocking reactions of adducts of 4,4'-diphenyl methane diisocyanate (MDI) with caprolactam, methyl ethyl ketoxime and benzotriazole using differential scanning calorimetry (d.s.c.) and infrared (i.r.) spectroscopy. Table 3.2 represents the results obtained by performing the deblocking reaction of pure MDI/blocking agent adduct and the same in the presence of polyethylene glycol with a molecular weight of 400, which is able to undergo reaction with the isocyanate group. The start of the deblocking was noticed to be the start of the endotherm at the d.s.c. scan for pure MDI adducts and the start of the exotherm where polyethylene glycol is present. It has been found that deblocking temperatures in all cases were lower when hydroxyl containing material is present in the system. The authors suggest that this can be explained by a bimolecular reaction between the primary hydroxyl group on the blocked site or by a solvating effect on the adduct itself. They have also concluded that the addition of catalyst does not alter the data obtained with uncatalysed systems very much.

The same influence of the hydroxyl groups on the deblocking reaction, manifested through the start of the curing process at a lower temperature compared to the deblocking temperature of the phenylurethane alone, was noticed earlier by Sal'nikova, Shmidt and Voladorskaya [133]. They studied the kinetics of the reaction of n-hexanol with a series of substituted N-phenylurethanes and proposed a mechanism involving the formation of an

Table 3.2 Initial and maximum temperatures in d.s.c. scans of blocked MDI and blocked MDI/polyethylene glycol systems (Adapted from Ref. 132)

Blocking agent	Blocked MDI		Blocked MDI + polyethylene glycol	
	Initial (°C)	Maximum (°C)	Initial (°C)	Maximum (°C)
ε-Caprolactam	158	162	153	155
Methyl ethyl ketoxime	120	130–139	107	not defined
Benzotriazole	214	222	179	184

intermediate complex between the alcohol and the urethane followed by a bimolecular nucleophilic displacement.

Mirgel and Nachtkamp [134] suggest that the mechanisms do not exclude each other and that both proceed at a rate dependent on the reaction conditions and types of reactants. They propose the following overall mechanism for crosslinking of blocked isocyanates:

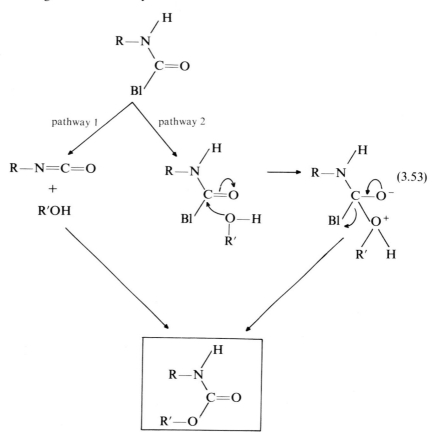

According to this mechanism the regeneration of the isocyanate group is not absolutely necessary for crosslinking. This is represented by the alternative mechanism (pathway 2) according to which the crosslinking is a result of addition of the polyol to the blocked isocyanate group followed by elimination of the blocking agent. In this addition–elimination reaction which is analogous to an ester interchange no isocyanate is involved. The well-known gelation of the system of blocked polyisocyanate/polyamine, even at room temperature [135, 136], can be considered as a support for this mechanism. However, there is not a direct confirmation that the deblocking step does not take place in this reaction; namely it can be supposed that because of extremely high reactivity of the amines towards isocyanates, the deblocking equilibrium is strongly affected by the very rapid consumption of the isocyanate groups:

$$\underset{\underset{Bl}{|}}{-N-C=O} \rightleftharpoons -N=C=O + HBl \xrightarrow{polyamine} polyureas \quad (3.54)$$
$$\,\,\,H$$

All compounds bearing an active hydrogen atom may be considered for use as blocking agents for isocyanate groups. In two excellent articles Wicks [137, 138] gives a review of different blocked isocyanates either used or having a potential use in the coating industry. The following classification is suggested by Mirgel and Nachtkamp for a variety of compounds that can be used for blocking the isocyanate group [134]:

X—H	Protonic acids
—O—H	Alcohols, phenols
—S—H	Thioalcohols, thiophenols
=N—OH	Oximes, hydroxamic acid esters
>N—H	Amines, amides, imides, lactams
—CO—C—H	β-Dicarbonylic compounds

Next to this, the heterocyclic ring opening reactions can be considered as a route to isocyanate formation, although some of these heterocyclic compounds are not strictly speaking blocked isocyanates since they are not synthesized from isocyanates. They can be more properly classified as isocyanate precursors. The fact that no volatile reaction products are formed during the curing process is certainly an advantage of these compounds.

The ease with which a blocked isocyanate RNHCOBl will dissociate will depend to a large extent upon the strength of the bond between Bl and the carbonyl group. Different blocking agents used in practice to prevent premature reaction between the isocyanate group and the functional group containing active hydrogen exhibit different deblocking temperatures with the same isocyanate. The deblocking temperature also depends on the type of isocyanate. Aromatic isocyanates undergo deblocking at lower temperatures compared to aliphatic isocyanates when blocked with the same blocking agent. Aliphatic isocyanates, for example, blocked with aliphatic alcohols undergo deblocking at about 250°C, while the same isocyanates blocked with phenolic compounds have a deblocking temperature that is approximately 50°C lower. On the other hand, the deblocking temperatures of aromatic isocyanates blocked with aliphatic alcohols and phenolic compounds are in the region of 180 and 120°C respectively [139].

The dissociation process of the aromatic isocyanates is significantly affected by the substituents on the aromatic ring. Electron-withdrawing groups decrease stability of the substituted isocyanates and consequently lower the dissociation temperature, while electron-releasing groups have an opposite effect [133, 140].

The deblocking temperature of isocyanate adducts is largely determined by the acidity of the blocking agent and the isocyanate, or their pK_a value. An increase in the acidity of the blocking agent or isocyanate usually has the effect of reducing the required curing temperature. Aliphatic alcohols, lactams and phenols have pK_a values of ca. 18, 17 and 10 respectively, while the pK_a values of the aliphatic and aromatic amines from which the corresponding isocyanates can be derived are 30 and 25 [141]. The order of the deblocking temperatures—aliphatic alcohols > lactams > phenols and aliphatic isocyanates > aromatic isocyanates—fits with the order of the pK_a values of the corresponding compounds.

A large number of compounds have the potential to be used as blocking agents, but only a few of them have achieved technical importance in practice. In the case of powder coatings caprolactam and methyl ethyl ketoxime are used on a commercial scale although many different compounds are disclosed in numerous patents. The structure of these blocking agents and the structure of the blocked isocyanate groups are as follows:

Caprolactam   Methyl ethyl ketoxime

The use of internally blocked isocyanates in the form of uretidiones (urethane dimers) is becoming more popular because of the obvious reason of having no by-products during the curing.

Wicks [137] has investigated the deblocking reaction of caprolactam blocked phenyl isocyanate. He found that the reaction is reversible with an equilibrium dependent on the temperature at which the deblocking is carried out. At 160°C, for example, in a solution of chlorobenzene, the equilibrium is reached within 5 minutes, when about 40% of the blocked adduct has dissociated. From the activation parameters of the reaction he suggests that a cyclic transition state is involved:

THERMOSETTING POWDER COATINGS

[Chemical structure: caprolactam-blocked isocyanate adduct showing R—N(H)—C(=O)—N ring with CH₂ groups]

McLafferty, Figlioti and Camilleri [142] have studied the influence of the carboxyl groups normally present in the hydroxyl functional polyesters used for powder coatings on the curing behaviour of a system based on caprolactam blocked isophorone diisocyanate oligomer and polyester resin. They have drawn the interesting conclusion that the balance of hydroxyl and carboxyl end groups in urethane powder coatings affects significantly the curing characteristics and performance of the coatings, especially in the case of aliphatic isocyanates.

The deblocking temperatures of the oxime blocked isocyanates are lower compared to caprolactam adducts. Methyl ethyl ketoxime blocked MDI undergoes deblocking at a temperature that is 38°C below the deblocking temperature of MDI/ε-caprolactam (Table 3.2)

Carlson *et al.* [143] have used thermal–mechanical analysis (t.m.a.) and Fourier transform infrared spectrometry (f.t.–i.r.) for cure kinetics characterization of trimer of isophorone diisocyanate blocked with methyl ethyl ketoxime, 2-ethyl-4-methylimidazole, phenol and caprolactam. The results obtained, showing the relatively free NCO concentration at various temperatures, are depicted in Figure 3.1. They have found that in the case of methyl ethyl ketoxime the first hint of deblocking appears at 120°C, with significant levels above 140°C. The deblocking reactions for caprolactam and phenol take place above 160°C. In contrast, 2-ethyl-4-methyl imidazole blocked IPDI trimer undergoes deblocking slightly above room temperature. In all cases they have found first-order kinetics. The presence of typical catalysts such as 1,4-diaza[2.2.2]bicyclooctane (DABCO), tetrabutylammonium chloride (TBAC) and dibutyltin dilaurate (DBTDL) did not change the reaction order and none catalysed the deblocking as reported previously by Wicks [138]. The results for DABCO are difficult to interpret due to the side reactions. However, all of them (especially DBTDL) have shown a catalytic effect towards the curing reaction.

Interesting blocking agents with attractive low deblocking temperatures are described by Cooray and Spencer [144]. They belong to the group of pyrazole derivatives. Pyrazole blocked aliphatic isocyanates remain stable up to 80°C, but begin to dissociate at 120°C at a considerable rate due to the proton transfer process illustrated below:

[Chemical structure: pyrazole-blocked isocyanate showing proton transfer mechanism with R—N, N=C—R₁, C—R₂, C=O, and R₃ substituents]

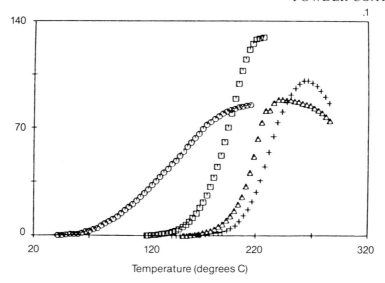

Figure 3.1 Effect of blocking agent on deblocking reaction [143] (Reproduced from *Advanced Urethane, Sci. Techn.*, Vol. 9, 1984, p. 9, Figure 4. © Technomic Publishing Co., Inc., 1984)

The proton attached to the urea nitrogen atom forms the fifth member ring. Similarly to the mechanism proposed earlier by Wicks [137], Cooray and Spencer found a ring opening reorganization for the deblocking mechanism as the temperature is raised above 100°C. Below this temperature, proton transfer to the pyrazole is almost eliminated by intermolecular hydrogen bonding between adjacent urea groups in the blocked urethane.

An interesting list of blocking agents that undergo deblocking in a relatively low temperature range has been given by Guise, Freeland and Smith [145]. As well as the oximes, compounds such as hydroxylamines, hydroxypyridines, hydroxyquinolines and pyrozalinones are considered as suitable blocking agents offering a good compromise between short curing cycles at temperatures down to 100°C and acceptable storage stability. All of them possess a common structural feature, namely an —OH group (or a group capable of enolization) and a basic nitrogen function.

Relatively low deblocking and curing temperatures are observed in the case of acylurethanes; their use as curing agents for powder coatings has been described by Brindoepke [141]. These compounds can be obtained retrosynthetically from aliphatic carboxylic acids and carbamates as illustrated below:

$$\text{RCOOH} + \text{H}_2\text{N}=\text{C}-\text{Bl} \longrightarrow \text{R}-\text{C}-\text{NH}-\text{C}-\text{Bl} \qquad (3.55)$$

In principle they have the structure of normal blocked isocyanates. The lower

THERMOSETTING POWDER COATINGS 67

deblocking temperature compared to that of the isocyanate blocked with the same blocking agent is due to the reduction of the basicity by the acylation.

Isocyanate dimers or the so-called uretidiones used as blocked isocyanates for curing powder coatings are obtained by dimerization of diisocyanates in the presence of phosphine catalysts. This is a reversible process which yields monomeric isocyanates under mild heating:

$$2R-N=C=O \rightleftharpoons R-N\underset{\underset{O}{\overset{\|}{C}}}{\overset{\overset{O}{\overset{\|}{C}}}{\diamond}}N-R \quad (3.56)$$

Uretidione formation is facilitated particularly when the reactivity of the isocyanate groups in the diisocyanate is not equal, as in the case of toluene diisocyanate or isophorone diisocyanate. There is little published work on the ring opening reactions of uretidiones, although they are used in practice as hardeners for powder coatings. Arnold, Nelson and Verbanc [146] have followed the ring opening reactions of aromatic isocyanates with amines and found that biuret is the normal product of the reaction. Under forced conditions, the reaction proceeds to the formation of two moles of substituted urea:

$$\text{Ar}-N\underset{\underset{O}{\overset{\|}{C}}}{\overset{\overset{O}{\overset{\|}{C}}}{\diamond}}N-\text{Ar} + R_2NH \longrightarrow \text{Ar}-\overset{H}{N}-\overset{\overset{O}{\|}}{C}-\overset{\text{Ar}}{\underset{}{N}}-\overset{\overset{O}{\|}}{C}-\overset{H}{N}-R_2 + R_2NH \longrightarrow$$

$$2\,\text{Ar}-\overset{H}{N}-\overset{\overset{O}{\|}}{C}-NR_2 \qquad (3.57)$$

On the other hand, the same reaction with alcohols leads to the formation of the corresponding allophonate [147]:

$$RO-\overset{\overset{O}{\|}}{C}-\overset{H}{N}-\text{Ar}-N\underset{\underset{O}{\overset{\|}{C}}}{\overset{\overset{O}{\overset{\|}{C}}}{\diamond}}N-\text{Ar}-\overset{H}{N}-\overset{\overset{O}{\|}}{C}-OR + R'OH \longrightarrow$$

$$\text{RO—}\underset{\parallel}{\text{C}}\text{—NH—Ar—}\underset{\underset{\underset{\text{OR}'}{|}}{\underset{\text{C=O}}{|}}}{\text{N}}\text{—}\underset{\parallel}{\text{C}}\text{—NH—Ar—NH—}\underset{\parallel}{\text{C}}\text{—OR} \qquad (3.58)$$

O   O   O

Although the uretidione obtained from IPDI is becoming important as a crosslinker for powder coatings, no work has been published dealing with the mechanism of the curing process.

3.1.7 CURING WITH AMINO RESINS

The chemistry of amino resins as curing agents for coatings has been a matter of interest and investigation for more than 50 years. The enormous number of articles dealing with this matter have tried to elucidate the curing mechanism of this system which despite its long history seems still to be not well understood.

There is almost no work published with respect to the use of amino resins as crosslinking agents for powder coatings. The published patent applications understandably do not deal with the chemistry of the process to the same extent that would be expected from a scientific paper. Therefore, the material presented below refers to the chemistry of curing with amino resins in solvent based coatings. Disregarding the differences emanating from the absence of solvents, the curing principles and the curing reactions should also be valid to a great extent for powder coatings.

There are four different types of alkylated amino formaldehyde resins that are used in the coating industry: urea, melamine, benzoguanimine and glycoluryl formaldehyde resins. All of them contain the same type of reactive groups that will be discussed in further detail. However, the reactivity of these groups differs considerably. Urea formaldehyde resins are the most reactive crosslinking agents, while benzoguanimine resins are at the bottom of the scale regarding reactivity. Since the coatings containing urea formaldehyde resins as crosslinkers have low humidity and corrosion resistances, they have never been considered as suitable in application areas that are typical for powder coatings. The chances of the other three gaining wider use in the powder coatings world are rather high. Benzoguanimine resins provide superior chemical resistance properties, humidity, corrosion and detergent resistances, and despite the inferior exterior durability they have a big potential for indoor applications. For many years melamine resins have acquired a solid reputation as crosslinkers for stoving coatings with good outdoor durability. The reports on glycoluryl formaldehyde resins indicate that they can be compared with the melamine resins with respect to their exterior durability [148].

The following functional groups can exist in the amino resins [149]:

$$\mathrm{-N} \begin{cases} CH_2OR \\ CH_2OR \end{cases}$$
F_1

$$\mathrm{-N} \begin{cases} CH_2OR \\ CH_2OH \end{cases}$$
F_2

$$\mathrm{-N} \begin{cases} CH_2OH \\ CH_2OH \end{cases}$$
F_3

$$\mathrm{-N} \begin{cases} CH_2OR \\ H \end{cases}$$
F_4

$$\mathrm{-N} \begin{cases} CH_2OH \\ H \end{cases}$$
F_5

$$\mathrm{-N} \begin{cases} H \\ H \end{cases}$$
F_6

$$\mathrm{-N} \begin{cases} CH_2OCH_2OR \\ CH_2OR \end{cases}$$
F_7

Depending on the method of preparation, all of the above mentioned reactive species can be present in the amino resins. It is possible to prepare resins in which almost all functional groups are of type F_1, but resins containing exclusively one of the groups from F_2 to F_6 do not exist. Partially alkylated commercial products contain combinations of F_1, F_2 and F_4 functionality. Potential candidates for powder coating resins are highly alkylated types containing predominantly F_1, F_7 and small levels of F_2 functionality.

The network formation during the curing process is highly influenced by the presence of polymeric material in the amino crosslinker since the self-crosslinking of the triazine rings during the synthesis of amino resins increases their average functionality. All commercial types of melamine resins contain variable levels of self-condensed triazine rings which are connected by methylene ($-N-CH_2-N-$) or methylene ether groups ($-N-CH_2-O-CH_2-N-$). Keeping in mind the structure of the functional groups given above, a variety of combinations of functional groups attached to two nitrogen atoms bridged by methylene or methylene ether groups are possible. Blank [149] describes twelve possible different functional groups in melamine resin bridges. Even the methylene ether bridge can be considered as a functional group able to undergo a normal condensation reaction with an ether group. This illustrates how complicated the structure of the amino resin is and how difficult it is to describe exactly the curing process where amino resins are used as crosslinkers.

The methods used for characterization of the functional groups in the amino resins and following the curing process, which are described in the excellent article by Christensen [150] and many other papers, include wet chemical analysis [151–153], elemental analysis [154], analysis of the volatile products produced during curing [149, 155, 156], infrared (i.r.) spectroscopy [154, 157–162], ^{13}C-n.m.r. spectroscopy [163–166], proton n.m.r. spectroscopy [149, 164, 167], size exclusion chromatography (s.e.c.) [168, 169], high pressure liquid chromatography (h.p.l.c.) [153, 164, 169–171], gel permeation chromatography (g.p.c.) [154], swelling of the crosslinked films [172] and dynamic mechanical analysis (d.m.a.) [158, 172–175]. The reaction between polymers containing hydroxyl functional groups and HMMM (hexamethoxymethyl melamine) is transetherification. This reaction proceeds when HMMM is heated with higher alcohol in the presence of acid catalyst followed by liberation of methanol:

$$-\text{N}\begin{matrix}\diagup \text{CH}_2\text{OCH}_3\\ \diagdown \text{CH}_2\text{OCH}_3\end{matrix} + \text{OH-polymer} \xrightarrow{\text{H}^+} -\text{N}\begin{matrix}\diagup \text{CH}_2\text{O-polymer}\\ \diagdown \text{CH}_2\text{OCH}_3\end{matrix} + \text{CH}_3\text{OH}$$

One of the complications that occurs in attempts to quantify the curing process is the self-condensation of the amino resins which proceeds to a different extent depending on the curing conditions and the catalyst used. Saxon and Lestienne [158] proposed the following self-crosslinking mechanism in the case of HMMM catalysed by p-toluene sulphonic acid:

(a) Initiation:

$$-\text{N}\begin{matrix}\diagup \text{CH}_2\text{OCH}_3\\ \diagdown \text{CH}_2\text{OCH}_3\end{matrix} \xrightarrow{\text{H}^+} -\text{N}\begin{matrix}\diagup \text{CH}_2\text{OCH}_3\\ \diagdown \text{CH}_2\text{OCH}_3\end{matrix}\text{H}^+ \longrightarrow$$

$$-\text{N}\begin{matrix}\diagup \text{CH}_2\text{OCH}_3\\ \diagdown \text{CH}_2{}^+\end{matrix} + \text{CH}_3\text{OH} \longrightarrow$$

(b) Propagation:

$$-\text{N}\begin{matrix}\diagup \text{CH}_2\text{OCH}_3\\ \diagdown \text{CH}_2{}^+\end{matrix} \xrightarrow[\text{on N}]{\text{attack}} -\text{N}\begin{matrix}\diagup \text{CH}_2\text{OCH}_3\\ \diagdown \text{CH}_2-\text{N}\begin{matrix}\diagup\\ \diagdown \text{CH}_2\text{OCH}_3\end{matrix}\end{matrix} + [{}^+\text{CH}_2\text{OCH}_3]$$

THERMOSETTING POWDER COATINGS

$$-N\begin{matrix}CH_2OCH_3\\ \\ CH_2OCH_3\end{matrix} \xrightarrow[\text{attack on O}]{[^+CH_2OCH_3]} -N\begin{matrix}CH_2OCH_3\\ \\ CH_2^+\end{matrix} + CH_3-O-CH_2OCH_3$$

According to this scheme, the reaction is initiated by protonation of the ether oxygen of the methoxymethyl group, evolution of methanol and formation of the carbonium ion. Propagation involves an alternate attack on nitrogen and oxygen. The initial carbonium ion displaces a methoxymethyl group from the nitrogen of a second HMMM molecule, thus forming a methylene crosslink, and the displaced methoxymethyl ion by reaction with a third molecule of HMMM liberates methylal and regenerates a triazine carbonium ion to continue the reaction. The analytical data disclosing the composition of the reaction products of HMMM cured at 150°C in the presence of paratoluene sulphonic acid, being 83% methylal and 17% methanol, support the above mechanism.

However, the same authors have found that in crosslinking an hydroxyl functional acrylic resin with HMMM under the same curing conditions, the ratio of methylal to methanol decreases from 83:17 to 5:95, indicating that crosslinking of the acrylic polymer is about twenty times as rapid as HMMM self-condensation under the selected experimental conditions.

Somewhat different results were obtained by Koral and Petropoulos [176]. They found that under dry nitrogen the self-condensation of HMMM proceeds by evolution of methylal as the only reaction product.

Wohnsiedler [177] and Blank [149] propose the following reactions leading to the formation of volatile products:

$$>NCH_2OH \rightleftharpoons\ >NH + HCHO \qquad (3.59)$$
$$>NCH_2OH\ +\ >NH \longrightarrow\ >N-CH_2-N<\ +\ H_2O \qquad (3.60)$$
$$>NCH_2OH + ROH \rightleftharpoons\ >NCH_2OR + H_2O \qquad (3.61)$$
$$>NCH_2OCH_3 + ROH \rightleftharpoons\ >NCH_2OR + CH_3OH \qquad (3.62)$$
$$2>NCH_2OCH_3 + H_2O \longrightarrow\ >NCH_2N<\ +\ HCHO + 2CH_3OH \qquad (3.63)$$
$$>NCH_2OCH_3\ +\ >NH \longrightarrow\ >NCH_2N<\ +\ CH_3OH \qquad (3.64)$$
$$2>NCH_2OCH_3 \longrightarrow\ >NCH_2N<\ +\ CH_3OCH_2OCH_3 \qquad (3.65)$$
$$2>NCH_2OH \longrightarrow\ >NCH_2OCH_2N>\ +\ H_2O \qquad (3.66)$$
$$>NCH_2OCH_3\ +\ >NCH_2OH \longrightarrow\ >NCH_2OCH_2N<\ +\ CH_3OH \qquad (3.67)$$
$$>NCH_2OCH_3 + ROH \longrightarrow\ >NCH_2OR\ +\ >NCH_2OH \qquad (3.68)$$
$$>NCH_2OCH_3 + H_2O \longrightarrow\ >NCH_2OH + CH_3OH \qquad (3.69)$$

Table 3.3 presents the results of the analysis of volatile products during the

Table 3.3 Volatile products developed during curing of HMMM/polyol blend for 20 minutes at 150°C. Catalyst dodecylbenzene sulphonic acid (0.84% by weight) (Adapted from Ref. 149)

Polyol/HMMM ratio	75:25	62.5:37.5	50:50	62.5:37.5 Dry	Wet
OH/triazine	4.48:1	2.49:1	1.49:1	2.49:1	
Molecular volatiles/triazine					
HCHO	0.2	0.52	0.33	0.35	0.83
MeOH	5.0	3.07	2.4	2.64	3.5
Methylal	—	—	0.09	—	—
H$_2$O	1.02	0.72	0.5		

curing of HMMM with polyols at three different ratios obtained by Blank [149]: 75:24, 62.5:37.5 and 50:50.

If only reaction (3.62) between the hydroxyl groups of the polyol and methoxymethyl groups of HMMM proceeds during the curing, the molar amount of methanol should be equivalent to the molar amount of the hydroxyl groups. However, the amount of methanol, as can be seen in Table 3.3, is in all cases higher, which is an indication that the self-condensation (reaction 3.63) or hydrolysis of the melamine resin (reaction 3.69) take place. On other hand, the amount of water collected is larger than should be formed due to reactions (3.60), (3.61) and (3.66) reduced for the amount spent due to reaction (3.63). Blank assumes that this comes from the water absorption of the uncured film.

The increase in the level of amino resin results in higher levels of methanol than those calculated. Some traces of methylal are noticed at a melamine/polyol ratio of 50:50, accompanied by significantly higher levels of methanol. This differs from the results obtained during the self-condensation of HMMM by Saxon and Koral [158, 176], in which methylal was the main or the only volatile reaction product. Curing performed in normal air and in air saturated with moisture (reported as dry and wet in Table 3.3) gives significantly different results. This shows that the diffusion of the moisture present in the air into the film makes a substantial impact on hydrolysis, demethylation and self-condensation reactions.

The experiments with different catalyst levels and different temperatures have shown that up to 100°C the level of catalysis does not play a significant role in the composition of the volatile reaction products. However, at 120°C, higher amounts of catalyst cause increased rates of self-condensation and demethylation, leading to considerably higher amounts of methanol in the volatile reaction products.

Different papers dealing with the mechanism of crosslinking with amino resin agree to a large extent with the general scheme presented below [149, 158, 169, 170, 175, 178, 179]:

$$-N{\overset{CH_2OCH_3}{\underset{CH_2OCH_3}{\diagup\!\!\!\diagdown}}} \quad \underset{}{\overset{H^+}{\rightleftharpoons}} \quad -N{\overset{CH_2OCH_3}{\underset{\underset{H}{CH_2\overset{+}{O}CH_3}}{\diagup\!\!\!\diagdown}}} \qquad (3.70)$$

$$-N{\overset{CH_2OCH_3}{\underset{CH_2\overset{+}{O}CH_3}{\diagup\!\!\!\diagdown}}} \quad \rightleftharpoons \quad -N{\overset{CH_2OCH_3}{\underset{CH_2^+}{\diagup\!\!\!\diagdown}}} \quad + \text{ CH}_3\text{OH} \qquad (3.71)$$

$$-N{\overset{CH_2OCH_3}{\underset{CH_2^+}{\diagup\!\!\!\diagdown}}} + \text{ ROH} \rightleftharpoons -N{\overset{\overset{+}{CH_2OCH_3}}{\underset{\underset{H}{CH_2\overset{+}{O}R}}{\diagup\!\!\!\diagdown}}} \qquad (3.72)$$

$$-N{\overset{CH_2OCH_3}{\underset{\underset{H}{CH_2\overset{+}{O}R}}{\diagup\!\!\!\diagdown}}} \rightleftharpoons -N{\overset{CH_2OCH_3}{\underset{CH_2OR}{\diagup\!\!\!\diagdown}}} + \text{ H}^+ \qquad (3.73)$$

Berge, Gudmundsen and Ugelstad [151, 152] postulate that the formation of the carbonium ion (reaction 3.71) is the slowest step in determining the maximum overall reaction rate of curing that can be obtained. This reaction mechanism is valid for a curing reaction catalysed by strong acids such as *p*-toluene sulphonic acid, dodecylbenzene sulphonic acid, naphthalene sulphonic acid, etc.

In the case of partially methylated amino resins when weak acid catalysts are used, for example the residual carboxyl groups of the polyester or alkyd resins or carboxyl groups introduced in acrylic copolymers by acrylic acid, a different curing mechanism has been proposed:

$$-N{\overset{CH_2OCH_3}{\underset{CH_2OH}{\diagup\!\!\!\diagdown}}} \rightleftharpoons -N{\overset{CH_2OCH_3}{\underset{H}{\diagup\!\!\!\diagdown}}} + \text{ HCHO} \qquad (3.74)$$

$$-N{\overset{CH_2OCH_3}{\underset{H}{\diagup\!\!\!\diagdown}}} \quad \overset{HA}{\rightleftharpoons} \quad -N{\overset{\overset{+}{CH_2OCH_3}\;H}{\underset{H}{\diagup\!\!\!\diagdown}}} + \text{ A}^- \qquad (3.75)$$

$$-\text{N}\begin{smallmatrix}\diagup \text{CH}_2\text{OCH}_3\\ \text{H}\\ \diagdown \text{H}\end{smallmatrix} + \text{A}^- \rightleftharpoons -\text{N}=\text{CH}_2 + \text{ROH} + \text{HA} \quad (3.76)$$

$$-\text{N}=\text{CH}_2 + \text{ROH} \rightleftharpoons -\text{N}\begin{smallmatrix}\diagup \text{CH}_2\text{OR}\\ \\ \diagdown \text{H}\end{smallmatrix} \quad (3.77)$$

The rate determining step in this so-called general acid catalysis is the abstraction of the hydrogen atom at the nitrogen to which the methoxymethyl group is attached (reaction 3.76). In the case of HMMM, which lacks this atom, this scheme is obviously not applicable. None of the experiments with HMMM reveal any general acid catalysis [152]. Yamamoto, Nakamichi and Ohe [154] have determined the change of viscosity in a system of acrylic resin/HMMM catalysed with p-toluene sulphonic acid as a strong acid catalyst and monodecyl phthalate as a weak acid catalyst. They have used two types of acrylic polymers

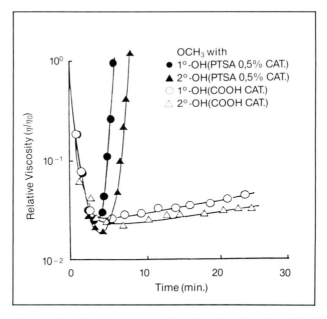

Figure 3.2 Relative viscosity changes as a function of curing time baked at 140°C for catalytic crosslinking reaction systems which consist of two types of acrylic polyols (1 primary OH and 2°—secondary OH groups) and HMMM. Initial viscosity is designated as ○ [154] (Reproduced by permission of Federation of Societies for Coatings Technology)

having the same hydroxyl value, acid value, molecular weight and glass transition temperature, being different only with respect to the nature of the hydroxyl groups. In one case the hydroxyl functionality in the acrylic resin was introduced through 2-hydroxyethylmethacrylate (primary hydroxyl groups) and in the other case through 2-hydroxypropylmethacrylate (secondary hydroxyl groups). The change in relative viscosity during the curing process at 140°C is shown in Figure 3.2 and at different curing temperatures in Figures 3.3 and 3.4. In the strong acid catalyst case with *p*-toluene sulphonic acid added, the relative viscosity rises rapidly after only four minutes of curing. However, in the case of monodecyl phthalate catalyst, both viscosities of primary and secondary hydroxyl systems hardly change with time. In the case of weak acid catalysis, the change in viscosity differs between the primary and secondary hydroxyl acrylic systems at different temperatures, indicating the higher reactivity of the primary hydroxyl groups. Although the curing reaction proceeds at higher temperatures, even at 180°C, 17–20 minutes are necessary to compensate for the viscosity drop with temperature and to reach the initial viscosity.

The influence of different catalysts on the curing rate is quite obvious. Especially in the case of fully methylated amino resins it seems that the use of

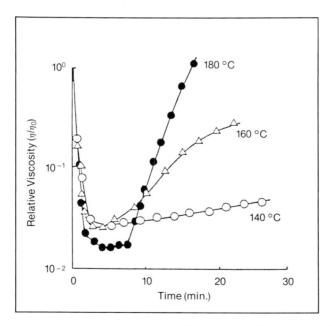

Figure 3.3 Relative viscosity changes as a function of curing time baked at 140, 160 and 180°C respectively (acrylic polyol with primary hydroxyl groups) catalysed with monodecyl phthalate [154] (Reproduced by permission of Federation of Societies for Coatings Technology)

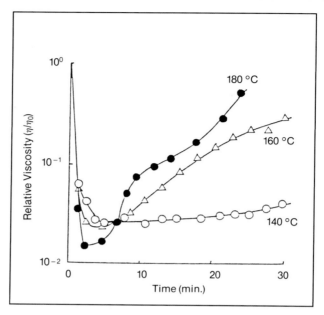

Figure 3.4 Relative viscosity changes as a function of curing time baked at 140, 160 and 180°C respectively (acrylic polyol with secondary hydroxyl groups) catalysed with monodecyl phthalate [154] (Reproduced by permission of Federation of Societies for Coatings Technology)

strong acid catalysts for obtaining acceptable curing time at relatively low curing temperatures is inevitable. However, coatings that have been cured with strong catalysts such as *p*-toluene sulphonic acid are sensitive to moisture that diffuses across the air/film interface and may cause blistering. Water resistance can be improved by substitution of *p*-toluene sulphonic acid with more hydrophobic acid such as dinonylnaphthalene disulphonic acid [180].

Considerable work has been done with respect to the kinetics of the curing process with amino resins, predominantly with HMMM. The already presented reactions which may take place during curing show the complexity of the system as well as any difficulties that may occur in attempts to make a quantitative description of the curing. Therefore, to a first approximation, many authors describe the cure of fully alkylated melamines by a single reaction. This is the formation of polymer–melamine crosslinks via the reaction of alkoxy groups on the melamine with hydroxy groups on the polymer. This approximation can be considered to be valid to a great extent at moderate curing temperatures and normal catalyst levels.

Holmberg [181] concluded that the principal cure reaction between the hydroxyl groups and alkoxymethyl groups in Sn2, the rate being proportional to the concentration of both the alkoxymethyl and hydroxyl groups. Lazzara has found a good fit of the kinetics data he has obtained to second-order kinetics [156]. The same reaction order has been proposed by Yamamoto, Nakamichi and Ohe [154] in the case of weak acid catalysed systems. However, according to them, the self-condensation reaction, which occurs even when weak acid catalyst is used, does not follow second-order kinetics, but rather that of pseudo third order (first order for OH and second order for OCH_3), in spite of the non-disappearance of the hydroxyl groups. Although they were not able to provide experimental evidence, they postulate that the residual polar hydroxyl groups play the role of activator on a carboxylic acid catalyst.

In contrast to Holmberg, Berge, Gudmundsen and Ugelstad [152] and Blank [182] have proposed an Sn1 type reaction on the basis that formation of the carbonium ion by the acid catalyst is a rate-determining step. Blank has also cautioned against drawing conclusions as to the mechanism from systems that are highly viscous and approaching gelation.

Using the same curing mechanism, which postulates that formation of the carbonium ion is the slowest step that determines the overall kinetics of the curing process, Bauer and Budde [167] have developed a kinetic model for crosslinking fully methylated melamines which accurately predicts the observed dependence of the extent of reaction on bake time and melamine concentration. Using the model and the extent of reaction, a rate constant for the reaction has been determined and found to be proportional to the catalyst concentration and to the square root of the acid strength of the catalyst. From a closer examination of the implications of the accepted curing mechanism they have concluded that the reaction is first order only if the methanol concentration in the film during the curing process is zero, which obviously is not the case in practice.

Alkoxy groups also react with other functional groups, including carboxylic acids which are typically present in the powder coating resins. Very often these groups are incorporated in the polymer on purpose to promote cure. The relative reactivity of the carboxyl groups depends on the level of catalyst and the type of amino resin. Bauer and Dickie found that the hydroxyl groups are more reactive with fully alkylated melamines in the presence of strong acid catalysis, while carboxyl groups are more reactive with partially alkylated melamines [161, 162].

A combination of the kinetic model fitting to a great extent the experimental results and theory of Miller and Macosko [183] of non-linear polymerization was used by Bauer and coworkers [161, 162, 167, 184, 185] to develop a model describing network formation during the crosslinking process. The mathematics involved in the derivation of the model is rather complex and beyond the scope of this book. One advantage of this model is that the data required as input can be obtained from practical coatings.

REFERENCES

1. Govarkier, V. R., in *Polymer Science*, John Wiley & Sons, New York, 1976, p. 98.
2. Batzer, H., and Lang, H., *Makromol. Chem.*, **15**, 211, 1955.
3. Carothers, W. H., *Trans. Faraday Soc.*, **32**, 39, 1936.
4. Flory, P. J., in *Principles of Polymer Chemistry*, Ch. 9, Cornell University Press, Ithaca, New York, 1953.
5. Stockmayer, W. H., *J. Appl. Polym. Sci.*, **9**, 69, 1952.
6. Gordon, M., *Proc. Roy. Soc. London*, A268, 240, 1962.
7. Gordon, M., and Malcom, G. N., *Proc. Roy. Soc. London*, A295, 29, 1966.
8. Macosko, C. W., and Miller, D. R., *Macromolecules*, **9**, 199, 1976.
9. Miller, D. R., and Macosko, C. W., *Macromolecules*, **9**, 206, 1976.
10. Tiemersma-Thoone, G. P. J. M., Scholtens, B. J. R., and Dusek, K., *Proceedings of the First International Conference on Industrial and Applied Mathematics*, Paris-La Vilette, 29 June–3 July 1987.
11. Good, I. J., *Proc. Camb. Phil. Soc.*, **45**, 360, 1948.
12. Good. I. J., *Proc. Camb. Phil. Soc.*, **56**, 367, 1960.
13. Good., I. J., *Proc. Roy. Soc. London*, A272, 54, 1963.
14. Gordon, M., and Scantlebury, G. R., *Trans. Faraday Soc.*, **60**, 605, 1964.
15. Gordon, M., Ward, T. C., and Whitney, R. S., *Polymer Networks*, Eds. A. S. Chompff and S. Newman, Plenum Press, New York, 1970.
16. Saito, O., *Polym. Eng. Sci.*, **19** (4), 234, 1979.
17. Ilavsky, M., and Dusek, K., *Polymer*, **24**, 981, 1983.
18. Dusek, K., Ilavsky, M., and Matejka, L., *Polym. Bull.* **12**, 33, 1984.
19. Ilavsky, M., Bogdanova, L. M., and Dusek, K., *J. Polym. Sci., Polym. Phys. Ed.*, **22**, 265, 1984.
20. Dusek, K., and Vojta, V., *Br. Polym. J.*, **9**, 164, 1977.
21. Dusek, K., and Ilavsky, M., *J. Polym. Sci., Polym. Phys. Ed.* **21**, 1323, 1983.
22. Dusek, K., *Makromol. Chem. Suppl.*, **2** (180), 35, 1979.
23. Flory, P. J., *Macromolecules*, **15**, 99, 1982.
24. Dusek, K., Ilavsky, M., and Lunak, S., *J. Polym. Sci. Symp.*, **53**, 29, 1975.
25. Dusek, K., Gordon, M., and Ross-Murphy, S. B., *Macromolecules*, **11**, 236, 1978.
26. Dusek, K., Bleha, M., and Lunak, S., *J. Polym. Sci., Polym. Chem. Ed.*, **15**, 2393, 1977.
27. Charlesworth, J. M., *J. Polym. Sci., Polym. Phys. Ed.*, **17**, 1557, 1571, 1979.
28. Topolkaraev, V. A., Zhorina, L.A., Vladimirov, L. V., Berlin, Al.Al., Zelenetskii, A. N., Prut, E. V., and Enikolopyan, N. S., *Vysokomol. Soed., Sec. A*, **A21**, 1655, 1979.
29. Allen, G., Holmes, P. A., and Walsh, D. J., *Faraday Disc.*, **1974**, 57.
30. Allen, G., Booth, C., and Price, C., *Polymer*, **8**, 397, 1967.
31. Shechter, L., and Wynstra, J., *Ind. Eng. Chem.*, **48**, 86, 1956.
32. Shechter, L., Wynstra, J., and Kurkjy, R. E., *Ind. Eng. Chem.*, **48**, 94, 1956.
33. Shechter, L., Wynstra, J., and Kurkjy, R. E., *Ind. Eng. Chem.*, **49**, 1107, 1957.
34. Malkemus, J. D., and Swan, J. D., *J. Am. Oil Chemists' Soc.*, **34**, 342, 1957.
35. Wrigley, A. N., Smith, D. F., and Stirton, A. J., *J. Am. Oil Chemists' Soc.*, **36**, 34, 1959.
36. Kakiuchi, H., and Tanaka, Y., *J. Org. Chem.*, **31**, 1559, 1966.
37. Tanaka, Y., and Takeuchi, H., *Tetrahedron*, **24**, 6433, 1968.
38. Tanaka, Y., *J. Org. Chem.*, **32**, 2405, 1967.
39. Tanaka, Y., Okada, A., and Suzuki, M., *Can. J. Chem.*, **48**, 3258, 1970.
40. Isaacs, N. S., *Tetrahedron Letters*, **1965**, 4549.
41. Albert, A., and Phillips, J. N., *J. Chem. Soc.*, **1956**, 129.
42. Brown, H. C., and Mihm, X. R., *J. Am. Chem. Soc.*, **77**, 1723, 1955.
43. Brown, H. C., and Gintis, D., *J. Am. Chem. Soc..*, **78**, 5387, 1956.

44. Brown, H. C., McDaniel, D. H., and Haflinger, O., *Determination of Organic Structures by Physical Methods*, Eds. F. C. Nochod and W. D. Phillips, Vol. I, Academic Press, New York, 1962.
45. Linde, R. van der, and Belder, E. G., *VIIth International Conference in Organic Coatings Science and Technology*, Athens, July 1981, p. 359.
46. Luston, J., Manasek, Z., and Kulickova, M., *J. Macromol Sci. Chem.*, **A12**, 995, 1978.
47. Matejka, L., Lovy, J., Pokorny, S., Bouchal, K., and Dusek, K., *J. Polym. Sci., Polym. Chem. Ed.*, **21**, 2873, 1983.
48. Tanaka, Y., and Mika, T. F., in *Epoxy Resins—Chemistry and Technology*, Eds. C. A. May and Y. Tanaka, Marcel Dekker, Inc., New York, 1973, p. 179.
49. Henig, A., Jath, J., and Mohler, H., *Farbe und Lack*, **86**, 313, 1980.
50. Fisch, W., and Hofmann, W., *J. Polym. Sci.*, **12**, 497, 1954.
51. Fisch, W., and Hofmann, W., *Makromol. Chem.*, **44–46**, 8, 1961.
52. Fisch, W., Hofmann, W., and Koskikallio, J., *J. Appl. Chem.*, **6**, 429, 1950.
53. Dearborn, E. C., Fuoss, R. M., and White, A. F., *J. Polym. Sci.*, **16**, 201, 1955.
54. Tanaka, Y., and Kakiuchi, H., *J. Appl. Polym. Sci.*, **7**, 1063, 1963.
55. Tanaka, Y., and Kakiuchi, H., *J. Polym. Sci.*, **2A**, 3405, 1964.
56. Newey, H. A., *Gordon Res. Conf. on Polymers*, New London, Connecticut, July 1955.
57. Kaplan, S. I., *Polym. Eng. Sci.*, **6**, 65, 1966.
58. Fedtke, M., and Domaratius, F., *Polym. Bull.*, **15**, 13, 1986.
59. Fischer, R. F., *J. Polym. Sci.*, **44**, 155, 1960.
60. Matejka, L., Lovy, J., Pokorny, S., Bouchal, K., and Dusek, K., *J. Polym. Sci., Polym. Chem. Ed.*, **21**, 2873, 1983.
61. Tanaka, Y., and Kakiuchi, H., *J. Makromol. Chem.*, **1**, 307, 1966.
62. Tzirkin, M. Z., *Soviet Plastics*, **7**, 17, 1963.
63. Fedke, M., and Tarnow, M., *Plaste und Kautchuk*, **30**, 70, 1983.
64. Kushch, P. P., Lagodzinskaya, G. W., Komarov, B. A., and Rozenberg, B. A., *Visokomol. Soed.*, **B21**, 708, 1979.
65. Fischer, R. F., *J. Polym. Sci.*, **44**, 155, 1966.
66. Feltzin, J., *ACS Meeting Chicago 1964, Division of Organic Coatings Plastics*, Paper 40.
67. Kakurai, T., and Noguchi, T., *J. Soc. Org. Synth. Chem. Japan*, **18**,, 485, 1960.
68. Dannenberg, H., *SPE Trans.*, **3**, 78, 1963.
69. Gross, A., Kollek, H., and Brockmann, H., *Int. J. Adhesion and Adhesives*, **8** (4), 225, 1988.
70. Gross, A., Brockmann, H., and Kollek, H., *Int. J. Adhesion and Adhesives*, **7** (1), 33, 1987.
71. Gross, A., Kollek, H., Schormann, A., and Brockmann, H., *Int. J. Adhesion and Adhesives*, **8** (3), 147, 1988.
72. Kwei, T. K., *J. Polym. Sci.*, **1A**, 2985, 1963.
73. Dusek, K., *Makromol. Chem., Macromol. Symp.*, **7**, 37, 1987.
74. Dusek, K., *Adv. Polym. Sci.*, **78**, 1, 1986.
75. Smith, T. I., *Polymer*, **2**, 95, 1961.
76. Gaylord, N. G., *Polyethers, Part 1*, Interscience, New York, 1963.
77. Ishii, Y., and Sakai, S., in *Ring Opening Polymerisation*, Eds. K. C. Frisch and S. L. Reegen, Marcel Dekker, New York, 1969.
78. Lidarik, M., Stary, S., and Mleziva, I., *Visokomol. Soed.*, **5**, 1748, 1963.
79. Tanaka, Y., Tomoi, M., and Kakiuchi, H., *J. Macromol. Sci.*, **1**, 471, 1967.
80. Steiner, E. C., Pelletier, R. R., and Trucks, R. O., *J. Am. Chem. Soc.*, **86**, 4678, 1964.
81. Farkas, A., and Ströhm, P., *J. Appl. Polym. Sci.*, **12**, 159, 1968.
82. Dearlove, T., *J. Appl. Polym. Sci.*, **14**, 1615, 1970.

83. Barton, J., and Scheperd, P., *Makromol. Chem.*, **176**, 919, 1975.
84. Ricciardi, F., Romanchik, W., and Jouille, M., *J. Appl. Polym. Sci.*, **21**, 1475, 1983.
85. Berger, J., and Lohse, F., *J. Appl. Polym. Sci.*, **30**, 531, 1985.
86. Berger, J., and Lohse, F., *Eur. Polym. J.*, **21**, 435, 1985.
87. Yamake, T., and Fukui, K., *Bull. Chem. Soc. Japan*, **42**, 2112, 1969.
88. Markevich, M. A., and Irzhak, V. I., *Visokomol. Soed.*, **28**, 60, 1986.
89. Bokare, V. M., and Ghandi, K. S., *J. Polym. Sci., Polym. Chem. Ed.*, **18**, 587, 1980.
90. Dusek, K., *Polym. Bull.*, **13**, 321, 1985.
91. Narracott, E. S., *Br. Plastics*, **26**, 120, 1953.
92. Bressers, H. J. L., and Goumans, L., *Crosslinked Epoxies*, Walter Gruyter and Co., Berlin, 1987, p. 233.
93. David, D. J., and Staley, H. B., *Analytical Chemistry of Polyurethanes, High Polymer Series*, Vol. XVI, Pt III, Wiley Interscience, New York, 1969.
94. Kresta, J. E., and Hsieh, K. H., *Macromol. Chem.*, **179**, 2779, 1978.
95. Flynn, K. G., and Nenortas, D. R., *J. Org. Chem.*, **28**, 3527, 1963.
96. Abbate, F. W., and Urlich, H., *J. Appl. Polym. Sci.*, **13**, 1929, 1969.
97. Naegele, W., and Wendisch, D., *Org. Magn. Reson.*, **2**, 439, 1970.
98. Butler, G. B., and Corfield, G. C., *J. Macromol. Sci. Chem.*, **A5**, 37, 1971.
99. Nguyen, A., and Marechal, E., *J. Macromol. Sci. Chem.*, **A16**, 881, 1981.
100. Entelis, S. G., and Nesterov, O. V., *Kinet. Katal.*, **7**, 464, 1966.
101. Weij van der, F. W., *J. Polym. Sci., Polym. Chem. Ed.*, **19**, 381, 3063, 1981.
102. Usmani, A. M., *J. Coat. Technol.*, **56** (716), 99, 1984.
103. Ono, H. K., Jones, F. N., and Pappas, S. P., *J. Polym. Sci., Polym. Lett. Ed.*, **23**, 509, 1985.
104. Wong, S. W., and Frisch, K. C., *J. Polym. Sci., Polym. Chem. Ed.*, **24**, 2867, 1986.
105. Wong, S. W., and Frisch, K. C., *J. Polym. Sci., Polym. Chem. Ed.*, **24**, 2877, 1986.
106. Dusek, K., *Br. Polym. J.*, **17** (2), 185, 1985.
107. Shy, L. Y., and Eichinger, B. E., *Br. Polym. J.*, **17** (2), 200, 1985.
108. Bauer, D. R., and Dickie, R. A., *ACS Pol. Mat. Sci. Eng.*, **52**, 550, 1985.
109. Nelen, P. J. C., *Proceedings of XVII Fatipec*, Lugano, 1984, p. 283.
110. Wu, D. T., *ACS Pol. Mat. Sci. Eng.*, **52**, 458, 1985.
111. Squiller, E. P., and Rosthauser, J. W., *Proceedings of the ACS Division of Polymeric Material: Science and Engineering*, Preprints 55, 1986, p. 640.
112. Kogon, I. C., *J. Org. Chem.*, **24**, 438, 1959.
113. Burkus, J., *J. Org. Chem.*, **27**, 474, 1962.
114. Okada, H., and Iwakura, Y., *Makromol. Chem.*, **66**, 91, 1963.
115. Rand, L., Thir, B., Reegen, S. L., and Frisch, K. C., *J. Appl. Polym. Sci.*, **9**, 1787, 1965.
116. Okuto, H., *Makromol. Chem.*, **98**, 148, 1966.
117. Rossmy, G. R., Kollmeier, H. J., Lidy, W., Schator, H., and Wiemann, M., *J. Cell. Plast.*, **13**, 26, 1977.
118. Furukawa, M., and Yokoyama, T., *J. Polym. Sci., Polym. Lett. Ed.*, **17**, 175, 1979.
119. Bechara, I. S., and Mascioli, R. L., *J. Cell. Plast.*, **15**, 321, 1979.
120. Rossmy, G. R., Kollmeier, H. J., Lidy, W., Schator, H., and Wiemann, M., *J. Cell. Plast.*, **17**, 319, 1981.
121. Bailey Jr, F. E., and Critchfield, F. E., *J. Cell. Plast.*, **17**, 333, 1981.
122. Potter, T. A., and Williams, J. L., *J. Coat. Technol.*, **59** (749), 63, 1987.
123. Alzner, B. G., and Frisch, K. C., *Ind. Eng. Chem.*, **51**, 715, 1959.
124. Farkas, A., Mills, G. A., Erner, W. E., and Maerker, J. B., *Ind. Eng. Chem.*, **51**, 1299, 1959.
125. Burkus, J., *J. Org. Chem.*, **26**, 779, 1961.
126. Britain, J. W., and Gemeinhardt, P. G., *J. Appl. Polym. Sci.*, **4**, 207, 1960.

127. Hüls Technical Bulletin 22 ME 377–7.
128. Lorenz, O., Decker, H., and Rose, G., *Angew, Makromol. Chem.,* **122**, 83, 1984.
129. Jonson, P. C., in *Advances in Polyurethane Technology*, Eds. J. M. Buist and H. Gudheon, McLaren and Sons Ltd, London, 1968, p. 19.
130. Gaylord, N. G., and Sroog, C. E., *J. Org. Chem.,* **18**, 1362, 1953.
131. Griffin, G. R., and Willwerth, J. L., *I & EC Product Research and Dev.,* **1** (4), 265, 1962.
132. Anagnostou, T. J., and Jaul, E., *J. Coat. Technol.,* **53** (673), 35, 1981.
133. Sal'nikova, G. A., Shmidt, Yu. A., and Voladorskaya, Yu. I., *Dokl. Akad. Nauk. SSSR,* **181**, 669, 1968.
134. Mirgel, V., and Nachtkamp, K., *Polym. Paint Colour J.,* **176** (4163), 200, 1986.
135. Hudson, G. A., Hixenburgh, J. C., Wells, E. R., Saunders, J. H., and Hardy, E. E., *Off. Dig.,* **32**, 213, 1960.
136. Furuya, Y., Goto, S., Itoko, K., Uraski, I., and Morita, A., *Tetrahedron,* **24**, 2367, 1968.
137. Wicks Jr, Z. W., *Progress Org. Coat.,* **9**, 3, 1981.
138. Wicks Jr, Z. W., *Progress Org. Coat.,* **3**, 73, 1975.
139. Myers, R. R., and Long, S. S., *Treatise on Coatings*, Book 1, Marcel Dekker, New York, 1967, p. 470.
140. Lateef, A. B., Reeder, J. A., and Rand, L., *J. Org. Chem.,* **36**, 2295, 1971.
141. Brindoepke, G., *Proceedings of XII International Conference on Organic Coatings Science Technology*, Athens, 1986, p. 45.
142. McLafferty, J. J., Figlioti, P. A., and Camilleri, L. T., *J. Coat. Technol.,* **58**, 733, 23, 1986.
143. Carlson, G. M., Neag, C. M., Kuo, C., and Provder, T., *Advanced Urethane, Sci. Technol.,* **9**, 47, 1984.
144. Cooray, B., and Spencer, R., *Paint and Resin*, October 1988, p. 18.
145. Guise, G. B., Freeland, G. N., and Smith, G. C., *J. Appl. Polym. Sci.,* **23**, 353, 1979.
146. Arnold, R. G., Nelson, J. A., and Verbanc, J. J., *Chem. Rev.,* **57**, 47, 1957.
147. Singh, P., and Boivin, J. L., *Can. J. Chem.,* **40**, 935, 1962.
148. Parekh, G. G., *J. Coat. Technol.,* **51** (658), 101, 1979.
149. Blank, W. J., *J. Coat. Technol.,* **51** (656), 61, 1979.
150. Christensen, G., *Prog. Org. Coat.,* **8**, 211, 1980.
151. Berge, A., Gudmundsen, S., and Ugelstad, J., *Eur. Polym. J.,* **5**, 171, 1969.
152. Berge, A., Gudmundsen, S., and Ugelstad, J., *Eur. Polym. J.,* **6**, 981, 1970.
153. Chiavarini, M., Del Fanti, N., and Bigatto, R., *Angew. Makromol. Chem.,* **56**, 15, 1976.
154. Yamamoto, T., Nakamichi, T., and Ohe, O., *J. Coat. Technol.,* **60** (762), 51, 1988.
155. Kooistra, M. F., *J. Oil Col. Chem. Assoc.,* **62**, 432, 1979.
156. Lazzara, M. G., *J. Coat. Technol.,* **56** (710), 19, 1984.
157. Dannenbeg, H., Forbes, J. W., and Jones, A. C., *Anal. Chem.,* **32**, 365, 1960.
158. Saxon, R., and Lestienne, F. C., *J. Appl. Polym. Sci.,* **8**, 475, 1964.
159. Hornung, K. H., and Biethan, U., *Farbe und Lack,* **76**, 461, 1970.
160. Dorffel, J., and Biethan, U., *Farbe und Lacke,* **82**, 1017, 1976.
161. Bauer, D. R., and Dickie, R. A., *J. Polym. Sci., Polym. Phys.,* **18**, 1997, 1980.
162. Bauer, D. R., and Dickie, R. A., *J. Polym. Sci., Polym. Phys.,* **18**, 2015, 1980.
163. Tomita, B., and Hatono, S., *J. Polym. Sci., Polym. Chem. Ed.,* **16**, 2509, 1978.
164. Dijk van, J. H., Brakel van, A. S., Dankelman, W., and Groenenboom, C. J., *XV Fatipec Congress*, Book II, Amsterdam, 1980, p. 326.
165. Bauer, D. R., Dickie, R. A., and Koenig, J. L., *Ind. Eng. Chem. Prod. Res. Div.,* **24**, 121, 1985.
166. Bauer, D. R., Dickie, R. A., and Koenig, J. L., *J. Polym. Sci., Polym. Phys. Ed.,* **22**, 2009, 1984.

167. Bauer, D. R., and Budde, G. F., *J. Appl. Polym. Sci.*, **28**, 253, 1983.
168. Aldersley, J. W., *Br. Polym. J.*, **1**, 101, 1969.
169. Anderson, D. G., Netzel, D. A., and Tessari, D. J., *J. Appl. Polym. Sci.*, **14**, 3021, 1970.
170. Feurer, B., and Gourdenne, A., *Bull. Soc. Chim. Fr.*, **12**, 2845, 1974.
171. Schindlbauer, H., and Anderer, J., *Angew. Makromol. Chem.*, **79**, 157, 1979.
172. Collete, J. W., Corcoran, P., Tannenbaum, H, P., and Zimmt, W. S., *J. Appl. Polym. Sci.*, **32**, 4209, 1986.
173. Hill, L. W., and Kozlowski, K., *J. Coat. Technol.*, **59** (751), 63, 1987.
174. Oshikubo, T., Yoshida, T., and Tanaka, S., *Proceedings of Xth International Conference on Organic Coatings Science Technology*, Athens, Greece, 1984, p. 317.
175. Nakamichi, T., *Prog. Org. Coat.*, **14**, 23, 1986.
176. Koral, J. N., and Petropoulos, J. C., *J. Paint Technol.*, **38**, 501, 600, 1966.
177. Wohnsiedler, H. P., *ACS Div. Org. Coat. Plastics Chem.*, **20** (2), 52, 1960.
178. Santer, J. O., *Prog. Org. Coat.*, **12**, 309, 1984.
179. Bauer, D. R., *Prog. Org. Coat.*, **14**, 193, 1986.
180. Calbo, *J. Coat. Technol.*, **52**, 75, 1980.
181. Holmberg, K., *J. Oil Colour Chem. Assoc.*, **61**, 359, 1978.
182. Blank, W., *J. Coat. Technol.*, **54**, 26, 1982.
183. Miller, D. R., and Macosko, C. W., *Macromolecules*, **9**, 206, 1976.
184. Bauer, D. R., and Budde, G. F., *Ind. Eng. Chem., Prod. Res. Dev.*, **20**, 674, 1981.
185. Bauer, D. R., and Dickie, R. A., *J. Coat. Technol.*, **54**, 57, 1982.

3.2 MONITORING THE CURING PROCESS

The characteristics of the cured film of thermosetting powder coatings are dependent on the degree of curing. Not only mechanical properties such as hardness, flexibility and impact resistance, but also the appearance of the film surface, flow properties during the film formation, chemical resistance of the cured film, gloss retention and the service life of powder coatings are directly related to the curing process. The curing schedules in practice vary considerably depending on the characteristics of the stoving line. Therefore it is of essential importance for the powder coating chemist to be able to follow the curing process by a suitable technique, to interpret the gathered data correctly and to use them as elements in adjusting the powder coating formulation to the demands of the end user.

A variety of techniques have been developed to evaluate the degree of curing or to follow the curing process at specific time sequences or continuously. These methods and techniques include infrared (i.r.) spectroscopy, nuclear magnetic resonance (n.m.r.), chemical analysis, dielectric analysis, measurement of certain mechanical properties of the film such as hardness, flexibility and impact resistance, determination of the solvent resistance and degree of swelling of the cured film, thermal analysis such as differential scanning calorimetry (d.s.c.) and thermogravimetry (t.g.a.) and thermal mechanical analyses such as torsional braid analysis (t.b.a.) or dynamic mechanical analysis (d.m.a.).

THERMOSETTING POWDER COATINGS

Most of these techniques are suitable for characterization of the curing process in the case of powder coatings. Unfortunately many of them supply either insufficient and not always relevant data for quantitative interpretation or data that cannot be directly related to the phenomena of film formation. Spectroscopical methods, which are probably the best analytical tools for characterization of the curing mechanism from a chemical point of view, do not say much about film formation which is a rather complex physical–chemical process.

In practice, the mechanism of the curing reactions for existing powder coating systems is relatively well known. Therefore, techniques that provide data that can be easily related to the film formation process and used for formulating coatings are of the greatest importance to the powder paint chemist. For this purpose, the thermal and thermal mechanical analytical techniques probably provide the most useful data, although they do not give any direct information about the chemical changes that take place. However, by these methods, one can measure the changes in the fundamental properties of the coating during curing which can be related to the structure of the material.

3.2.1 DIFFERENTIAL SCANNING CALORIMETRY (d.s.c.)

Differential scanning calorimetry (d.s.c.) belongs to a group of thermal analytical methods that are based on detection of the thermal effects in the sample resulting from physical and chemical changes as a function of temperature. The latter is usually programmed at a linear rate. The principle on which this technique operates is based on determining the energy required to keep the sample and the blank reference at the same temperature. The d.s.c. technique could be considered as a follow-up of the differential thermal analysis (d.t.a.) which was developed much earlier with the basic principles established by Roberts-Austen at the end of the last century. Contrary to d.s.c. in d.t.a. the difference in temperature between the sample and the reference is observed while both are being heated (or cooled) through a preselected temperature programme, which, as in the case of d.s.c., is mostly linear.

There are two types of d.s.c. instruments that have been widely used for calorimetric measurements: the heat-flux d.s.c and the power-compensational d.s.c.

The principal scheme of the heat-flux d.s.c. instrument is presented in Figure 3.5. Sample holder S and reference R (usually an empty holder) are placed on a disc D which serves for heat transfer to the sample and the reference. Heat supplied by the furnace F is transferred from the disc up to the sample and the reference via the sample pans. In the case of a dynamic run, the furnace temperature T_c increases or decreases at a linear rate. The differential heat flow to the sample and reference is monitored by thermocouple wires which are connected to the underside of the disc. The heat flow dH/dt to the sample is equal

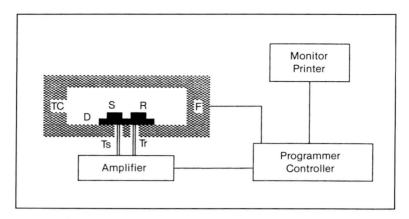

Figure 3.5 Scheme of heat-flow d.s.c. apparatus

to the difference between dQ_s/dt and dQ_r/dt representing the total heat flow to the sample holder plus the sample and reference holder:

$$\frac{dH}{dt} = \frac{dQ_s}{dt} - \frac{dQ_r}{dt} \qquad (3.78)$$

According to the thermal analogue of Ohm's law the heat flow is proportional to the driving force (temperature difference between the furnace and the sample or reference) and inversely proportional to the temperature resistance R_{th}. Consequently, for the d.s.c. cell, equation (3.78) can be written as follows:

$$\frac{dH}{dt} = \frac{T_c - T_s}{R_{th}} - \frac{T_c - T_r}{R_{th}} \qquad (3.79)$$

Due to the symmetric construction of the instrument, T_c and R_{th} are equal on both sides [1], which leads to the following equation:

$$\frac{dH}{dt} = -\frac{T_s - T_r}{R_{th}} \qquad (3.80)$$

For known thermal characteristics of the instrument, the thermal effect of the process in terms of the energy consumed or emitted by the sample can be determined by measuring the temperatures of the sample and reference. The inherent problem emanating from the fact that the thermal characteristics of the instrument are dependent on the actual temperature and change permanently during the dynamic run is solved in different ways depending on the producer of the instrument.

Figure 3.6 represents schematically the power-compensational d.s.c. apparatus. The main element of the instrument is the furnace in which the sample (S) and

THERMOSETTING POWDER COATINGS

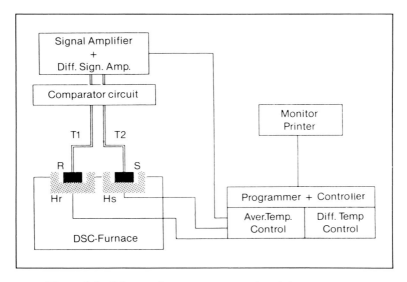

Figure 3.6 Scheme of power-compensational d.s.c. apparatus

the reference (R) are heated separately by individually controlled heating elements (Hs and Hr). The temperatures of the sample and reference are measured by two thermocouples T1 and T2 and the amplified signals are transmitted to the computer as input data for two separate 'control loops', the first for average temperature control and the second for differential temperature control. In the average temperature loop a programmer provides an electric signal that is proportional to the desired temperature of the sample and reference holders. This signal is compared with the corresponding one received from the thermocouples. The same signals measured by the pair of thermocouples before entering the differential temperature control loop are fed to a differential temperature amplifier via a comparator circuit. The output of this circuit determines the magnitude and direction of the power increment to the required heater to correct any temperature difference.

For the powder paint chemist, d.s.c. has a considerable advantage over d.t.a. This emanates from fundamental differences between the techniques. In d.s.c. acquisition of the calorimetric data resulting from any thermal effects is determined directly from the energy input in order to compensate for the difference in temperature between the sample and the reference. With d.t.a. the calorimetric information is obtained indirectly from thermometric data. Not only is interpretation of the results obtained from d.t.a. more complicated but the reliability of the technique is in many cases doubtful for quantitative interpretation. In other words, since the thermal conductivity of the sample can change during the transition, the proportionality between the temperature differences and the energy changes is in general unknown. In such cases, d.t.a. can provide more or less qualitative results. In contrast, in d.s.c. the peak area is always a true measure of the energy input required to keep the sample and reference temperatures equal, independent of the changes in the thermal behaviour of the sample or the thermal characteristics of the instrument itself. With proper design the performances of the d.t.a. apparatus can get close to those of d.s.c. However, the ease of handling and processing the acquired data and their reliability made d.s.c. the most common instrument used for executing thermal analysis in powder coating laboratories.

Next to the kinetic parameters of the curing process, which can be obtained by the d.s.c. technique, several additional facts relevant to powder coating can be gathered. These include glass transition temperatures of the powder coating and the components used in the coating composition, the crystalline melting point of the polymers (especially in the case of thermoplasts) and the melting point of the crosslinkers, the onset flow temperature of the powder coating and the onset cure temperature. Indirectly, from the kinetic parameters, the conversion of the curing reaction with time at a certain temperature can be easily calculated. This gives an indication of the necessary curing time at a given temperature or vice versa, which can easily be related to common practical problems.

Three different methods can be used to obtain kinetic information from dynamic thermal experiments. The easiest method developed by Borchardt and Daniels [2] which is widely used to examine the cure behaviour of powder coatings involves analysis of only one exotherm. A deficiency of this method is its lack of consistent accuracy in measuring kinetic parameters [3]. Much more reliable results concerning the activation energy and frequency factor for different types of reactions can be obtained by the method based on variations in the peak exotherm temperature with heating rate [4]. The last method also uses experiments at different heating rates and analyses the dependence of the temperature to reach a constant conversion on the heating rate [5, 6]. Although the last two methods provide much more reliable results, the method that acquires the kinetic parameters from analysis of only one exoterm seems to be the most attractive. In fact its accuracy is quite acceptable in practice for most cases dealing with powder coatings.

Figure 3.7 represents a typical d.s.c. scan of a powder coating in which the carboxyl functional polyester binder is cured with bisphenol A epoxy resin. This graph will be used to explain the methods that use the data obtained from d.s.c. for kinetic analysis of the curing process.

The characteristic points of the curve marked by T_g and T_{of} will be discussed later. The exothermal reaction of curing starts at a considerable rate at temperature T_{oc}, a designation for the temperature onset cure. As already stated, the advantage of the d.s.c. technique compared to d.t.a. is in the basic principle of the technique of measuring the power required to prevent the temperature change in the sample compared to the reference due to the thermal effects during

THERMOSETTING POWDER COATINGS

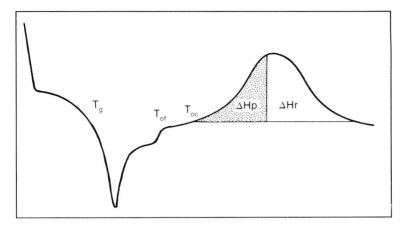

Figure 3.7 Typical d.s.c. curve of a polyester/epoxy powder coating

the process that is monitored. This means that dH/dt, the time rate of change of the enthalpy of the curing reaction, can be directly followed, which makes the kinetic analysis very simple since the thermal energy effects measured directly with the instrument can be correlated to the reaction speed.

In kinetic calculations based on data acquired by d.s.c. the rate of change in certain materials accompanied by thermal effects has to be related to the temperature. The heat evolved or absorbed should be proportional to the number of moles reacting, as expressed by

$$\frac{dH}{dt} = -\frac{(\Delta H/N_0)dN}{dt} \tag{3.81}$$

where N_0 is the initial number of moles of the reactant, $-dN/dt$ is the loss of moles per unit time, dH/dt is the corresponding enthalpy change and ΔH is the enthalpy of the reaction.

The area under the thermal peak in Figure 3.7 is a measure of the enthalpy of the reaction and it can be assumed that the total area under the peak (ΔH_t) corresponds to the total thermal effect of the process. Since the degree of conversion is proportional to the respective enthalpy change, the relation between the initial number of moles of reactant (N_0), the number of moles at time t, the part of the peak area (ΔH_p) at time t, the residual peak area (ΔH_r) and the total area under the peak $(\Delta H_t = \Delta H_p + \Delta H_r)$ can be written as follows:

$$\frac{N}{N_0} = \frac{\Delta H_r}{\Delta H_t} \tag{3.82}$$

Consequently the relation between the d.s.c. signal and the conversion of the reaction with respect to the initial sample α will be given by

$$\alpha = \frac{\Delta H_p}{\Delta H_t} \tag{3.83}$$

The differential equation

$$\frac{d\alpha}{dt} = k(1-\alpha)^n \tag{3.84}$$

in which k represents the specific rate constant and n the reaction order can be written for many reactions at a given temperature [7]. The rearrangement and integration of equation (3.83) gives the following results:

$$\frac{d\alpha}{dt} = k[kt(n-1)+1]^{n/(1-n)} \tag{3.85}$$

$$\alpha = 1 - [kt(n-1)+1]^{1/(1-n)} \tag{3.86}$$

For a first-order reaction, the following equations can be obtained:

$$\frac{d\alpha}{dt} = k\exp(-kt) \tag{3.87}$$

$$\alpha = 1 - \exp(-kt) \tag{3.88}$$

The reaction rate constant is generally assumed to have a temperature dependence according to the well-known Arrhenius equation:

$$k = k_0 \exp\left(\frac{-E_a}{RT}\right) \tag{3.89}$$

Here k_0 is the preexponential constant or the so-called frequency factor, E_a is the activation energy of the reaction, R is the gas constant and T is the absolute temperature in kelvin.

Combining equations (3.84) and (3.89) gives

$$\frac{d\alpha}{dt} = k_0 \exp\left(\frac{-E_a}{RT}\right)(1-\alpha)^n \tag{3.90}$$

The logarithmic form of equation (3.90) is very suitable for further analysis:

$$\ln\left(\frac{d\alpha}{dt}\right) = \ln k_0 - \frac{E_a}{RT} + n\ln(1-\alpha) \tag{3.91}$$

Combining equations (3.81), (3.82) and (3.83) gives the following relation between the degree of conversion and reaction enthalpy:

$$\frac{dH}{dt\,\Delta H_t} = \frac{d\alpha}{dt} \tag{3.92}$$

By proper substitution in (3.91) the following equation, which includes data that can be easily acquired by a single dynamic run, is obtained:

$$\ln\left(\frac{dH}{dt\,\Delta H_t}\right) = \ln k_0 - \frac{E_a}{RT} + n\ln\left(\frac{\Delta H_r}{\Delta H_t}\right) \quad (3.93)$$

The three unknowns in equation (3.93), k_0, E_a and n, can be obtained using three of the many points of the d.s.c. curve. Contemporary d.s.c instruments perform multilinear regression analysis using $\ln(dH/dt\,\Delta H_t)$, $1/T$ and $\ln(\Delta H_r/\Delta H_t)$ as variables that have been evaluated from the d.s.c. data. In this way the accuracy of the results is increased since any number of curve points can be evaluated. Knowing the activation energy, the preexponential constant and the reaction order, conversion of the reaction with time can easily be calculated at any given temperature using equations (3.86) or (3.88). An example of such applied kinetic calculations obtained on a Mettler TA3000 d.s.c. system for a curing reaction of polyester/TGIC powder coating is presented in Figure 3.8.

The d.s.c. method has been widely used to examine the curing behaviour of powder coatings. The interested reader can consult a number of published papers

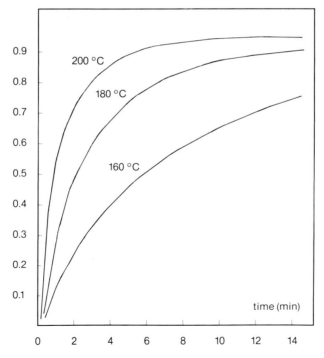

Figure 3.8 Degree of curing as a function of time at 160, 180 and 200°C for polyester/TGIC powder coating

in which the curing kinetics of different powder coating systems, curing mechanism or activity of different catalysts have been studied by the d.s.c. technique [8–17]. Apart from the kinetic data, which can be obtained by analysis of the d.s.c. curve, other relevant data can also be obtained from a single d.s.c. run. The glass transition temperature of the powder coating, whose importance will be discussed later in Chapter 4, is the temperature at which there is a 'sudden' increase in the specific heat (C_p) of the powder coating. This is manifested by a shift in the baseline of the d.s.c. curve (Figure 3.7, point T_g). The International Confederation of Thermal Analysis (ICTA) proposes an evaluation procedure to be used to determine the glass transition temperature (Figure 3.9) [18, 19]. According to this procedure three regression lines R_1, R_2 and R_3 are applied to the d.s.c. curve: the regression line before (R_1) and after (R_3) the event and the regression line at the inflection point (R_2). These three lines define the temperature T_{g1} as the first significant deviation from R_1, T_{g2} as the intersection between R_1 and R_2 and T_{g3} as a mid point between the two intersection points. T_{g2} is considered to be the relevant value for the glass transition temperature.

It should be noted that the values for the T_g obtained by d.s.c. are dependent on the heating rate chosen during the experiment. This is not an unexpected

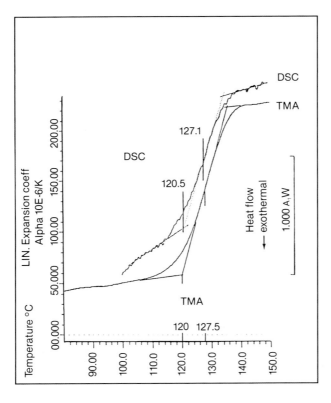

Figure 3.9 Evaluation of the characteristic glass transition temperatures by the ICTA procedure for d.s.c. curves [18] (Reproduced by permission of Dr R. Riesen, Mettler Instruments AG, CH-8606 Greifensee)

THERMOSETTING POWDER COATINGS

phenomenon since, as will be seen later, the glass transition temperature is related to the mobility of the polymer chains. This is a kinetic dependent process and the rate of the experiment will influence the results obtained. Values of T_g determined by the mid-point method will increase with a increasing heating rate. Gargallo and Russo, for example, using the d.s.c. technique, have obtained T_g values of 26, 29, 34 and 39°C for heating rates of 4, 8, 16 and 32°C/minute respectively for the case of poly(neopentyl) methacrylate [20]. In order to avoid confusion, the T_g values obtained by d.s.c. measurements should always be accompanied by the data concerning the heating rate.

Other useful information that can be obtained by the d.s.c. technique is the temperature at which the powder coating begins to melt. The larger contact surface between the sample and the walls of the sample holder after melting

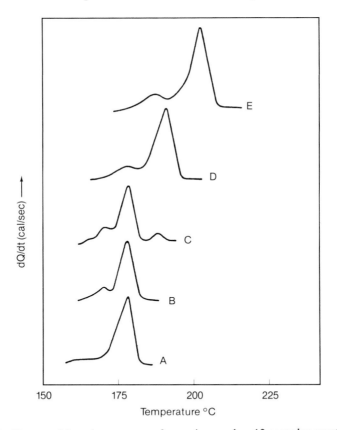

Figure 3.10 D.s.c. melting thermograms for various nylon-12 samples crystallized for 16 h under a pressure of 4.9 kbar at different temperatures: (A) original sample; (B) crystallization at 260°C; (C) crystallization at 250°C; (D) crystallization at 240°C; (E) crystallization at 240°C for 48 h [21] [Reproduced by permission of the publishers, Butterworth & Co. (Publishers) Ltd. ©]

influences the heat transfer characteristics of the system. This is manifested by the shift in the baseline. The temperature at which this shift is observed is called the temperature onset flow (point T_{of} in Figure 3.7) [14]. Depending on the reactivity of the system, the exothermal peak is observed to start at the so-called temperature onset cure (T_{oc} in Figure 3.7). The distance between T_{of} and T_{oc} gives an indication of the time available for the powder coating to flow. The size of the plateau between the melting of the powder and the start of the exothermal reaction on the d.s.c. curve cannot be quantitatively correlated to the flow properties of the powder coating. Relations between the factors influencing the flow are rather complex and taking only the time available for flow as a single element is an oversimplification of the problem. However, it gives an indication of the activity of the catalysts used in the system, and especially an indication about their latency, which is very often a desired characteristic of a good catalyst.

The d.s.c. technique can be used to determine the melting temperature of the crystalline polymers, which is an important parameter for thermoplastic powder coatings. Since the melting of the crystallites is an endothermal process, it is easily registered by d.s.c. as a sharp endothermal peak with its minimum corresponding to the melting temperature. Stamhuis and Pennings have investigated the crystallization of nylon-12 [21]. Thermograms of nylon-12 samples crystallized under various conditions are shown in Figure 3.10.

3.2.2 THERMOGRAVIMETRIC ANALYSIS (t.g.a.)

The d.s.c. technique can be successfully used in the cases of curing where the thermal effect of the curing reaction is not accompanied by the other phenomena characterized by emission or absorption of energy. In principle the cure kinetics of processes involving reactions between pairs of functional groups such as acid/epoxy, acid anhydride/epoxy, acid anhydride/hydroxyl, epoxy/amine and epoxy/phenolic can be successfully followed by d.s.c. However, in cases where the curing of powder coatings involves simultaneously several chemical or physical transformations, each having its own thermal effect, the interpretation of the results obtained by d.s.c. is rather complex and in many cases impossible. An example of such a curing process is the crosslinking of hydroxyl functional polymers with blocked isocyanates in polyurethane powder coatings. The curing process starts with deblocking of the polyisocyanate. This reaction, as well as evaporation of the blocking agent from the systems, is endothermal. The crosslinking between hydroxyl groups and isocyanate in the next step is an exothermal reaction. Therefore, the d.s.c. curve obtained as a result of these three processes cannot supply data relevant and reliable for kinetic analysis. A similar example is the curing of hydroxyl functional polymers with amino resins, where methanol or other higher alcohol is a by-product of the curing reaction.

In such cases, thermogravimetry is a technique that can be used with partial success for acquiring useful data. Thermogravimetric analysis (t.g.a.) is a rather old thermal analysis technique. Although the first thermobalances had been constructed at the beginning of this century, the first commercial instrument appeared in 1945, based on the work of Chevenard, Wache and De la Tullaye [22]. This thermobalance was used for the first time in the polymer field in 1948 by Duval [23]. Contemporary instruments with a sensitivity of 0.1 μg and a precision of one part in 100 000 of the mass change were developed in the 1960s by several producers [24]. The t.g.a. instrument is designed to produce a continuous record of weight as a function of temperature. The sample can usually be subjected to a linear temperature change or may be isothermally maintained at a given temperature. Figure 3.11 presents a schematic diagram of a typical t.g.a. instrument. It consists of the following main parts: furnace (F), furnace temperature programmer, recording balance, a part of which is the sample holder with a temperature sensor (S), and recorder.

The most important component of the t.g.a. instrument is the recording balance. The wide variety of recording balances can be divided into three main groups based on their mode of operation: (a) deflection-type instruments, (b) null-type instruments and (c) those based on changes in resonance frequency [24, p. 10]. Deflection balances involve the conversion of balance beam deflections into an electrically recorded signal by means of an appropriate displacement measuring transducer. Four different types of deflection balance are presented in Figure 3.12 [7, p. 282].

The automatic null-type balance incorporates a sensing element that detects a deviation of the balance beam from its null position. A restoring force of either electrical or mechanical mass loading is then applied to the beam through an appropriate electronic or mechanical linkage, restoring it to the null position. The principle of the null-type balance is presented in Figure 3.13. The null-type balance principle is now used in almost all commercially available t.g.a. instruments [24]. They can convert the analog signal output to the derivative of the mass change or digitize it by an analog-to-digital converter for processing by digital computers.

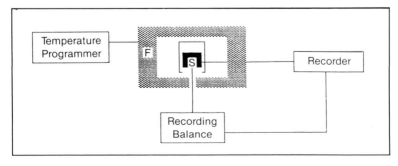

Figure 3.11 Schematic diagram of t.g.a. instrument

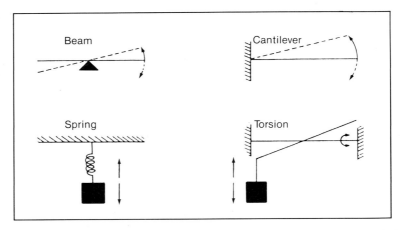

Figure 3.12 Deflection-type balance principles [24] (Reproduced by permission of Academic Press)

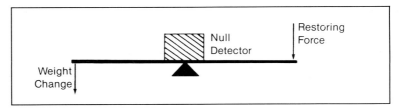

Figure 3.13 Null-type balance principle [7] (Reprinted from [7] by courtesy of Marcel Dekker, Inc.)

The use of the t.g.a. for evaluation of the kinetic parameters of the curing process of powder coatings makes sense only when the deblocking mechanism is the slowest step determining the overall curing rate and when the evaporation of the blocking agent is fast.enough and thus does not affect the equilibrium of the curing reaction. The second assumption can be considered as valid for methoxymethyl amino resins, since the evaporation of the methanol which is the by-product of the curing reaction at the curing temperature is rather high. However, even in such a case, Bauer and Dickie have shown that although very short, the residence time of the free methanol in the film during the curing process affects the reaction rate in the case of solvent based acrylic coatings cured with HMMM [25]. In the case of polyurethane powder coatings the curing rate should not be affected by the presence of caprolactam above the deblocking temperature, assuming that the curing mechanism is a two-step process in which the urethane formation is preceded by unimolecular thermal dissociation of the blocked isocyanate to free isocyanate. As already discussed, it is quite possible that in the presence of compounds containing active hydrogen, the curing mechanism can also be a direct bimolecular displacement reaction. Therefore it is difficult to predict how reliable the kinetic parameters can be when evaluated from data acquired by thermogravimetric analysis of the cure of powder coatings. Although t.g.a. has been widely used in determining the kinetics of polymer systems, it is very difficult to find any paper referring to the use of t.g.a. for following the curing process of powder coatings.

The method of Freeman and Carrol [26], in spite of its limited precision [27], is quite convenient for processing the acquired t.g.a. data, since it belongs to the so-called derivative methods.

The weight lost per unit time during the t.g.a. experiment can be written as

$$-\frac{dw}{dt} = kf(w) \qquad (3.94)$$

where w is the weight of the reactive portion of the sample, which means the actual weight at time t minus the final weight of the sample at the end of the experiment, and k is the reaction rate constant.

Assuming that $f(w) = w^n$, then

$$-\frac{dw}{dt} = Z \exp\left(-\frac{E_a}{RT}\right) w^n \qquad (3.95)$$

The above equation can be written in terms of temperature, introducing the linear heating rate F:

$$-\frac{dw}{dT} = \left(\frac{Z}{F}\right) \exp\left(-\frac{E_a}{RT}\right) w^n \qquad (3.96)$$

or in logarithmic form:

$$\ln\left(-\frac{dw}{dt}\right) = \ln\left(\frac{Z}{F}\right) - \frac{E_a}{RT} + n \ln w \qquad (3.97)$$

Differentiating equation (3.97) and then integrating under the assumption that the frequency factor Z is temperature independent, the following equation suitable for graphical determination of E_a and n is obtained:

$$\Delta \ln\left(-\frac{dw}{dt}\right) = -\left(\frac{E_a}{R}\right) \Delta\left(\frac{1}{T}\right) + n \Delta \ln w \qquad (3.98)$$

Freeman and Carrol suggest the following form of equation (3.98):

$$\frac{\Delta \log(-dw/dt)}{\Delta \log w} = \left[\frac{(-E_a/R)\Delta(1/T)}{2.303 \Delta \log w}\right] + n \qquad (3.99)$$

Plotting the values for $\Delta \log(-dw/dt)/(\Delta \log w)$ versus $\Delta(1/T)/(\Delta \log w)$, the values of E_a and n can be obtained from the slope and the intercept of the curve.

It has been found convenient to have the logarithm of the experimental rates plotted initially against $1/T$, so that the obtained curve may be used to select a set of values for constant increments in $1/T$. Keeping $1/T$ constant a subsequent plot of $\Delta \log(-dw/dt)/(\Delta \log w)$ versus $1/(\Delta \log w)$ leads to a value of n as the intercept and E_a as the slope of the curve.

Several other derivative methods include the method of multiple heating rates [28, 29], the method of maximal rates suitable only for determination of the activation energy [7, p. 320] and the method which assumes a prior correct analytical form of the kinetic equation for determination of the kinetic parameters [30].

A variety of integral methods have been developed for evaluation of kinetic data obtained in t.g.a. experiments. Most of them are based on the best fit of curve obtained from the assumed analytical form of the equation expressing the kinetic of the process with the experimental t.g.a. trace. The equation usually selected is similar to that used in the d.s.c. experiments:

$$f(\alpha) = (1-\alpha)^n \qquad (3.100)$$

Here the meaning of α is the same as before, i.e. the degree of completion of the reaction with respect to the initial sample, and n represents the reaction order.

From the several integral methods proposed by different authors [31–35] the one developed by Coats and Redfern [31] seems to be the most suitable from a practical point of view. This method assumes the validity of equation (3.100) and takes into account the equation describing the temperature dependence of the degree of conversion in a linear temperature programme during the execution of the t.g.a. experiment equivalent to equation (3.96):

$$\frac{d\alpha}{dT} = \left(\frac{Z}{F}\right) f(\alpha) \exp\left(-\frac{E_a}{RT}\right) \qquad (3.101)$$

The solution of the differential equation (3.101) for the boundary conditions $T = T_0 = 0$, $\alpha = 0$ and $T = T$, $\alpha = \alpha$ gives the following expressions:

$$\frac{(1-\alpha)^{1-n} - 1}{1-n} = \left(\frac{ZE_a}{FR}\right) p(x). \quad \text{for } n \neq 1 \qquad (3.102)$$

$$\ln(1-\alpha) = \left(\frac{ZE_a}{RT}\right) p(x). \quad n = 1 \qquad (3.103)$$

where $p(x)$ is a polynomial solution of the integral appearing in the solution of the equation (3.101) having the following form:

$$p(x) = \left(\frac{e^x}{x^2}\right)\left[1 + \left(\frac{2!}{x}\right) + \left(\frac{3!}{x^2}\right) + \ldots\right]\ldots \qquad (3.104)$$

where $x = E_a/RT$.

THERMOSETTING POWDER COATINGS

The assumption $T_0 = 0$ can be considered as valid since the reaction rate at low temperature is negligible. The method of Coats and Redfern takes into account the first three terms of the asymptotic approximation of equation (3.104), thus giving the following final form of expression (3.102):

$$\frac{(1-\alpha)^{1-n}-1}{1-n} = \left(\frac{ZRT^2}{E_a F}\right)\left[1-\left(\frac{2RT}{E_a}\right)\right]\exp\left(\frac{-E_a}{RT}\right) \qquad (3.105)$$

or in logarithmic form:

$$\log\left[\frac{1-(1-\alpha)^{1-n}}{T^2(1-n)}\right] = \log\left\{\left(\frac{ZR}{E_a F}\right)\left[1-\left(\frac{2RT}{E_a}\right)\right]\right\}\frac{-E_a}{2.3RT} \quad \text{for } n \neq 1$$

$$(3.106)$$

$$\log\left[-\log\left(\frac{1-\alpha}{T^2}\right)\right] = \log\left(\frac{ZR}{E_a F}\right)\frac{-E_a}{2.3RT} \qquad \text{for } n = 1 \qquad (3.107)$$

Thus the plot of either

$$\log\left[\frac{1-(1-\alpha)^{1-n}}{T^2(1-n)}\right] \text{ versus } \frac{1}{T} \quad \text{for } n \neq 1$$

or

$$\log\left[-\log\left(\frac{1-\alpha}{T^2}\right)\right] \text{ versus } \frac{1}{T} \quad \text{for } n = 1$$

should result in a straight line with a slope $-E_a/2.3R$ and intercept reading $\log[(ZR/E_a F)(1-2RT/E_a)]$ or $\log(ZR/E_a F)$ for a correct value of n.

3.2.3 THERMAL MECHANICAL ANALYSIS (t.m.a.)

The film formation of thermosetting powder coatings is a complex process. It starts with melting and coalescence of the individual powder particles into a continuous film followed by crosslinking of the low molecular weight binder. This results in a permanent change in the physical–chemical properties of the material, until the final film properties are obtained by completion of the curing reaction to a desired degree.

For following the degree of conversion of the functional groups involved in the curing reactions d.s.c. and t.g.a. are helpful techniques. As already discussed, by using these techniques, kinetic data relevant for predicting the curing behaviour of the system can be obtained. Using these kinetic parameters, the degree of conversion of the functional groups during curing at different temperatures can be calculated as a function of time or vice versa. However, d.s.c. and t.g.a. do not provide information with respect to the changes of mechanical and rheological properties of the coating during the film forming process. Such information can be gathered by dynamic mechanical analysis (d.m.a.). They can be correlated with the flow behaviour and mechanical properties of the powder coating as a function of the resin/crosslinker ratio and resin structure. Combined with the kinetic data obtained by d.s.c. or t.g.a., mechanical properties can be followed as a function of the degree of curing. As well as d.s.c., the d.m.a. technique can be used to determine the physical transitions in the coating, temperature onset flow and temperature onset cure.

The dynamic mechanical analysis (d.m.a.) as a technique for characterization of polymeric materials exists for a relatively long time. However, the use of d.m.a. for characterization of cured coatings and for following the progress of curing reactions is not widely described in the literature typical for the coating field. Therefore a short introduction of the theory backing up the d.m.a. seems to be useful before starting with a description of the technique.

According to Hook's law, which assumes perfect elasticity of a material body, the relation between the tensile stress (σ) and strain (ε) is described by

$$\sigma = E\varepsilon \qquad (3.108)$$

where E is defined as Young's modulus, serving as a measure of the stiffness of the material. This means that materials with a higher modulus are more resistant to being stretched. If the sample initial length is L_0 and after being subjected to stress σ the final length is L, then the strain ε is defined as (Figure 3.14)

$$\varepsilon = \frac{L - L_0}{L_0} \qquad (3.109)$$

It is obvious that the stress should be defined in terms of force per unit area.

In another type of deformation the sample can be subjected to a shearing or twisting. In the case of a perfect elastic body, a cube of material fixed with one side to an immobile surface subjected to the action of a force parallel to the surface will

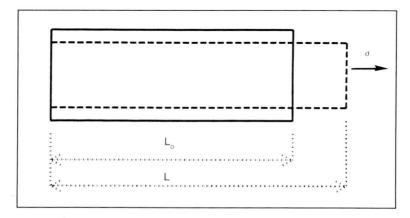

Figure 3.14 Behaviour of a perfect elastic body subjected to stress σ

THERMOSETTING POWDER COATINGS

take another position, as presented in Figure 3.15. As in the case of tensile deformation, the shear stress and shear rate will be connected through the shear modulus G:

$$\sigma = G\varepsilon \tag{3.110}$$

The above equation should be compared to the one used to define a perfect liquid or so-called Newtonian liquid:

$$f = \mu\left(\frac{ds}{dt}\right) \tag{3.111}$$

where f and s represent the shear stress and strain respectively, t is the time and μ is the viscosity of the liquid as a measure of the resistance of the liquid to flow. This equation is valid for many low molecular materials above their melting point.

Equation (3.111) defines that, in an ideal Newtonian liquid, the ratio of the force that opposes the flow to the rate of flow is constant. On the other hand, in an ideal Hookean solid (equation 3.110) the ratio of the force that opposes deformation to the extent of deformation is constant. While neither equation (3.110) nor (3.111) accurately describes polymer behaviour, they represent two important limiting cases.

Powder coatings, as is the case with most of the polymeric materials, neither obey equation (3.110) describing the behaviour of a perfect elastic body under stress nor equation (3.111) describing the behaviour of a perfect liquid under the same conditions. They are defined as viscoelastic materials exhibiting some of the properties of solids and some of the properties of liquids. This means that under an applied force they undergo deformation as solid materials, but also undergo flow in the manner of liquids. Depending on the magnitude of these phenomena powder coatings can be identified as viscoelastic liquids immediately after

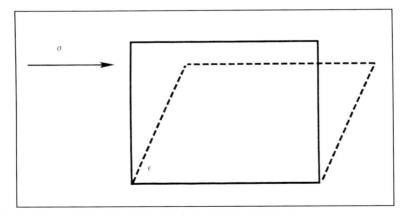

Figure 3.15 Shear type of deformation of an elastic body

melting, when the flow is the usual response to the applied force and as viscoelastic solids as the curing progresses, when the response of the cured film to the applied force is a deformation.

The quantities E and G in equations (3.108) and (3.110) refer to measurements not dependent on time, or in other words to static measurements (the definition of a static measurement here is debatable, since the experiment is performed within a certain time scale anyway). In the case of dynamic mechanical measurements, the sample in question is subjected to cyclic or repetitive motions. In this type of experiment, time as a factor influencing the results should be introduced.

Figure 3.16 represents schematically the principal construction of the d.m.a. instrument.

It is common practice in dynamic mechanical measurements for the applied force to vary sinusoidally with time, by a rate usually specified by the frequency f

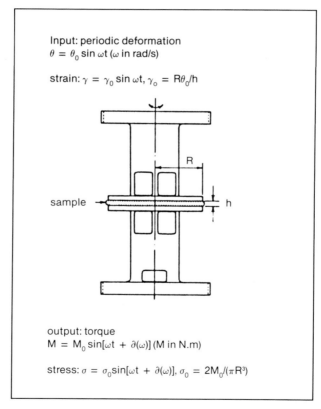

Figure 3.16 Measuring principle of the mechanical spectrometer [36] (Reproduced by permission of DSM Resins BV)

THERMOSETTING POWDER COATINGS

in cycles per second or $\omega = 2\pi f$ in radians per second. For Hookean solids having the behaviour of a perfect elastic body the strain will follow the same pattern. In the case of powder coatings, when such a sinusoidal excitation is applied to the sample, the strain will alternate sinusoidally also. However, depending on the temperature of the sample and the degree of curing, the strain will be out of phase with the stress. This phase lag results from the time necessary to make molecular rearrangements and is associated with relaxation phenomena:

$$\sigma = \sigma_0 \sin(\omega t + \delta) \quad (3.112)$$

$$\varepsilon = \varepsilon_0 \sin \omega t \quad (3.113)$$

In equations (3.112) and (3.113), δ is the phase angle, σ_0 is the peak stress and ε_0 is the peak strain (Figure 3.17).

Equation (3.112) can be expanded according to the well-known trigonometrical transformation to

$$\sigma = \sigma_0 \sin \omega t \cos \delta + \sigma_0 \cos \omega t \sin \delta \quad (3.114)$$

In d.m.a. experiments the stress is resolved into two components, one in phase with the strain $\sigma_{(in)}$ and the other out of phase with the strain $\sigma_{(out)}$, each given by

$$\sigma_{(in)} = \sigma_0 \cos \delta \quad (3.115)$$

$$\sigma_{(out)} = \sigma_0 \sin \delta \quad (3.116)$$

The values of in-phase and out-of-phase stress divided by the strain separate the modulus into in-phase (G') and out-of-phase (G'') components:

$$G' = \left(\frac{\sigma_0}{\varepsilon_0}\right) \cos \delta \quad (3.117)$$

$$G'' = \left(\frac{\sigma_0}{\varepsilon_0}\right) \sin \delta \quad (3.118)$$

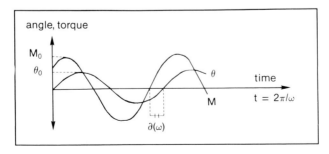

Figure 3.17 Stress and strain as a function of time in sinusoidal excitation of a viscoelastic specimen

Consequently the value of the stress from equation (3.114) can be expressed in the following way:

$$\sigma = \varepsilon_0 G' \sin \omega t + \varepsilon_0 G'' \cos \omega t \qquad (3.119)$$

The ratio between the peak stress and peak strain is called the dynamic modulus G_d:

$$G_d = \frac{\sigma_0}{\varepsilon_0} \qquad (3.120)$$

The modulus can also be expressed in complex form as follows:

$$\sigma = \sigma_0 \exp i(\omega t + \delta) \qquad (3.121)$$

$$\varepsilon = \varepsilon_0 \exp i\omega t \qquad (3.122)$$

$$\frac{\sigma}{\varepsilon} = G_d \exp i\delta = G_d(\cos \delta + \sin \delta) \qquad (3.123)$$

$$G^* = \frac{\sigma}{\varepsilon} = G' + iG'' \qquad (3.124)$$

The relations between the different components of the modulus are summarized in the so-called Argand diagram in Figure 3.18.

The in-phase component G' is called the real part of the modulus or the storage modulus because it is related to the energy stored in the specimen as a potential energy which is released in the periodic deformation. The out-of-phase component G'', also called the imaginary part of the modulus, is associated with the viscous loss of energy, or the energy which is dissipated as heat when the specimen is deformed.

The ratio between the loss and the storage component of the modulus equals the tangent of the phase shift (the Argand diagram, Figure 3.18)

$$\tan \delta = \frac{G''}{G'} \qquad (3.125)$$

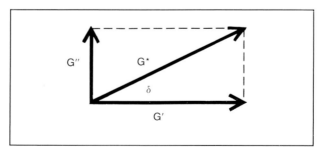

Figure 3.18 Definition of the storage G' and loss G'' moduli and tan δ under application of a sinusoidal strain to a viscoelastic specimen

This so-called loss tangent is a dimensionless term representing the ratio of the energy lost (dissipated as heat) per cycle to the energy stored and hence recovered per cycle. The loss tangent is temperature dependent and is at a maximum near the glass transition temperature. Thus, by an isofrequent scan at different temperatures, the glass transition temperature of the polymers can be located at the maximum of the curve representing tan δ versus temperature. As in the case of determination of T_g by the d.s.c. technique, here again it should be stressed that the T_g determined by d.m.a. is the frequency dependent value, as presented in Figure 3.19 [37].

When equation (3.111) is related to the dynamic mechanical measurements, the first derivation of the strain with time can be replaced by the oscillation frequency. Thus, the determination of the dynamic modulus of the specimen at a certain frequency enables us to obtain the value of the dynamic viscosity of the system:

$$\mu = \frac{G_d}{w} \qquad (3.126)$$

According to the theory of rubber elasticity [38, 39] the degree of crosslinking above the glass transition temperature can be expressed through the concentration of the elastically active network chains, V_e:

$$V_e = \frac{G'_{(0)}}{RT} \qquad (3.127)$$

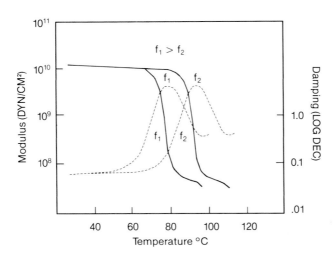

Figure 3.19 Schematic diagram of the effect of frequency on modulus and damping temperature curves [37]

where $G'_{(0)}$ is the value of the storage modulus at zero frequency, $R = 8.31$ J/(mol K) and T is expressed in kelvin.

In principle, powder coating can be scanned over a wide frequency range or a wide temperature range in order to study motional transitions and changes in the mechanical and rheological properties during curing. While the frequency scans at constant temperature are suitable for the theoretical study of the relaxation phenomena, the temperature scans at constant frequency provide more relevant data for following the progress of the curing reactions.

Scholtens, van der Linde and Tiemersma-Thoone [36] suggest the following experiment design in dynamic mechanical analysis to acquire data that are relevant for the curing process of powder coatings.

A compression moulded sample (about 10 minutes at 90°C for polyester/TGIC systems) is placed between two concentric plates of the instrument. Firstly, an oscillation run, starting at $\log w = -0.2$ up to $+2.0$ with steps of $\log w = 0.2$, is performed at 110–120°C. At this temperature the curing of the powder coating can be neglected. This oscillation run provides information on the melt viscosity of the powder coating. Figure 3.20 shows the d.m.a. isotherms of unpigmented polyester/epoxy hybrid powder coating catalysed with different amounts of dimethyl benzyl amine. The phase angle and dynamic modulus are presented as functions of the frequency. A remarkable increase in dynamic modulus and consequently in viscosity can be observed in the system with ten times higher catalyst concentration. This is an indication that some prereaction has taken

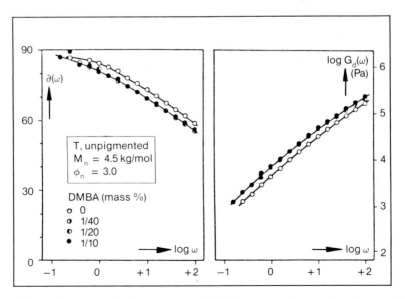

Figure 3.20 The d.m.a. isotherms at 110°C for an unpigmented polyester/TGIC powder coating at four concentrations of DMBA catalyst [36] (Reproduced by permission of DSM Resins BV)

place during the mixing of the components in the extruder or during the measuring stage. It is obvious that powder coating behaves as a viscoelastic material since the phase angle dependent on the catalyst concentration and applied frequency lies between 35 and 90°. Otherwise a purely viscous material is characterized by $\delta = 90°$ and $G_d/w = $ constant (equal to the zero shear viscosity μ_0) and a purely elastic material by $\delta = 0°$ and $G_d = $ constant (equal to the equilibrium modulus G_{d0}).

For monitoring the curing of the coating, the oven of the instrument is heated to 200°C linearly at a heating rate of 20°C/min. The crosslinking is followed by a period of 10 minutes at a constant frequency of log $w = 0.2$ by determining the values of the storage modulus. Using equations (3.117) and (3.126) the dynamic viscosity of the system can be easily calculated. This series of measurements shows the initial decrease in viscosity as a result of the heating and subsequent onset of network formation and network build-up as the curing proceeds, as presented in Figure 3.21.

Determination of the dynamic modulus after completion of the curing process makes it possible to calculate the degree of crosslinking in the completely cured film. For this purpose the final oscillation run should be performed at a different oscillation rate at constant temperature which is higher than the T_g of the system. From an extrapolation of G' to zero frequency, the equilibrium modulus $G'_{(0)}$ can be obtained and the crosslinking density of the fully cured film can be calculated

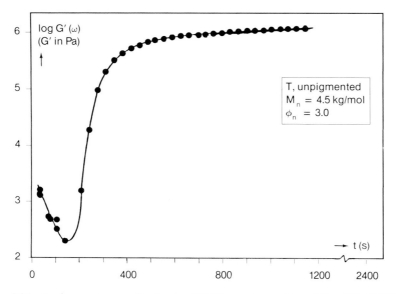

Figure 3.21 Isofrequent runs of polyester/TGIC powder coating at $T = 110–200°C$ at a heating rate of 20°C/min. (Adapted from Ref. [36]) (Reproduced by permission of DSM Resins BV)

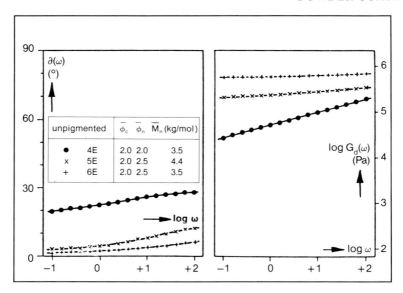

Figure 3.22 The d.m.a. isotherms at 200°C for polyester/epoxy hybrid powder coating after completion of the curing reaction [36] (Reproduced by permission of DSM Resins BV)

using equation (3.127). Figure 3.22 represents d.m.a. isotherms at 200°C for three polyester/epoxy hybrid powder coatings prepared from polyester resins that differ in functionality ($F_n = 2.0$ and 2.5) and molecular weight ($M_n = 3500$ and 4400). The relatively high value of δ and the frequency dependence of G_d indicates that the reaction between linear polyester and linear epoxy resin (4E) does not lead to a permanent network. However, the increase of the functionality to 2.5 causes the formation of the permanent networks (5E and 6E). The lower molecular weight of polyester 6E results in a higher value for the dynamic modulus, indicating that the molecular weight between the crosslinking points (M_c) in this case is lower.

The use of d.m.a. to study the influence of the ratio between the functional groups and the catalyst concentration in the powder coating composition, discussed later in Chapter 4, is another indication of the potential use of this technique in powder coating.

REFERENCES

1. Widmann, G., Proceedings of a Symposium in *Polymerteknisk Selskab*, 7 May 1980.
2. Borchardt, H. J., and Daniels, F., *J. Am. Chem. Soc.*, **79**, 41, 1956.
3. Prime, R. B., in *Thermal Characterisation of Polymeric Materials*, Ed. E. A. Turi, Academic Press, New York, 1981, p. 532.

4. Ozawa, T., *J. Therm. Anal.*, **2**, 301, 1970.
5. Fava, R. A., *Polymer*, **9**, 137, 1968.
6. Barton, J. M., *J. Macromol. Sci. Chem.*, **A8** (1), 53, 1974.
7. Manche, E. P., and Carrol, B., in *Physical Methods in Macromolecular Chemistry*, Ed. B. Carrol, Vol. 2, Marcel Dekker, Inc., New York, 1972, p. 309.
8. Mohler, H., *Farbe und Lack*, **86**, 211, 1980.
9. Henig, A., Jath, M., and Mohler, H., *Farbe und Lack*, **86**, 313, 1980.
10. Zicherman, J. B., and Holsworth, R. M., *J. Paint. Technol.*, **46**, 55, 1974.
11. Peyser, P., and Bescom, N. D., *J. Appl. Polym. Sci.*, **21**, 2359, 1977.
12. Olcese, T., Spelta, O., and Vargiu, S., *J. Polym. Sci. Symp.*, **N53**, 1975, p. 113.
13. Gabriel, S., *J. Oil Col. Chem. Assoc.*, **58**, 52, 1975.
14. Klaren, C. H. J., *J. Oil Col. Chem. Assoc.*, **60**, 205, 1977.
15. Linde, R. van der, and Belder, E. G., *Proceedings of VIIth International Conference on Organic Coating Science Technology*, Athens, Greece, July 1981, p. 359.
16. Ghijssels, A., and Klaren, C. H. J., *Proceedings of XIIth Fatipec Congress*, 1974, p. 269.
17. Franiau, R. P., *Paintindia*, September 1987, p. 33.
18. Riesen, R., and Wyden, H., *Mettler TA3000 Seminar*, 5–7 October 1982.
19. Riesen, R., *STK Meeting*, Basel, Switzerland, 30 September 1981.
20. Gargallo, L., and Russo, M., *Makromol. Chem.*, **176**, 2735, 1975.
21. Stamhuis, J. E., and Pennings, A. J., *Polymer*, **18**, 667, 1977.
22. Chevenard, P., Wache, X., and De la Tullaye, R., *Bull. Soc. Chim. Fr.*, **11**, 48, 1944.
23. Duval, C., *Anal. Chem. Acta*, **2**, 92, 1948.
24. Wendlandt, W. W., and Gallagher, P. K., in *Thermal Characterisation of Polymeric Materials*, Ed. E. A. Tuti, Academic Press, New York, 1981, p. 4.
25. Bauer, D. R., and Dickie, R. A., *J. Coat. Technol.*, **54**, 57, 1982.
26. Freeman, E. S., and Carrol, B., *J. Phys. Chem.*, **62**, 394, 1958.
27. Scharp, J. H., and Wentworth, S. A., *Anal. Chem.*, **41**, 2060, 1969.
28. Reich, L., and Levi, W., in *Macromolecular Reviews*, Eds. A. Peterlin, M. Goodman, S. Okamura, B. H. Zimm, and H. F. Mark, Vol. 1, Wiley–Interscience, New York, 1963, p. 173.
29. Friedman, H. L., *J. Macromol. Sci. Chem.*, **A1**, 57, 1967.
30. Sharp, J. H., and Wentworth, S. A., *Anal. Chem.*, **41**, 2060, 1969.
31. Coats, A. W., and Redfern, J. P., *Nature*, **201**, 68, 1964.
32. Flynn, J. F., and Wall, L. A., *J. Res. Natl. But. Std.*, **70**, A487, 1966.
33. Zsako, J., *J. Phys. Chem.*, **72**, 2406, 1968.
34. Cameron, G. G., and Fortune, J. D., *Eur. Polym. J.*, **4**, 333, 1968.
35. MacCallum, J. R., and Tanner, J., *Eur. Pol. J.*, **6**, 1033, 1970.
36. Scholtens, B. J. R., Linde, R. van der, and Tiemersma-Thoone, P. J. M., *DSM Resins Powder Resins Symposium, Nordwijk*, 1988.
37. Adapted from Skrovanek and Schoff, *Prog. Org. Coat.*, **16**, 135, 1988.
38. Treloar, L. R. G., in *The Physics of Rubber Elasticity*, 3rd ed., Clarendon, Oxford, 1975.
39. Ikeda, S., *Prog. Org. Coat.*, **1**, 205, 1973.

3.3 CROSSLINKERS FOR POWDER COATINGS

After the binders, the crosslinkers used in thermosetting powder coatings are the most important part of the coating composition. They exert an influence on the production, storage, application and exploitation properties of the coatings, such as extrusion temperature, storage stability, curing time, curing temperature, flow behaviour, impact resistance and hardness, weathering and chemical resistance, etc.

It is sometimes very difficult to distinguish between the crosslinker and the binder, because both constituents at the last instance make the complete binding system. Very often the low molecular weight compound present in a minor amount and participating in the curing reaction is considered as a crosslinker. However, in some cases, as in the polyester/epoxy hybrids, it is a question of what is the crosslinker and what is the binder, especially in the 50:50 systems. In such a case the choice of the right word is more or less a semantic problem. Therefore bisphenol A epoxy resins and acid functional polyesters, which, depending on someone's preference, can be classified in a group of the crosslinkers, will be omitted in this section, since they will be discussed in Section 3.4 as binders for powder coatings.

3.3.1 CROSSLINKERS OF THE EPOXY TYPE

3.3.1.1 Triglycidyl Isocyanurate (TGIC)

Triglycidyl isocyanurate is one of the most important curing agents for carboxyl functional binders, specifically for carboxyl terminated polyesters. In Europe this is certainly the most popular crosslinker for polyester powder coatings for outdoor use. The formula of TGIC is given below:

$$\text{CH}_2-\text{CH}-\text{CH}_2-\text{N}\underset{\underset{\text{CH}_2-\text{CH}-\text{CH}_2}{|}}{\overset{\overset{\text{O}}{\|}}{\underset{\text{O}=\text{C}}{\overset{\text{C}}{\diagup}}\underset{\text{N}}{\diagdown}\overset{\text{C}=\text{O}}{\diagup}}}\text{N}-\text{CH}_2-\text{CH}-\text{CH}_2$$

Different methods of preparation of TGIC such as isomerization of triglycidyl cyanurate obtained from cyanuric chloride and glycidol or transesterification of cyanuric tosylate with glycidol are described in the patent literature. However, only the epichlorohydrine route has found practical importance.

The industrial preparation of TGIC from cyanuric acid and epichlorohydrin is extensively described in the patent literature [1–3]. The synthesis is performed by heating cyanuric acid in a large excess of epichlorohydrin in the presence of ammonium or phosphonium halide as catalyst. The subsequent dehydrohalogenation of the formed halohydrin by the base gives TGIC as a product, while the excess of epichlorohydrin is distilled off. The obtained crude product is purified by recrystallization:

$$\underset{\underset{C=O}{HN\diagup\diagdown NH}}{O=C\diagdown\diagup C=O} \quad \longleftrightarrow \quad \underset{\underset{C-OH}{N\diagup\diagdown N}}{HO-C\diagdown\diagup C-OH} \quad \xrightarrow[NaOH]{CH_2-CH-CH_2Cl \text{ (epoxide)}}$$

[Structure of TGIC: triazine ring with three N-CH$_2$-CH(epoxide)-CH$_2$ glycidyl groups]

Because of the racemic blend of epichlorohydrin used in the syntheses, by fixing one glycidyl group per molecule, a product that is a 50:50 blend of the two stereoisomers should be obtained. Since in the case of TGIC three glycidyl groups are fixed on one molecule, eight different diastereoisomers can be obtained. Because of the symmetry of the molecule of third order three pairs of diastereoisomers cannot be distinguished, thus leading to the following four isomeric forms: RRR/SSS and three pairs of RRS/SSR, RSR/SRS and SRR/RSS. Statistically, the composition with respect to the ratio of different pairs will be 1:3 or 25:75. It has been found that TGIC produced in the way described above is a blend containing 75% of so-called form A with a melting point of 103–104.5°C and 25% of form B with a melting point of 156–157.5°C [4, 5]. Merck and Loutz have found that the amount of form B with a higher melting point in commercially available TGIC from different suppliers varies between 13 and 29% [6]. They have isolated and purified forms A and B and have shown that they behave very differently when cured with carboxyl functional compounds. Form B is more reactive, and with an equivalent amount of adipic acid it has a gel time at 180°C of 38 seconds, in comparison with 73 seconds for form A. On the other hand, in a coating formulation with carboxyl functional polyester, form B shows a longer gel time, a slower viscosity increase during curing and requires a longer curing time to obtain good mechanical properties than form A. At first sight these are contradictory results, explained by the particularly low solubility of form B in organic media even at 200°C, which considerably influences the kinetics of the curing process. This could also be an explanation for the surface and transparency anomalies that are frequently observed in TGIC/polyester powder coatings, especially in the case of dark coloured paint formulations.

The presence of the B form with a higher melting point in the commercial types of TGIC can considerably influence the curing behaviour of the powder coatings. It will be emphasized even more in low TGIC containing powder formulations, which are in principle more critical. For the time being, there is no strong evidence that TGIC will isomerize itself during storage to thermodynamically more stable, but unfortunately not desirable, form B.

The theoretical epoxy equivalent weight of TGIC is 99, but in practice, due to partial self-condensation during its production and certain hydrolysis during storage, the epoxy equivalent value is somewhat higher, being in the range of 105–110. This low epoxy equivalent value is a rather important characteristic of TGIC. Low molecular weight materials in powder coating compositions dramatically decrease the glass transition temperature of the system. This adversely affects the storage stability of the coating. Because of its low equivalent weight, the concentration of TGIC in powder coating formulations is relatively low, usually being no higher than 10%. Therefore TGIC affects the T_g of the coating composition to a level that is acceptable in practice. However, it should be kept in mind that every weight percentage of TGIC in the powder formulation decreases the T_g of the coating by approximately 2°C.

Compared to the bisphenol A types of epoxy crosslinkers TGIC is much more reactive. Powder coating compositions containing TGIC as a crosslinker do not contain curing catalysts in many cases. If a fast curing system has to be formulated, the type and amount of catalyst should be chosen very carefully in order to avoid prereaction during the extrusion of the coating or problems connected with the chemical stability of the coating during storage.

Coatings in which TGIC is used as a crosslinker exhibit outstanding outdoor durability, and with a good choice of the binders and pigments, systems with a service time longer than ten years can be obtained.

3.3.2 POLYISOCYANATES

From a number of methods available for preparation of isocyanates, only the phosgenation of amines has found real commercial importance:

$$RNH_2 + COCl_2 \longrightarrow RNCO + 2\,HCl$$

Polyisocyanates used as crosslinkers for powder coatings can be of the aromatic or aliphatic types. Aromatic isocyanates, especially the derivatives of toluene diisocyanate, are very attractive from a price point of view. On the other hand, the main drawback associated with the use of aromatic isocyanates is their tendency to yellow on exposure to sunlight. This was assumed to be due to the oxidation of some terminal aromatic amino groups derived from these isocyanates [7].

The isocyanate groups in the crosslinkers used for the manufacture of polyurethane powder coatings must be protected in order to avoid the crosslinking reaction at room temperature and to provide good storage stability

THERMOSETTING POWDER COATINGS

of the coating. Another reason for this protection is to reduce the health and safety hazards to a level that allows relatively safe handling of these materials during production and use of the coatings. These crosslinkers are synthesized by the reaction of diisocyanates or their derivatives with protective species, commonly referred to as blocking agents to produce isocyanate-free adducts that are stable in the presence of moisture and in blends with polymers containing reactive hydroxyl groups at normal storage conditions.

Although there are many different combinations between the considerable number of available isocyanates and different blocking agents, only the derivatives of isophorone diisocyanate blocked with caprolactam are of real commercial importance. Despite its low price, toluene diisocyanate has found limited applicability only in cheap polyurethane coatings for indoor use because of its low yellowing resistance and light stability.

Commercially available types of IPDI derivatives will be presented in this section, together with isocyanates which are not yet commonly used in practice for powder coating production, but which have potentials to be commercialized in the near future.

3.3.2.1 Caprolactam Blocked IPDI Derivatives

Caprolactam blocked isophorone diisocyanate derivatives are the most widely used crosslinkers for polyurethane powder coatings. They possess a suitable melting range to be used as hardeners for powder coating manufacture by the standard extrusion procedure. This is combined with an acceptable deblocking temperature and curing rate which is suitable for the common curing schedules when the system is catalysed properly.

There are three important commercial IPDI caprolactam blocked adducts on the market produced by Chemische Werke Hüls AG: crosslinkers B989, B1065 and B1530, the last one being the most popular. The NCO content of this crosslinker is 15% with less than 1% of free NCO groups. All of them have melting temperatures in the range between 75 and 90°C and a glass transition temperature of about 50°C. The deblocking reaction takes place at 175°C for B989 and B1065, and somewhat lower (160–170°C) for B1530.

The chemical structure of these crosslinkers is as follows:

where D is a residue of a diol.

Different patents cover the production of caprolactam blocked IPDI oligomers suitable as crosslinkers for polyurethane powder coatings. In a patent of Hüls [8] 1 mole of ethylene glycol is reacted with 4 moles of IPDI followed by blocking with 6 moles of caprolactam. The product has a melting point of 70–80°C and T_g of 35–46°C.

Instead of diols, diamines can be used to extend the molecular weight of the IPDI blocked oligomers, in this way introducing urea groups into the oligomer structure [9]. Oligomer obtained by blocking IPDI with caprolactam followed by reaction with isophorone diamine is characterized by a melting point of 80–84°C and 16.3% of a blocked NCO content.

Isocyanates with a higher functionality than two can be obtained by trimerization of diisocyanates to isocyanurates [10]:

$$3\ OCN-R-NCO \xrightarrow{\text{strong bases}} \begin{array}{c} \text{isocyanurate ring} \end{array}$$

By the correct choice of isocyanate, such as 2-methylpentane-1,5-diisocyanate, 2-ethylbutane-1,4-diisocyanate or 3-isocyanatomethyl-3,5,5-trimethylcyclohexanediisocyanate, separately or in a mixture, a suitable catalyst for trimerization, such as triethylenediamine-propylene oxide, and caprolactam as a blocking agent [11], a curing agent containing 10–16% blocked NCO groups with a melting range between 60 and 93°C can be obtained.

Oligomers of IPDI with a higher functionality than two can be obtained by using blends of diols and triols in the preparation of the final product. A patent of UCB [12] discloses a two-stage process for the production of caprolactam blocked IPDI oligomer. In the first step, 1 mole of IPDI reacts with 1 mole of caprolactam at 70–90°C in the presence of Sn-octoate as catalyst, giving a partly blocked intermediate product. The latter reacts in the second step at 90–140°C with the equivalent quantity of blend comprising 30:70 to 50:50 diol/triol mixture until full consumption of the isocyanate groups.

An interesting hardener which is also a derivative of isophorone diisocyanate is described in a patent assigned to Bayer [13]. The hardener is a condensation product of isophorone diisocyanate, caprolactam and 2,2-bis-(hydroxymethyl)-propionic acid in a molar ratio of 2:2:1. The condensation reaction performed in toluene gives a crystalline colourless product with an acid value of 70 and melting point between 120 and 128°C. It is claimed that in combination with TGIC and hydroxyl functional polyesters as binders, this hardener is suitable for producing matt powder coatings.

3.3.2.2 Uretidione of IPDI

Blocking-agent-free polyisocyanate hardener suitable for producing polyurethane powder is internally blocked IPDI obtained by a self-condensation process in basic media. The basic structure of the IPDI uretidione is the following:

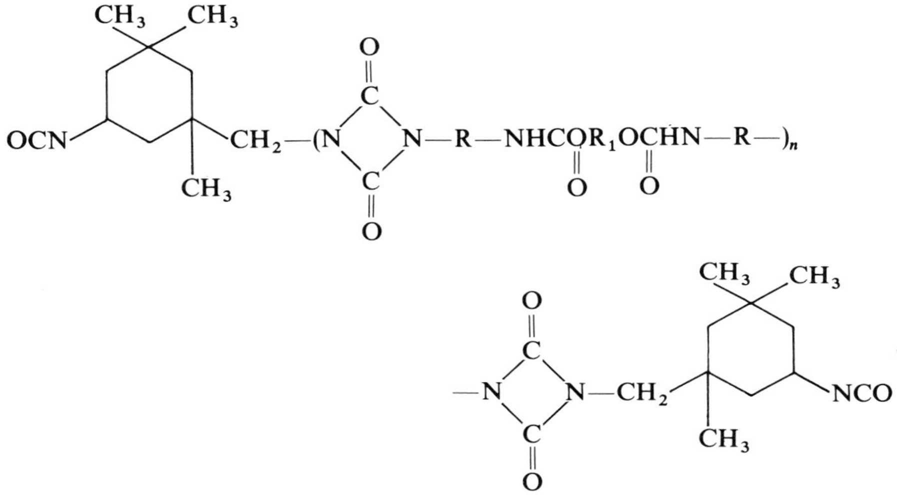

where R is the residue from IPDI, n is at least 1 and R_1 is a divalent aliphatic, cycloaliphatic, arylaliphatic or aromatic hydrocarbon group from a diol. The process according to Ref. [14] proceeds in two steps. In the first step a dimerization of IPDI is performed at room temperature in the presence of tridimethylamine phosphine catalyst to produce an uretidione of IPDI of the following formula:

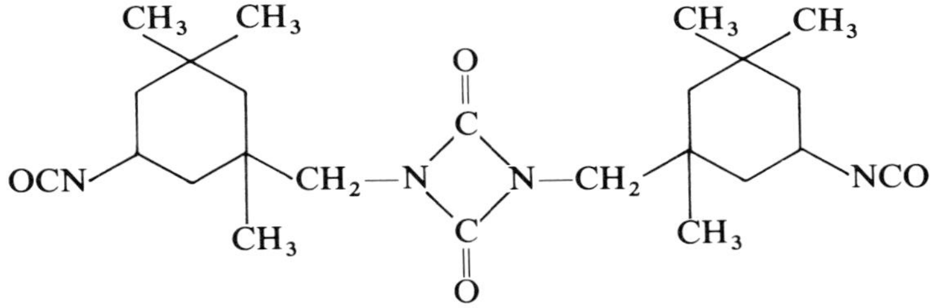

The intermediate uretidione in the second step is reacted with butane diol to produce the final product. Dependent on the IPDI-uretidione/butanediol ratio, a range of crosslinkers can be obtained with molecular weights between 500 and 4000, melting points between 70 and 130°C and less than 8% free NCO content. When heated to the stoving temperature of the coating, 98% of the material converts back to isophorone diisocyanate.

Other patents of Chemische Werke Hüls AG [15] disclose the possibility of producing practically isocyanate free uretidione of IPDI, using monoalcohols or monoamines as blocking agents for the free-end isocyanate groups. In one of the examples a polymeric IPDI uretidione obtained in the way described above is blocked with 2-ethyl hexanol to reduce the free isocyanate content below 1%. During the curing, 98% of the uretidione converts back to IPDI, which serves as a crosslinker for hydroxyl functional polymers.

A commercial product of this type is introduced and sold on the market by Hüls AG under the name of IPDI-BF 1540. The content of free NCO groups is very low (below 1%) while the total NCO content is 15.4%. The melting point of the crosslinker is in the range between 105 and 115°C. Since the usual extrusion temperature does not exceed 120°C, practically no prereaction takes place while the melt is compounding with the binder.

An obvious advantage of IPDI uretidione compared to blocked isocyanates is the absence of a side product which leaves the system during curing.

3.3.2.3 TDI Derivatives

Toluene diisocyanate is the cheapest diisocyanate available on the market. The commercial product is a colourless liquid with a boiling point of 120°C at 10 mmHg; it is an 80:20 blend of the 2,4 and 2,6 isomers:

2,4 Isomer   2,6 Isomer

The NCO group in position 4 is 8 to 10 times as reactive as that in position 2 at room temperature. However, with increasing temperature the reactivity of the ortho group increases at a greater rate than that of the para group. At 100°C the reactivity of both isocyanate groups is the same.

The difference in the rates of reactivity between the isocyanate groups is a specific advantage of TDI since it allows preparation of blocked adduct with a much better defined structure than would be the case if the reactivities were equal. Namely in a reaction with an equimolar amount of caprolactam the main product will be half-blocked TDI with a predominantly free isocyanate group in the 2 position. This intermediate can then be used for condensation with different polyols or polyamines to produce caprolactam blocked TDI oligomers with a well-defined structure.

THERMOSETTING POWDER COATINGS

Although the price and its suitability for preparation of blocked oligomers with tailored structures and properties are strong points of TDI, its use in powder coatings is very limited because of its sensitivity to u.v. light and emphasized yellowing. Powder coatings for indoor application where the whiteness of the coating is not of primary importance are typical products where TDI blocked derivatives are used as crosslinkers.

3.3.2.4 TMXDI and other Polyisocyanates

Two new isocyanates, *m*- and *p*-tetramethylxylene diisocyanates (*m*- and *p*-TMXDI), developed by the American Cyanamid Company are potentially very interesting materials for making crosslinkers for powder coatings. They have characteristics typical of aliphatic diisocyanates, even though they contain aromatic rings. This is because the NCO groups are shielded from the ring by protective methyl groups [16]. Since they are made by a non-phosgenation process [17, 18], they offer the potential for much greater economy than the conventional aliphatic isocyanates.

The structural formulas of *m*-TMXDI and *p*-TMXDI are the following:

p-TMXDI   *m*-TMXDI

m-TMXDI is a liquid at room temperature with a melting point of $-10°C$ and boiling point of 150°C at 3 mmHg. *p*-TMXDI has the same boiling point, but it is a solid material at room temperature with a melting point of 72°C.

The reactivities of *m*- and *p*-TMXDI with hydroxyl groups are somewhat lower compared to the other aliphatic isocyanates.

A large number of catalysts are examined for improving the reactivity of the TMXDI isomers in the NCO—OH reaction. It is found that the amine catalysts are relatively ineffective as compared with metal salts such as dimethyltin dilaurate, lead octoate or tetrabutyl diacetyl distannoxane.

TMXDI isomers cannot be used as crosslinkers for powder coatings. They have to be blocked in order to prevent storage instability and viscosity increase during processing of the powder coatings. In order to increase the functionality and to decrease the vapour pressure of *m*-TMXDI it has been reacted with trimethyl propane to produce a crosslinker with a functionality of three. The use of the trimethylol propane adduct of *m*-TMXDI in two-component acrylic coatings has been reported by Fiori and Dexter [19]. Since the *m*-TMXDI/TMP adduct is relatively rigid, it is necessary to select binders that have a lower glass transition temperature and therefore are softer and more flexible, in order to compensate for the rigidity of the crosslinker.

An advantage of TMXDI isomers is their relatively lower toxicity compared to the conventional isocyanates. For example, it has been found that *m*-TMXDI does not produce pulmonary sensitization at inhalation doses that cause acute irritation effects [20]. This behaviour is assigned to the tertiary character of the isocyanate group in *m*-TMXDI in contrast to the primary and secondary isocyanate groups in the conventional isocyanates.

Although caprolactam blocked isocyanates are the most widely used for powder coatings, the possibility of using other blocking agents or other isocyanates is also described in the patent literature.

The patent of Takeda Yakuhin Kogy [21] describes the use of 1,3,5-tris(isocyanatomethyl)benzene or 1,3,5-tris(isocyanatomethyl)cyclohexane of the formulas given below blocked with methyl ethyl ketoxime:

$$\begin{array}{cc} \text{CH}_2\text{NCO} & \text{CH}_2\text{NCO} \\ | & | \\ \text{C}_6\text{H}_3-\text{CH}_2\text{NCO} & \text{C}_6\text{H}_6-\text{CH}_2\text{NCO} \\ | & | \\ \text{CH}_2\text{NCO} & \text{CH}_2\text{NCO} \end{array}$$

It is claimed that in powder composition using hydroxyl functional polyester as a binder, the product gives a coating film free of yellowing, blistering, combined with good outdoor durability, chemical resistance, mechanical properties and storage stability. Moreover, the low deblocking temperature of the ketoxime blocked isocyanates allows curing of the coating at 160°C.

An Eastman Kodak patent [22] discloses the use of adducts of 1,4-cyclohexane bismethylisocyanate with compounds containing active hydrogen as crosslinkers for powder coating compositions. Specifically, the patent claims the use of oximes such as acetone oxime, for example, which enables a powder coating composition to be made with a very low curing temperature of 140°C.

Cooray and Spencer [23] have reported a dissociation temperature of pyrazole blocked aliphatic isocyanates of 120°C. Although there is no evidence that these crosslinkers have been used in powder coatings, they could be potentially very interesting hardeners for producing low temperature curing powders.

A process of obtaining caprolactam blocked aromatic diisocyanate with deblocking temperature of 175°C is disclosed in a patent of Ciba–Geigy [24]. To prepare the diisocyanate, a toluene solution of diamine isomers being 38% 5-amino-1-(4'aminophenyl)-1,3,3-trimethylindane and 62% 6-amino-1-(4'aminophenyl)-1,3,3-trimethylindane is added to a toluene solution of phosgene. After reaction at 55°C, a solid isocyanate with a melting point of 60–80°C is obtained. The caprolactam blocked adduct obtained in a standard procedure has a melting range between 100 and 108°C.

THERMOSETTING POWDER COATINGS

A solid ketoxime blocked isocyanate being a reaction product of acetone oxime and 1,4-cyclohexane bis(methylisocyanate) suitable for manufacturing powder coatings with low curing temperatures is described in a patent of Kodak [22].

$$\mathrm{\underset{H_3C}{\overset{H_3C}{>}}C=N-OCONH-CH_2-C_6H_{10}-CH_2-NHOCO-N=C\underset{CH_3}{\overset{CH_3}{<}}}$$

It is reported that in combination with hydroxyl functional polyester a gel time of 18.5 minutes at 140°C is obtained. In the presence of tin catalyst the gel time is reduced to 7.5 minutes, which does not affect the stability of the system.

3.3.3 POLYAMINES

3.3.3.1 Dicyandiamide and its Derivatives

Dicyandiamide (DICY) or 1-cyanoguanidine is the most widely used crosslinking agent for epoxy powder coatings. It is a white solid crystalline compound which is produced from calcium cyanamide according to the following reaction:

$$\mathrm{CaNCN \xrightarrow[CO_2]{H_2O} \underset{H}{\overset{H}{N}}-\underset{}{\overset{NH}{\underset{\|}{C}}}-NH-CN}$$

The amino hydrogens present in the structure of DICY can react with the oxirane rings of epoxy resins. The curing mechanism of DICY/epoxy resin is complex involving addition reactions with both epoxide and hydroxyl groups. DICY also catalyses considerably the epoxide/hydroxyl reaction promoting the self-crosslinking of the epoxy resins. Therefore, in the case of DICY and its derivatives there is no need for strict stoichiometric proportions between the curing agent and the binder in the coating composition for obtaining infinite network formation, as is the case with most crosslinker/binder systems in powder coatings.

The melting point of DICY (211°C) is much higher than the extrusion temperature during the compounding process. On the other hand, its solubility in the binder is relatively low. Therefore, the reaction rate between the amino and epoxy groups during curing depends on the availability of reactive groups on the surface of the DICY particle. For this reason commercial grades of DICY are usually delivered in fine micronized form with the particle size below 75 μm [25, 26]. The lower availability of reactive groups because of the high melting point and low solubility of DICY must be compensated by a slight excess of hardener in the coating composition. Even the excess of DICY does not improve the reaction rate. For obtaining acceptable curing schedules, a proper catalyst has to be used. The catalysts are incorporated in the hardener during its manufacture. Most of the DICY producers supply the so-called 'accelerated dicyandiamides' providing different levels of reactivity. In this way curing schedules of 15 minutes at 155°C can be achieved [27] that do not affect the storage stability of the system very much; stability is only slightly reduced due to the insolubility of DICY in the binder.

Although the reaction rate can be controlled and improved by proper choice of a catalyst, several problems still can be addressed to the DICY/epoxy resin system. The insolubility of DICY creates inhomogeneity problems which are manifested in low gloss, poor flow and in extreme cases non-uniform appearance of the film after curing. DICY itself is a water soluble compound, and often increases the water sensitivity of the cured films. For these reasons, DICY has been chemically modified to produce a family of curing agents which give better flow, high gloss and very low water sensitivity. These so-called substituted dicyandiamide curing agents whose general formula is presented below, are characterized by good solubility in the epoxy resins, providing in this way a homogeneous powder coating composition that does not suffer from the typical disadvantages of pure DICY systems:

$$\begin{array}{c} H \\ \diagdown \\ N-C-NH-C-NH-\!\!\!\!\bigcirc\!\!\!-R \\ \diagup \\ H \end{array} \begin{array}{c} NH \\ \| \\ \end{array} \begin{array}{c} NH \\ \| \\ \end{array}$$

Since the solubility of the substituted DICY in the binder is much better than DICY itself, it will affect the storage stability of the system. Therefore, a good compromise has to be found between the desirable curing schedule and acceptable storage stability by choosing a suitable amount of the correct catalyst. In properly formulated coatings, curing schedules of 15 minutes at 160°C can be obtained combined with an acceptable storage stability.

3.3.3.2 Modified Aromatic and Aliphatic Polyamines

The majority of primary and secondary aliphatic and aromatic amines, even those that are solids at room temperature, cannot be used as hardeners for epoxy powder coatings because of the extremely high reactivity of the system. This creates considerable problems during the processing in the extruder, and dramatically decreases the storage stability of the coating. However, several types of amine derivatives have been developed which provide the necessary stability when combined with solid epoxy resins in a powder coating composition.

Certain aromatic diamines can be used as hardeners for epoxy powder coatings. Methylene dianiline, for example, the formula of which is presented below, has been claimed as a hardener for epoxy powder coatings [28]:

$$H_2N-C_6H_4-CH_2-C_6H_4-NH_2$$

It is solid compound at room temperature with a melting point of 90°C and an active hydrogen equivalent of 49.5%. Epoxy powder coatings formulated with this hardener possess excellent chemical resistance, particularly to mineral acids and bases, and good retention of the properties at elevated temperatures.

It is not common practice to use low molecular aromatic diamines as crosslinkers for powder coatings even when they are solid at room temperature, because of their relatively high vapour pressure at elevated temperatures combined with considerable toxicity. Therefore they are usually modified by a reaction with monoglycidyl phenolic ethers [7, p. 677] in order to increase their boiling point or by diglycidyl ethers making resinous adducts [29].

In general, aromatic polyamines and their derivatives give systems with excellent reactivity, extremely good chemical resistance, but with a very emphasized tendency towards yellowing. Another weak point of aromatic amines and their derivatives is their very questionable physiological activity because of toxicity. This makes them very inconvenient hardeners for a system that is claimed to be environmentally friendly.

A group of aromatic amine curing agents that are monosalts of aromatic polycarboxylic acids and 2-phenyl-2-imidazoline are described in several patents of Hüls [30, 31]. These hardeners are obtained by reaction of equimolar amounts of pyromellitic or trimellitic anhydride previously hydrolysed with water to obtain their acid and 2-phenyl-2-imidazoline. Combined with epoxy resins, pigments and flow agents in a powder coating composition, after curing they produce matt films with excellent flow and a very low gloss of only 9% (at an angle of 60°).

Aliphatic polyamines that differ from DICY produced by condensation of aliphatic polyvalent amine and bisphenol A epoxy compounds are described in the patent literature. For example, the condensation product of isophorone diamine and diglycidyl ether of bisphenol A with low molecular weight (Epicote 828) in the ratio of 2:1 with a melting point of 55–57°C is used in combination with high molecular weight epoxy resins as the curing agent [32]. It is claimed that the hardener can be mixed in the melt with an epoxy resin homogeneously, without gelation, providing extremely short curing times.

In the patent of Yuka Shell Epoxy [33] a method of producing dicarboxy-polymethylene diamine of formula $H_2N-(CH_2)_n-OCO-R-OCO-(CH_2)_n-NH_2$ is described involving esterification of aminododecanoic acid and aliphatic diols with two to six carbon atoms. A modification of the process also involves diols derived by condensation of low molecular weight epoxy resins with phenol followed by esterification of the formed secondary hydroxyl groups with amino acids. In this way, different diamines having a melting point in the range between 50 and 140°C can be obtained.

3.3.4 POLYPHENOLS

Conventional thermosetting phenolic resins of both resol and novolac type can be used in principle as hardeners for epoxy powder coatings. They are prepared by step-growth polymerization of phenols with aldehydes. This involves electrophilic substitution of the aromatic ring to form a phenol alcohol, followed by condensation reaction between the methylol groups leading to the formation of a high molecular compound. Phenol and formaldehyde are by far the most important raw materials for production of phenol-formaldehyde resins in general, but other types of phenolic compounds such as cresol and diphenylol propane are also reported to be used for the manufacture of phenolic resins for powder coatings [34, 35].

Although the condensation reaction between phenol and an aldehyde had already been described by Bayer in 1872 [36], the commercial production of these resins started in 1910 after the work of Baekeland who developed formulations and an economical method for manufacturing moulded parts using phenol formaldehyde resins as binders.

Two different types of product can be obtained by condensation of phenol with formaldehyde, depending on the reaction conditions. The reaction between phenol and formaldehyde in alkaline conditions or slightly acidic conditions at a pH above 5 results in a formation of mono, bis and tris alcohols, as well as in many other rather complicated compounds:

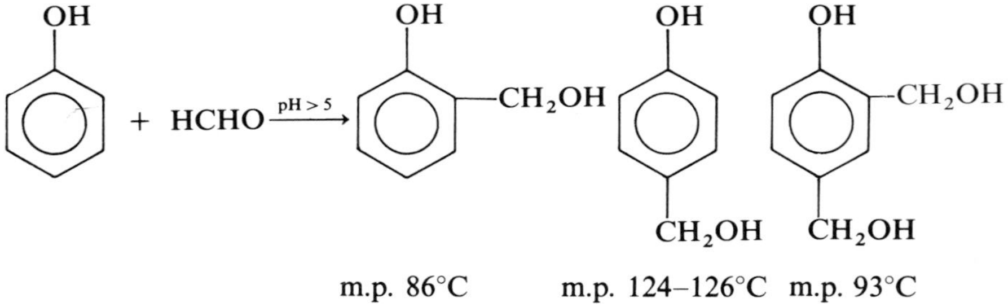

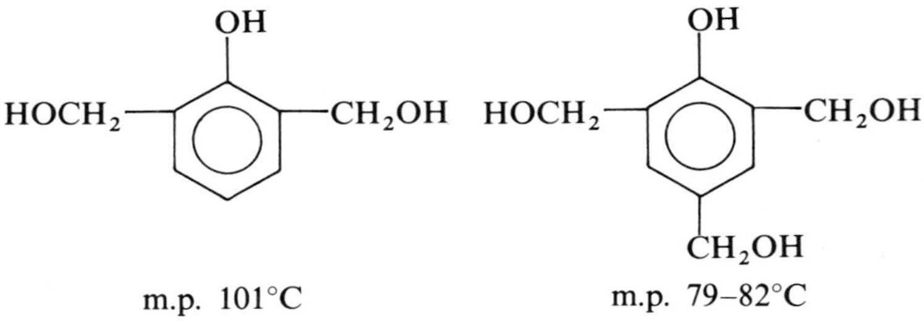

Under alkaline conditions, the formed methylol phenols undergo a condensation reaction forming a polymer. Based on the experiments with model compounds it has been proposed that the following two reactions dominate in this stage of the process [37, 38]:

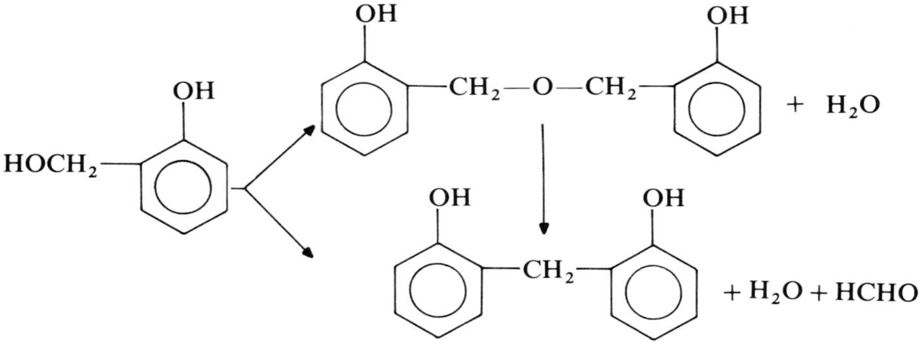

Yeddanapalli and Francis [39] have shown that under strong alkaline conditions the reaction yielding diphenylmethanes with splitting formaldehyde as a by-product is the dominating one. However, under weak acidic or neutral conditions, the condensation proceeds through a formation of dihydroxy benzyl ethers.

Under strongly acidic conditions, it has been proposed that the reaction between phenol and formaldehyde is an electrophilic substitution [36, p. 47]:

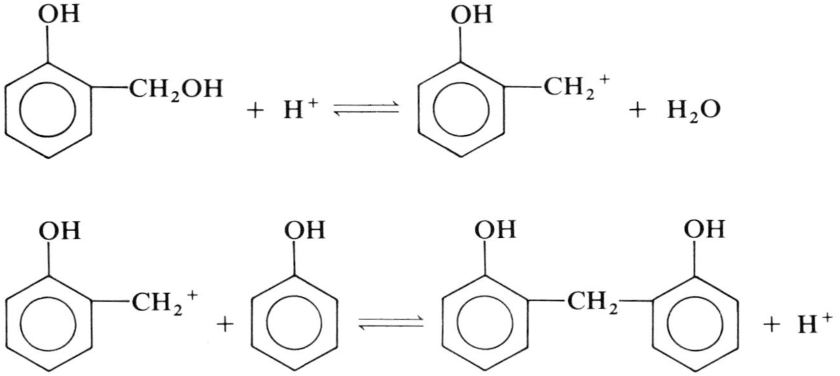

Depending on the molar ratio between the starting raw materials and the reaction conditions, two different types of phenol formaldehyde resins can be produced.

In an excess of formaldehyde and under alkaline conditions the products are branched resins commonly known under the name resoles. The mole ratio between formaldehyde and phenol is between 1:1 and 3:1. Catalysts used commercially include sodium hydroxide, sodium carbonate, alkaline earth oxides and hydroxides, ammonia and tertiary amines.

Under acidic conditions and a slight excess of phenol (formaldehyde/phenol mole ratios between 0.75 and 0.85) novolac types of linear phenol formaldehyde resins are obtained. Oxalic acid is the most widely used catalyst giving low coloured products.

The relatively cheap and well-established process of production of phenolic resins in combination with conventional raw materials used for their production, which belong to the typical commodity chemicals, makes them very attractive crosslinkers from an economical point of view. On the other hand, bisphenol A epoxy resins when crosslinked with phenolic hardeners produce coatings with exceptionally high chemical, solvent, detergent and boiling water resistance. These properties cannot be achieved to such an extent by any other binder/crosslinker system.

Several disadvantages are encountered when phenolic resins are used as crosslinkers for powder coatings. The yellowing resistance of the system is rather poor which makes them less suitable for white powder coatings. The presence of the aromatic ether group dramatically decreases their u.v. resistance. Therefore their use is exclusively limited to indoor application. On the other hand, the curing reaction between oxirane rings of the epoxy resin and phenolic hydroxy groups is rather slow, giving cured films that are in general brittle and have a poor impact resistance.

The latest developments in the novolac phenolic resins specially designed to be used as crosslinkers for powder coatings and the new catalysts for increasing the curing rate have overcome the problems mentioned above. As a result commercial crosslinkers have appeared that are characterized by the following properties:

— A high degree of compatibility with the bisphenol A epoxy resins.
— The same melting range as that of the epoxy resins, thus providing excellent processing properties of the coating.
— A reactivity that allows curing at low temperatures or the formulation of very fast curing systems (5 minutes at 180°C).
— Sufficient latency to ensure safe compounding at high temperature.
— Sufficient film flexibility to meet normal customer requests.

Crosslinkers of the phenolic type can be used for decorative purposes as well as for functional coatings, providing excellent corrosion, solvent and chemical resistance.

Crosslinkers containing phenolic functional groups can be obtained by condensation of epoxy resins with a molar excess of phenolic compounds. However, this is a rather expensive process of making phenolic crosslinkers. An interesting modification of such a process in order to make the system flexible by introducing aliphatic chains is described in a patent assigned to Shell Oil Company [40]. According to this patent phenolic curing agents specifically suitable for use in powder coatings are prepared by reacting dihydric phenol (diphenylol propane), diepoxide (low molecular weight diglycidyl ether of

THERMOSETTING POWDER COATINGS

diphenylol propane) and dicarboxylic acid (adipic for example). The phenolic functionality is in excess, exceeding the difference between the epoxy and acid functionality. In one example of the patent a mixture of 2 moles of bisphenol A epoxy resin, 2 moles of bisphenol A and 1 mole of adipic acid are reacted in the presence of tetramethyl ammonium chloride as catalyst for 1 hour at 140°C. In these conditions the whole amount of carboxyl groups has been consumed in the reaction with the oxirane rings. The second reaction between the epoxy and phenolic groups is then carried out at 160°C for 2 hours. A clear brittle product with a phenolic hydroxyl content of 230 meq/100 g with a melting range between 75 and 90°C is obtained.

Most of the epoxy resin producers have developed phenolic crosslinkers specially matching their resins. All of the commercially available phenolic crosslinkers have almost the same hydroxyl equivalent weight, ranging between 240 and 260 and relatively low softening point between 65 and 75°C [41–43]. They are characterized by a very low melt viscosity which is at 150°C usually below 1 Pa s, in most cases between 100 and 400 mPa s. The producer usually incorporates the curing catalyst during the manufacture of the phenolic crosslinkers. They are delivered to the market with different degrees of reactivity, very often offering extremely short curing schedules of 1 min at 230°C to 15 min at 150°C. Despite the high reactivity, it is possible to formulate powder coatings with acceptable storage stability.

3.3.5 ACID ANHYDRIDES

Acid anhydrides can be used as curing agents for powder coatings having an epoxy resin or hydroxyl functional polyester as binders. In the case of epoxy binders, acid anhydrides containing only one anhydride ring can lead to an infinite network formation, because after the ring opening esterification of the anhydride group, the resulting carboxyl group can react with the epoxy group of the binder. In the case of hydroxyl functional binders, only di- or polyanhydrides can be suitable curing agents.

In principle all low molecular weight acid anhydrides such as phthalic anhydride, maleic anhydride and trimellitic anhydride can be used for curing epoxy resins, or resins containing epoxy groups.

Trimellitic anhydride (TMA) is widely used as a hardener for epoxy composites. In applications such as powder coatings, powder particles, which are always present in the air surrounding the spraying booths, and the large open surface to be coated, which emits a considerable amount of trimellitic anhydride during curing, restrict the use of trimellitic anhydride as a curing agent because of its toxicity and irritancy. The same holds for the other potentially useful hardener pyromellitic dianhydride (PMDA).

Preparation of dianhydrides from trimellitic anhydride and polyols or amines, as curing agents for epoxy compounds, is described in a patent of the General

Electric Co. [44]. Resinous dianhydrides can be obtained, starting from trimellitic anhydride and diols or diamines. The process proceeds at sufficiently high temperatures (230–300°C) to remove water, resulting in an intermolecular transesterification. During the reaction a tetracarboxylic acid is formed as an intermediate product, which under vacuum at 260°C undergoes transformation to a dianhydride. A modification of this patent [45] involves a one-step procedure for obtaining the same products. Common commercially available glycols can be used as diols. The process can be performed at much lower temperatures up to 200°C, starting from dialkyl ester of different diols (ethyleneglycol diacetate, for example) and trimellitic anhydride. The formation of the desired compound goes through a process of acidolysis, taking place on the carboxylic acid radical, which leaves the anhydride radical intact [46].

A patent of Rhone Poulenc Ind. [47] describes a powder coating composition comprising polyester resin with terminal hydroxyl groups with an hydroxyl value of 150–250 and a softening point of 70–130°C. The crosslinking agent is a reaction product of trimellitic anhydride and diol, having a softening point between 60 and 130°C. The dianhydride curing agent is prepared by reacting trimellitic anhydride with C2–C10 aliphatic diols at 180–250°C, with a mole ratio for anhydride diol of 2:1. The curing process is claimed to be rather short, taking 5–30 minutes at 140–200°C.

It is reported that pure ethylene glycol bistrimellitate (EGBM) is a crystalline material with a melting point of 166°C [48] and therefore under normal extrusion conditions complete homogenization with polyester resins is not possible. Reduction of the melting point of EGBM can be achieved by making oligomers with a higher molecular weight. In this way the crystallinity of the product is destroyed, and a material with a melting range below 120°C can be obtained.

The high solubility of the amorphous oligomer of EGBM, however, creates another problem because of its high reactivity with the hydroxyl groups during processing. By careful control of the temperature below 120°C and short residence times in the extruder, relatively safe processing can be performed without gelation problems. It is reported that this material increases by roughly five times the viscosity of the powder coatings during 10 minutes of processing at 120°C compared to the initial viscosity. It is considered that this is an acceptable value for obtaining powder coating compositions that exhibit good flow [48].

Although the technical problems connected with the use of acid anhydrides as crosslinkers for powder coatings are solved, toxicity considerations still remain a permanent obstacle for wider acceptance of these systems by the market.

3.3.6 AMINO RESINS

Amino resins were used for a short period at the beginning of the 1970s as crosslinkers for polyester powder coatings. The appearance of TGIC/polyester systems swept them from the market, so there is no evidence that at this moment there are commercial systems based on amino resins as crosslinkers. In the solvent based coatings, however, amino resins play an extremely important role in formulating paints for industrial applications, which include the market for household equipment, general industrial applications, coil coating, can coating, tube coating and automotive coatings. These crosslinkers offer an outstanding price/performance ratio combined with relatively low toxicological and health hazard considerations. This is a good enough reason to include them in the group of crosslinkers for powder coatings, in order at least to stimulate powder coating chemists to consider them as a possibility, something which has been neglected for years.

Melamine, benzoguanimine and glycoluryl are the main raw materials for condensation with formaldehyde to produce crosslinkers with potential use in powder coatings:

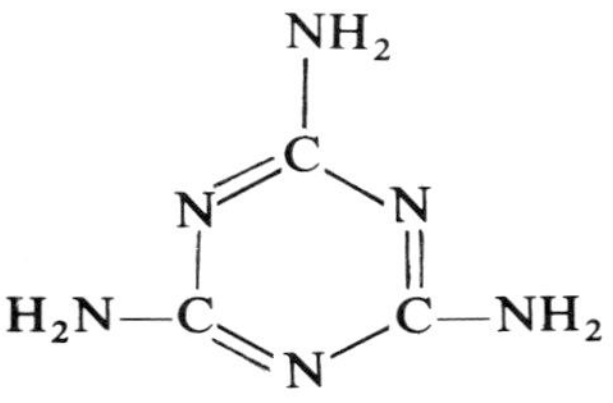

Melamine 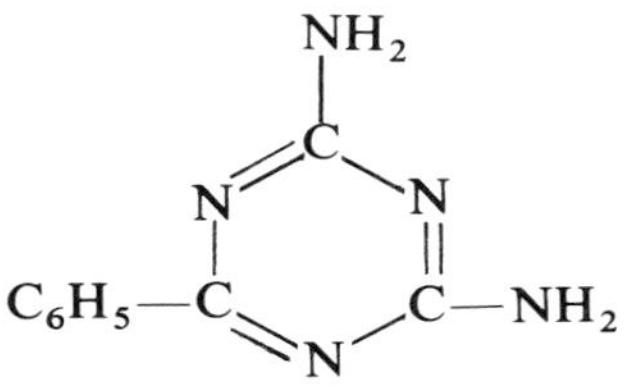   Benzoguanimine 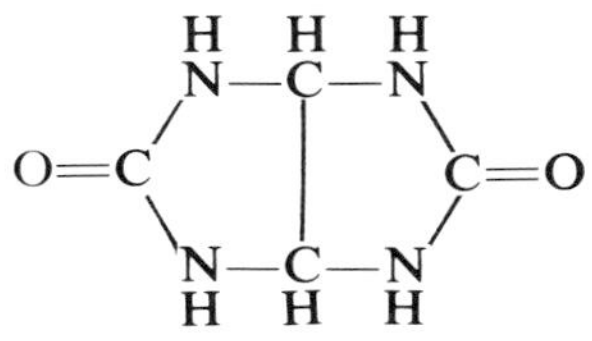   Glycoluryl

Two main reactions are involved in the formation and cure of amino resins. The first is a simple addition of formaldehyde to the amino compound forming a methylol intermediate according to the following scheme:

$$R-NH_2 + HCHO \longrightarrow RNH-CH_2OH$$

This reaction is also known as methylolation and may be catalysed by either acids or bases. The methylol compounds formed are relatively stable under neutral or basic conditions.

In the second step in the production of amino resins a condensation reaction between the methylol intermediate and amino compound proceeds, forming a methylene bridge and liberating a molecule of water as a by-product:

$$RNH-CH_2OH + H_2NR \longrightarrow RNH-CH_2-NHR + H_2O$$

Unlike the methylolation reaction, the condensation reaction is catalysed by acids only.

Because of their high reactivity leading to unstable products and generally poor compatibility with most of the binders used in the coating industry, methylol products of melamine, benzoguanimine, urea or glycoluryl are not used as crosslinkers. Etherification (or the so-called alkylation) with an aliphatic alcohol is the usual modification which overcomes the stability and compatibility problems:

$$RNH-CH_2-OH + R'OH \longrightarrow RNH-CH_2-OR' + H_2O$$

The reactions of methylolation, condensation and etherification are only an oversimplified presentation of the chemistry involved in the manufacture of amino resins. In fact it is a rather complicated process in which, in general, all of the three reactions proceed simultaneously, involving very complicated kinetics. Therefore, especially for melamine and benzoguanimine resins, it is very difficult to assign an exact mechanism of the resin formation. For details concerning the kinetics of the process, which determines to a great extent the type of resins obtained and their properties, a considerable number of published research papers can be consulted [49–68]. The overall reaction scheme of the process of preparation for alkylated melamine or benzoguanimine resins can be presented as shown on the facing page.

Depending on the mole ratio between the reactants and the reaction conditions, fully alkylated monomeric products or partially alkylated polymers can be obtained. Fully alkylated melamines or benzoguanimine monomers have a lower reactivity than the partially alkylated representatives of these resins, and they are potential candidates for crosslinkers for powder coatings, providing sufficient stability to the system.

Different alcohols may be used for alkylation of the amino resins. Butylated urea and melamine formaldehyde resins were among the first amino resins to be used as crosslinkers in the coating industry. In the early 1960s methylated melamine resins were introduced to the market as alternatives for butylated types, providing a higher solid content to the paint formulation at application viscosity. The evaluation of these products known under the common name of HMMM types of melamine resins (hexamethoxymethyl melamine) indicated a remarkable superiority over the butylated melamine. HMMM enamels have excellent gloss, without the characteristic haze of the butylated melamines, and in addition improved flexibility, adhesion, impact resistance with equal or better chemical resistance and exterior durability [69]. At the same time, HMMM resins impart equal film hardness at half the level of butylated melamine resins. Since melamine resins are in general low molecular compounds with a relatively low glass transition temperature, they decrease the T_g of the powder coating formulation remarkably. Much lower amounts of HMMM resins in the coating composition allow the formulation of powder coatings with a T_g high enough to provide good storage stability.

Fully methylolated grades of methylol melamine are produced on a commercial scale with a solids content higher than 98% and degree of polymerization ranging between 1.35 and 1.7 [69]. At room temperature they are highly viscous to waxy solid materials, which makes them rather unsuitable for use in a powder coating composition bearing in mind the present production technology.

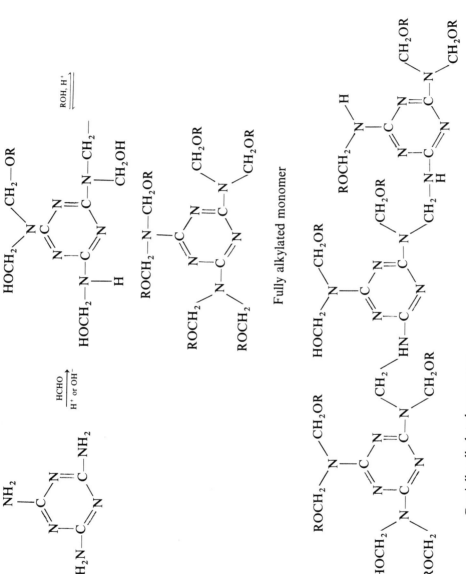

Another weak point of HMMM is the liberation of methanol during curing. Since the boiling point of methanol is rather low its rate of leaving the film during curing is very rapid. In thicker films this causes surface defects, usually manifested in the form of blisters. Therefore its use is limited to thin film applications.

An interesting way of obtaining solid melamine crosslinkers for powder coatings is described in a patent of Sherwin-Williams [70]. According to the patent, 1.8–2.2 moles of monohydroxyl phenol compounds are condensed with 1 mole of hexamethoxymethyl melamine in a presence of oxalic acid as catalyst. The methanol which is a by-product of the condensation process is removed by vacuum, after which a high viscous material is obtained having upon cooling a softening point of 106°C and a melting point of 116°C. It is claimed that such hardeners can be used in combination with hydroxyl containing binders without having surface defect problems in coating layers up to $25\,\mu$m. At curing temperatures ranging from 95 to 205°C films with good hardness, flexibility and gloss can be obtained.

Fully methylated monomeric benzoguanimine formaldehyde condensates are solid compounds at room temperature, with a melting point around 90°C depending on the purity. This makes them very suitable for incorporating in the powder coating composition using conventional extrusion procedures preceded with dry blending of the coating constituents. Benzoguanimine resins as a type of aminotriazine resins are very popular crosslinkers for can coatings, giving excellent sterilization resistance to the cured films. They are also widely used in the electrodeposition coatings. The benzoguanimine structure contributes superior detergent and chemical resistances. As well as splitting off methanol during curing, which could cause the same blistering problem as in the case of HMMM, another weak point of the benzoguanimine resins is their rather poor exterior durability, which restricts their use to indoor applications only.

Tetramethylmethoxy glycoluryl (TMMGU) with the formula presented below is another potential crosslinker for powder coatings [71–75]:

$$\begin{array}{c} CH_3OCH_2-N\text{---}CH\text{---}N-CH_2OCH_3 \\ | \qquad\qquad | \\ O=C \qquad\quad C=O \\ | \qquad\qquad | \\ CH_3OCH_2-N\text{---}CH\text{---}N-CH_2OCH_3 \end{array}$$

TMMGU is a solid at room temperature with a melting point of about 90°C. It has been reported that coatings containing glycoluryl resins have excellent mechanical properties, corrosion resistance, humidity resistance and good accelerated durability [71]. During baking, films with glycoluril/formaldehyde resins release much lower amounts of formaldehyde than melamine/formaldehyde containing films. Compared to melamine resins, glycoluril/formaldehyde resins require stronger acid catalysis to achieve a fast crosslinking reaction. At low stoving temperatures some of the titanium pigments, depending on the nature of their surface treatment, show inhibition in the cure response. This is especially the case of alumina-treated rutile titanium dioxide pigments. In the case of alumina silica zinc-treated rutile titanium dioxide pigments the inhibition of curing is not emphasized. By increasing the catalyst level it is possible to overcome the cure inhibition by the pigments, but, as can be expected, the increased catalyst level has an adverse effect on the humidity resistance of the coatings.

REFERENCES

1. CIBA Ltd, Br. Pat. 814 511, 1960.
2. Shell, Neth. Pat. 67 16 630, 1969.
3. CIBA, Fr. Pat. 1 570 826, 1970; Br. Pat. 1 250 840, 1972.
4. Budnowski, M., *Angew. Chem.,* **80**, 851, 1968.
5. Budnowski, M. *Kunstoffe,* **55**, 641, 1965.
6. Merck, Y., and Loutz, J. M., *Proceedings of XVe Congress AFTPV,* 1983, p. 183.
7. Paul, S., in *Surface Coatings, Science and Technology,* John Wiley & Sons, Chichester, 1985, p. 267.
8. Chemische Werke Hüls AG, Ger. Pat. 3 004 876, 1980.
9. Chemische Werke Hüls AG, Ger. Pat. 3 143 060, 1981.
10. Nicholas, L., and Gmitter, G. T., *J. Cell. Plast.* **1**, 85, 1965.
11. Chemische Werke Hüls AG, Ger. Pat. 3 322 718 and Ger. Pat. 3 322 719, 1983.
12. UCB, Ger. Pat. 2 542 191, 1974.
13. Bayer AG, Ger. Pat. 3 232 463, 1982.
14. Chemische Werke Hüls AG, Eur. Pat. 0 045 994, 1980.
15. Chemische Werke Hüls AG, Eur. Pat. 0 045 996 and Eur. Pat. 0 045 998, 1980.
16. Arendt, V. D., Logan, R. E., and Saxon, R., *J. Cell. Plast.,* November–December, 376, 1982.
17. Alexanian, V., Forgione, P. S., and Chang, L. W., to American Cyanamide Co., US Pat. 4 379 767, 1983.
18. Singh, B., Chang, L. W., and Forgione, P. S., to American Cyanamide Co., US Pat. 4 439 616, 1984.
19. Fiori, D. E., and Dexter, R. W., *Water-Borne and Higher Solids Coatings Symposium,* New Orleans, 5–7 February 1986.
20. *The Pharmacologist,* **28**, 3, 1986.
21. Takeda Yakuhin Kogy, GB Pat. 2 080 305, 1980.
22. Kodak Ltd, WP 8 300 328, 1981.
23. Cooray, B., and Spencer, R., *Paint and Resin,* October 1988, p. 18.
24. Ciba-Geigy Corp., US Pat. 4 393 002, 1982.
25. 'DEH 40 epoxy curing agent', EU 3566-E-680, *Dow Epoxy Resins in Power Coatings,* 1981 ed., Dow Chemicals.
26. 'Epicure DX-143', Epikote Technical Manual EP 1.4.16, 2nd ed., Shell Resins.
27. 'Epikote resin powder coatings', Epikote Technical Manual EP 2.10, 2nd ed., p. 8.
28. Hardener HT 972, Ciba-Geigy Corp. Brochure CR 8895M81, 1981.
29. Hardener HU 261, Ciba-Geigy Corp. Brochure CR 8615C90, 1981.
30. Chemische Werke Hüls AG, Eur. Pat. 0 044 030, 1080.
31. Chemische Werke Hüls AG, Ger. Pat. 3 311 404, 1983.
32. Nitto Electric Industry, Jap. Pat. 57 042 760, 1980.
33. Yuka Shell Epoxy, Jap. Pat. 58 118 818, 1982.
34. Sumitomo Durez, Jap. Pat. 59 226 066, 1983.
35. Shell Oil Company, US Pat. 4 214 068, 1980.
36. Knop, A., and Pilato, L. A., in *Phenolic Resins—Chemistry, Applications and Performance,* Springer-Verlag, Berlin, Heidelberg, 1985, p. 1.

37. Megson, N. J. L., *Phenolic Resin Chemistry*, Butterworths, London, 1958.
38. Martin, R. W., *The Chemistry of Phenolic Resins*, John Wiley and Sons, New York, 1956.
39. Yeddanapalli, L. M., and Francis, D. J., *Makromol. Chem.*, **55**, 74, 1962.
40. Shell Oil Company, US Pat. 4 214 068, 1980.
41. 'Dow epoxy resins in powder coatings', 1981 ed., EU 3568-E-680, EU 3569-E-680, EU 3570-E-680.
42. Shell Resins, 'Epicure 171 and Epikure DX-169', Epikote Technical Manual EP 1.4.19 and EP 1.4.20, 1983.
43. Ciba-Geigy Corp., Brochure CR 853A5C81, 1981.
44. General Electric Co., US Pat. 3 435 002, 1967.
45. General Electric Co., US Pat. 3 632 608, 1967.
46. General Electric Co., Br. Pat. 1 032 372, 1962.
47. Rhone Poulenc Ind., Ger. Pat. 2 558 295, 1974.
48. Walz, G., and Kraft, K., *Proceedings of 4th International Conference on Organic Coatings, Science and Technology*, Athens, 1978, p. 56.
49. Okano, M., and Ogata, Y., *J. Am. Chem. Cos.*, **74**, 5728, 1952.
50. Wohnsiedler, H. P., *Ind. Eng. Chem.*, **44**, 2679, 1952.
51. Ugelstad, J., and Jonge, J. de, *Rec. Trav. Chem.*, **76**, 919, 1957.
52. Gordon, M., Halliwell, A., and Wilson, T. SCI Monograph 20, Society of Chemical Industry, London, 1966, pp. 187–98.
53. Gordon, M., Halliwell, A., and Wilson, T., *J. Appl. Polym. Sci.*, **10**, 1153, 1966.
54. Sato, K., *Bull. Chem. Soc. Japan*, **40**, 2963, 1967.
55. Sato, K., *Bull. Chem. Soc. Japan*, **40**, 724, 1967.
56. Aldersley, J. W., Gordon, M., Halliwell, A., and Wilson, T., *Polymer*, **9**, 345, 1968.
57. Sato, K., *Bull. Chem. Soc. Japan.*, **41**, 7, 1968.
58. Anderson, I. H., Cawley, M., and Steedman, W., *Br. Polym. J.*, **1**, 24, 1969.
59. Berge, A., Gudmindsen, S., and Ugelstad, J., *Eur. Polym. J.*, **5**, 171, 1969.
60. Berge, A., Kraeven, B., and Ugelstad, J., *Eur. Polym. J.*, **6**, 981, 1970.
61. Sato, K., and Naito, T., *Polym. J. Japan*, **5**, 144, 1973.
62. Sato, K., and Abe, Y., *J. Polym. Sci., Chem. Ed.*, **13**, 263, 1975.
63. Tomita, B., *J. Polym. Sci.*, **15**, 2347, 1977.
64. Berge, A., *3rd International Conference on Organic Coatings, Science and Technology*, Athens, 1977.
65. Tashiro, T., and Oiwa, M., *J. Polym. Sci., Polym. Chem.*, **19** (3), 645, 1981.
66. Ubdegraff, I. H., and Suen, T. J., 'Condensations with formaldehye', Ch. 14 in *Polymerisation Processes*, Ed. C. E. Schidhenecht, Wiley–Interscience, New York, 1977.
67. Drumm, M. F., and Le Blank, J. R., 'The reactions of formaldehyde with phenols, melamine, aniline and urea', Ch. 5 in *Step Growth Polymerisation*, Ed. D. H. Solomon, Marcel Dekker, New York, 1972.
68. Brydson, J. A., *Plastic Materials*, 2nd ed., Chs. 19–23, Van Nostrand Reinhold, New York, 1970.
69. Kirsch, A. J., in *50 Years of Amino Coatings Resins*, The Twinchell Company, Philadelphia, 1986, p. 10.
70. The Sherwin-Williams Co., US Pat. 4 189 421, 1980.
71. Parekh, G. G., *J. Coat. Technol.*, **51** (658), 101, 1979.
72. Parekh, G. G., US Pat. 4 118 437, 1978.
73. SCM Corp., US Pat. 4 444 941, 1981.
74. SCM Corp., Eur. Pat. 0 021 770, 1979.
75. SCM Corp., US Pat. 4 271 277, 1979.

3.4 INDUSTRIAL THERMOSETTING POWDER COATINGS

3.4.1 EPOXY POWDER COATINGS

After the successful introduction of thermoplastic polyethylene powder coatings at the beginning of the 1950s, epoxy powder coatings were introduced in the market during the mid 1950s as the first thermosetting powder coatings.

The epoxy group can be introduced into the polymer backbone by different methods, but only a few of them are of commercial importance.

The epoxidation of unsaturated compounds by catalytic oxidation with oxygen is one of the most economical methods for production of ethylene oxide, but it has never found a real commercial application in the production of epoxy resins.

An easy and commercially widely used method for epoxidation of unsaturated compounds is that of using peroxy acids or esters of peroxy acids:

$$R_1R_2C=CR_3R_4 + R_5-COOOH \longrightarrow R_1R_2C\underset{O}{-}CR_3R_4 + R_5COOH$$

The ease of preparation of peroxy acids by treating the corresponding acids with hydrogen peroxide has led to the development of *in situ* epoxidation methods. In this way separate preparation of peroxy acids and the difficulties connected to their storage and loss of active oxygen due to decomposition are eliminated.

The method of epoxidation with peroxy acids is widely used on an industrial scale for obtaining epoxidized drying oils. Although it is not the typical method for production of epoxy resins used in the powder coating field, it is used for production of liquid epoxy resins of low molecular weight to solid epoxy resins of high molecular weight by epoxidation of polybutadiene, polyisoprene, polycyclopentadiene, poly(isoprene–styrene), poly(isoprene–isobutylene), unsaturated hydrocarbon resins, poly(cyclopentadiene–styrene–methylstyrene), etc. [1, 2].

The most important route for preparation of epoxy resins is conversion of compounds containing activated hydrogen to epoxides by the hydrohalogenation reaction with epichlorohydrin, followed by dehydrohalogenation with a suitable base:

$$RH + CH_2CH-CH_2Cl \longrightarrow R-CH_2-CH-CH_2Cl$$
$$\underset{O}{\diagdown\diagup}\underset{OH}{|}$$

$$R-CH_2-CH-CH_2Cl + NaOH \longrightarrow R-CH_2-CH-CH_2$$
$$\underset{OH}{|}\underset{O}{\diagdown\diagup}$$

The active hydrogen containing compounds used in this process which gives glycidyl types of epoxies are phenols, alcohols, carboxylic acids, amines, mercaptanes, silanols and certain phosphorus compounds.

By far the most important representatives of binders for powder coatings obtained in this way belong to the group of bisphenol A and novolac types of epoxy resins. Several very important low molecular weight epoxy crosslinkers are also obtained through the epichlorohydrin route as already described in Section 3.3.1.

Bisphenol A epoxy resins are produced by addition of epichlorohydrin to 2,2-bis(4-hydroxyphenyl) propane, commercially called bisphenol A, and subsequent dehydrohalogenation with sodium hydroxide.

It is believed that the first bisphenol A was ionized with an appropriate anion to form a phenoxy anion:

$$HO-C_6H_4-C(CH_3)_2-C_6H_4-OH + A^- \rightleftharpoons HO-C_6H_4-C(CH_3)_2-C_6H_4-O^- + HA$$

The phenoxy anion then attacks the oxirane ring of epichlorohydrin to form a chlorohydrin ether anion:

$$HO-C_6H_4-C(CH_3)_2-C_6H_4-O^- + CH_2-CH-CH_2Cl \longrightarrow$$
$$\underset{O}{\diagdown\diagup}$$

$$HO-C_6H_4-C(CH_3)_2-C_6H_4-O-CH_2-CH(O^-)-CH_2Cl$$

The formed chlorohydrin ether anion then abstracts a proton from another bisphenol A molecule forming another phenoxy anion:

$$HO-C_6H_4-C(CH_3)_2-C_6H_4-O-CH_2-CH(O^-)-CH_2Cl +$$

$$HO-C_6H_4-C(CH_3)_2-C_6H_4-OH \rightleftharpoons$$

THERMOSETTING POWDER COATINGS

HO—⌬—C(CH₃)(CH₃)—⌬—O—CH₂—CH(OH)—CH₂Cl + HO—⌬—C(CH₃)(CH₃)—⌬—O⁻

The regenerated phenoxy anion can then react with another epichlorohydrin producing chlorohydrin ether anion, etc. Finally, the chlorohydrin ether is dehydrochlorinated by the base to yield an epoxy resin:

HO—⌬—C(CH₃)(CH₃)—⌬—O—CH₂—CH(OH)—CH(Cl)—CH₂ + OH⁻ ⟶

HO—⌬—C(CH₃)(CH₃)—⌬—O—CH₂—CH—CH₂ (epoxide)

Since the hydroxyl anion is consumed in the last reaction, it is necessary to add the sodium hydroxide continuously, which means that all reactions mentioned above are carried out simultaneously. The overall reaction can be presented in a simplified way as follows:

2Cl—CH₂—CH(—O—)CH₂ + HO—⌬—C(CH₃)(CH₃)—⌬—OH ⟶

Cl—CH₂—CH(OH)—CH₂—O—⌬—C(CH₃)(CH₃)—⌬—O—CH₂—CH(OH)—CH₂—Cl +

2 NaOH ⟶ CH₂—CH(—O—)CH₂—O—⌬—C(CH₃)(CH₃)—⌬—O—CH₂—CH—CH₂ (epoxide) + 2 NaCl

However, because of the simultaneous nature of the process with respect to all of the reactions involved, the newly formed epoxy groups may be attacked by the phenoxy anions instead of epichlorohydrin, leading to the formation of oligomers with a higher molecular weight according the following overall scheme:

$$CH_2\!-\!CH\!-\!CH_2\!-\!O\!-\!\langle\!\bigcirc\!\rangle\!-\!\underset{\underset{CH_3}{|}}{\overset{\overset{CH_3}{|}}{C}}\!-\!\langle\!\bigcirc\!\rangle\!-\!O\!-\!CH_2\!-\!CH\!-\!CH_2\; +$$
$$\underset{O}{\diagdown\!\diagup}\qquad\qquad\qquad\qquad\qquad\qquad\underset{O}{\diagdown\!\diagup}$$

$$n\,HO\!-\!\langle\!\bigcirc\!\rangle\!-\!\underset{\underset{CH_3}{|}}{\overset{\overset{CH_3}{|}}{C}}\!-\!\langle\!\bigcirc\!\rangle\!-\!OH + n\,CH_2\!-\!CH\!-\!CH_2\!-\!Cl \longrightarrow$$
$$\qquad\qquad\qquad\qquad\qquad\qquad\qquad\underset{O}{\diagdown\!\diagup}$$

$$CH_2\!-\!CH\!-\!CH_2\!-\!O\!-\!\langle\!\bigcirc\!\rangle\!-\!\underset{\underset{CH_3}{|}}{\overset{\overset{CH_3}{|}}{C}}\!-\!\langle\!\bigcirc\!\rangle\![-O\!-\!CH_2\!-\!\underset{\underset{OH}{|}}{CH}\!-\!CH_2\!-\!O\!-$$
$$\underset{O}{\diagdown\!\diagup}$$

$$\langle\!\bigcirc\!\rangle\!-\!\underset{\underset{CH_3}{|}}{\overset{\overset{CH_3}{|}}{C}}\!-\!\langle\!\bigcirc\!\rangle\!-\!]_n\!-\!O\!-\!CH_2\!-\!CH\!-\!CH_2$$
$$\qquad\qquad\qquad\qquad\qquad\underset{O}{\diagdown\!\diagup}$$

This reaction cannot be avoided, but its extent can be minimized by using a high excess of epichlorohydrin over the stoichiometric ratio.

Another problem that occurs during the manufacture of bisphenol A epoxy resins is the hydrolysis of epichlorohydrin leading in the subsequent reaction with the sodium hydroxide to the formation of glycidol:

$$CH_2\!-\!CH\!-\!CH_2Cl + H_2O \longrightarrow CH_2\!-\!\underset{\underset{OH}{|}}{CH}\!-\!\underset{\underset{Cl}{|}}{CH_2} + NaOH \rightleftharpoons$$
$$\underset{O}{\diagdown\!\diagup}\qquad\qquad\qquad\qquad\underset{OH}{|}$$

$$\qquad\qquad\qquad\qquad\qquad\qquad CH_2\!-\!CH\!-\!\underset{\underset{OH}{|}}{CH_2} + NaCl$$
$$\qquad\qquad\qquad\qquad\qquad\qquad\underset{O}{\diagdown\!\diagup}$$

The glycidol formed can undergo further hydrolysis with water forming glycerine, reducing in this way the yield of the process with respect to epichlorohydrin. What is even worse, glycidol tends to react with bisphenol A to form a resin with two hydroxyl end groups. These end groups are not reactive when the resin is applied in a powder coating composition, thus adversely affecting the coating properties.

Diglycidyl ethers of bisphenol A are high viscous liquids at room temperature. With increasing molecular weight the melting point of the resins increases too, and those of interest for production of powder coatings usually have a molecular weight higher than 900, which corresponds to a melting point higher than 70°C.

THERMOSETTING POWDER COATINGS

The epoxy equivalent weight (e.e.w), defined as the weight of the resin containing one gram equivalent of epoxide, is an important parameter for the powder coating chemist. Very often terms such as weight per epoxide (w.p.e.) or epoxy molar mass (e.m.m.) are used interchangeably instead of e.e.w.

Although the first bisphenol A type of epoxy resin was synthesized in the laboratories of I.G. Farbenindustrie in 1934, the real commercialization of these new polymers began after the Second World War based on the work of Castan in Switzerland and Geenlee in the United States. A great number of patents issued in the more than forty years long industrial history of the epoxy resins has been reviewed several times in the literature [3–6].

From the reaction scheme of the synthesis of epoxy resins it is clear that the molecular mass of the resin can be adjusted by a proper ratio of bisphenol A/epichlorohydrin. Very often, however, especially when solid epoxy resins have to be obtained, a condensation of low molecular weight epoxy resin with bisphenol A is performed in the presence of a suitable catalyst:

$$CH_2\!-\!CH\!-\!CH_2\!-\!O\!-\!\bigcirc\!-\!\underset{\underset{CH_3}{|}}{\overset{\overset{CH_3}{|}}{C}}\!-\!\bigcirc\!-\!O\!-\!CH_2\!-\!CH\!-\!CH_2\ +$$

$$HO\!-\!\bigcirc\!-\!\underset{\underset{CH_3}{|}}{\overset{\overset{CH_3}{|}}{C}}\!-\!\bigcirc\!-\!OH = CH_2\!-\!CH\!-\!CH_2\!-\!O\!-\!\bigcirc\!-\!\underset{\underset{CH_3}{|}}{\overset{\overset{CH_3}{|}}{C}}\!-\!\bigcirc$$

$$[-O\!-\!CH_2\!-\!\underset{\underset{OH}{|}}{CH}\!-\!CH_2\!-\!O\!-\!\bigcirc\!-\!\underset{\underset{CH_3}{|}}{\overset{\overset{CH_3}{|}}{C}}\!-\!\bigcirc]_n\!-\!O\!-\!CH_2\!-\!CH\!-\!CH_2$$

Depending on the ratio between the low molecular weight liquid epoxy resin and bisphenol A, resins with a degree of polymerization up to 30 can be obtained. However, in the coating industry, epoxy resins with degrees of polymerization higher than 12 are not normally employed [7].

A modification of bisphenol A epoxy resin with an improved flexibility is described in the patent of Asahi Chemical Ind. [8] where bisphenol A epoxy in molar excess resin is reacted with a long-chain (C10–C25) aliphatic dicarboxylic acid of a saturated or unsaturated nature. It is claimed that the powder coating having these modified resins as binders, cured with DICY, imidazole derivatives, dihydrazides of dicarboxylic acids, solid aromatic amines or acid anhydrides, exhibits much better flexibility compared with that of the unmodified epoxy resins.

In the mid 1960s new multifunctional resins of the epoxy-novolac type were introduced in the market. The chemistry of these resins is very similar to that of bisphenol A type. Starting materials are novolac resins obtained by condensation of phenol, bisphenol A or cresol with formaldehyde in acidic media, followed by hydrohalogenation with epichlorohydrin and subsequent dehydrohalogenation with sodium hydroxide. The epichlorohydrin/phenolic hydroxyl mole ratio should be about 6 to assure complete epoxidation and to avoid crosslinking and possible gelation [9].

Aliphatic glycidyl ethers prepared from aliphatic alcohols are not used as sole binders for powder coatings because they are liquid at room temperature. The condensation of aliphatic polyols with epichlorohydrin is performed in two steps. The epichlorohydrin and the polyol are reacted in the presence of Lewis acid to produce the corresponding chlorohydrin. In the second step, the process proceeds with dehydrohalogenation by means of sodium aluminate or other base in a water immiscible solvent. Although, alone, they are not suitable as binders for powder coatings, they can be used in combination with other resins as additives which provide flexibility.

The principal difference between the bisphenol A epoxies and novolac epoxy resins is in their functionality. While bisphenol A epoxy resins have a functionality of 2 or in the case of low molecular representatives even 1, 9 [7], the functionality of the novolac epoxy resins can be easily adjusted by the resin manufacturer, and it is usually higher than two. This characteristics is very important in applications where a high density of reactive groups in a binder system is needed. Such cases are coatings with high solvent resistance combined with very high reactivity and high performances at elevated temperatures.

Epoxy powder coatings were used at the beginning for functional purposes only, but soon their potential as decorative coatings was recognized. Nowadays, functional and decorative epoxy powder coatings enjoy more than 40% of the thermosetting powder coatings market share, and more than 50% including the so-called hybrids using carboxyl functional polyesters as the main binder and epoxy resins as crosslinkers.

The curing agents commonly used in epoxy powder formulations are solid aliphatic amines, solid aromatic amines and their resinous adducts, solid acid anhydrides and their resinous adducts, solid phenolic resins, polybasic organic acids and acid functional polyester resins.

Dicyandiamide (DICY) is a typical representative among the aliphatic solid amines used as curing agents in the production of epoxy powder coatings. DICY has a reasonably high melting point of 211°C and during the extrusion of the powder coating it does not melt to give a homogeneous mixture with the epoxy resin on a molecular scale. Therefore, the commercial grades of DICY are finely ground powders with a particle size below 75 μm, which provide an acceptable homogeneous distribution of the fine particles of the hardener during the extrusion step. Since it is very difficult to obtain an optimum distribution of the hardener, in many cases it is necessary to use more DICY than stoichiometric quantities in order to ensure a full cure. Another reason for not having strict stoichiometric proportions of the resin and the curing agent of this class is the curing mechanism of DICY which is complex, involving addition reactions with both the epoxy and hydroxyl groups and catalysis of the epoxide/epoxide reaction. The optimum concentration is dependent on the types and concentrations of pigments present, the particle size of the curing agent, the degree of dispersion of DICY in the resin and the desired stoving schedule. The best level in a particular formulation therefore has to be established by trials.

The curing speed of powder coating with pure DICY is relatively low, even at comparatively high temperatures, so that stoving schedules of 30 minutes at 185°C or 20 minutes at 200°C are required. To obtain an acceptable curing time at temperatures that are normally used in the stoving lines in industry, use of an accelerator is necessary. Boron trifluoride/amine complexes have been suggested as latent catalysts for epoxy powders [10]. A wide range of decomposition temperatures between 70 and 300°C can be obtained dependent on the choice of amine complexing agent. Imidazoline derivatives such as 2-methylimidazoline [11, 12] are also reported as catalysts for epoxy/DICY systems. Tertiary amino compounds and quaternary ammonium salts are also very suitable catalysts for accelerating the curing reaction. Commonly the producers supply DICY already containing accelerator in the form of a dry blend. The curing speed of the powder coating can then be adjusted by incorporating in the powder formulation different ratios of accelerated and non-accelerated DICY. A typical formulation of this kind is presented in Table 3.4 [13].

Table 3.4 Epoxy powder coating formulations with dicyandiamide as curing agent [13] (Reproduced by permission of Dow Chemical Europe)

Material	Parts by weight	
	A	B
DER 663[a]	47.4	47.4
DER 673MF[b]	10.0	10.0
DICY DEH 41[c]	1.3	2.6
DICY	1.3	—
TiO$_2$—rutile	40.0	40.0
	100.0	100.0
Bake schedule (min at 180°C)	12–15	10–12

[a] Bisphenol A epoxy resin, trademark of Dow Chemical Company.
[b] Bisphenol A epoxy resin, trademark of Dow Chemical Company, containing 5% of flow additive.
[c] Accelerated DICY, trademark of Dow Chemical Company.

A patent of Hitachi Chemical [14] describes a method for obtaining the powder coating composition in with dicyandiamide is incorporated by dissolving it in novolac phenol formaldehyde resin. In this way the homogeneity of the system is improved, resulting in higher gloss.

Better gloss can be obtained with the substituted grades of DICY which have higher compatibility with the epoxy resin. Water resistance of the epoxy powders cured with substituted DICY is also improved, since these curing agents have lower solubility in water than DICY itself.

In general, it is difficult to obtain matt films with powder coatings. An interesting method to produce matt finishes has been proposed by Klaren [15]. He has found that by dry blending two epoxy/DICY powder coatings with different reactivities and high gloss, a gloss reduction can be obtained by mixing them in various ratios. For example, when 2 parts of an epoxy powder coating with DICY as the curing agent having a gel time of 6–30 minutes (dependent on the temperature) are dry blended with 1 part of powder coating having the same composition, but being catalysed with a tertiary amine, thus having a gel time of 1–6 minutes, a gloss of only 18% can be obtained after curing for 15 minutes at 180°C. Increasing the amount of the high reactive powder coating brings about a gloss increase of up to 68%, although the individually cured powders have a gloss of 100%. A proper combination of the fast and slow curing powder coatings gives a range of films from matt (20–25% gloss) to silk-like (60–56% gloss) that have at the same time good mechanical properties.

A combination of DICY with an adduct of abietic acid and maleic anhydride is described in a patent assigned to Dow Chemicals [16]. The curing agent comprises 50–80 wt% of the adduct and 50–20 wt% dicyandiamide. In one of the examples in which the ratio between the different curing agents is 80:20 with a bisphenol A epoxy resin with an epoxy equivalent weight of 785 as binder, a gloss of only 15% (60° angle) has been indicated.

Epoxy powder coatings cured with DICY types of curing agents show excellent chemical and solvent resistances. The resistance towards atmosphere containing sulphur dioxide is extremely good. Water sensitivity is substantially improved when substituted DICYs are used as crosslinkers. When properly formulated, these powder coatings offer outstanding impact, abrasion and scratch resistances. A very good point is the excellent adhesion without using any primer, which is in general a typical characteristic of most epoxy coatings. The availability of accelerated grades of DICY enables the powder coating formulator to adjust the curing behaviour of the coatings to various stoving conditions with considerable flexibility. Curing schedules of 8–15 minutes at 150–180°C are commonly used in practice.

Like all powder coatings based on bisphenol A epoxy resins as the main binder, coating formulations containing DICY suffer from sensitivity towards u.v. light. In the accelerated weathering tests they have shown only 7% of the initial gloss after 192 hours (Atlas Weather–Omether carbon-arc type), while at the same

THERMOSETTING POWDER COATINGS

conditions, acrylic, polyurethane and polyester/TGIC powder coatings retain 98% of the initial gloss [17]. The yellowing resistance of DICY cured epoxy powder coatings is rather poor. On a relative scale range where 10 indicates heavy and 0 no yellowing, epoxy/DICY systems are among the worst, having a mark of 10 [17].

Because of these two weak points of epoxy/DICY powder coatings, their application is restricted mainly to functional coatings and decorative coatings for indoor use, where the yellowing resistance is not a requested property. The main application of epoxy/DICY systems is certainly the coating of pipes for subsurface pipelines, where the epoxy powder coatings provide high corrosion resistance.

Amine derivatives containing primary or secondary amino groups are generally unsuitable as crosslinkers for powder coatings because of their high reactivity with epoxy resin. Even this is the case with those amines that are solids at normal temperature. This creates certain problems during processing of the coating or considerably reduces the storage time. However, aliphatic polyamines which differ from DICY are described in the patent literature, and some of them are commercially available. For example, the condensation product of isophorone diamine and diglycidyl ether of bisphenol A with a low molecular weight (Epicote 828) in the ratio of 2:1 with a melting point of 55–57°C can be used in combination with high molecular weight epoxy resins as the curing agent [18].

Aliphatic polyamine for preparation of corrosion resistant powder coatings is described as a reaction product of aminododecanoic acid and condensate of bisphenol A epoxy resin with an equivalent amount of phenol [19]. It is recommended as a hardener for epoxy powder coatings for pipes, drums, tanks, etc.

Powder coatings with amine derivatives as hardeners are characterized by extremely short curing cycles at relatively low temperature. The formulations presented in Table 3.5 [20] based on a commercially available amine curing

Table 3.5 Low curing powder coating with amine hardener [20] (Reproduced by permission of Shell International Chemical Company Ltd)

Material	Parts by weight	
	A	B
Epikote 1055[a]	100.0	100.0
Barytes	—	50
Modaflow	0.5	0.5
Epikure DX-137A[b]	2.5	2.5

[a] Bisphenol A epoxy resin, trademark of Shell Chemicals.
[b] Amine curing agent, trademark of Shell Chemicals.

agent from Shell Chemicals, Epikure DX-137A, refer to coatings with curing schedules of 15 min at 140°C, 20 min at 130°C and 30 min at 120°C. Since the proportion of curing agent is very small, the preblending is usually done by high-speed mixers in order to ensure efficient dispersion and homogeneity of the hardener throughout the powder in the next compounding step. It is interesting that in spite of the short curing schedules, the stability of the powder coating is satisfactory, having a reduction of the gel time at 150°C less than 20% after storage of the powder for one week at 50°C.

Epoxy powder coatings with phenolic resins as hardeners are also widely used in practice. In principle, common thermosetting solid phenolic resins can be used as crosslinkers for formulating epoxy powder coatings. The initial problems with respect to the dark colour of the phenolic resins, low reactivity and brittleness have been successfully overcome by the new generation of resinous phenolic hardeners. Coatings with high crosslinking density, high chemical, solvent and boiling water resistances, good mechanical and decorative properties can be obtained. The yellowing tendency is still an obstacle for universal use of these epoxy powder coatings, and like the other epoxy powder coatings they are suitable only for indoor application. In comparison with DICY systems, much shorter curing times can be obtained when suitable catalysts in proper amounts are used. In most cases, the catalyst is usually incorporated in the hardener, which provides better homogeneity of the system. Table 3.6 contains three guide formulations for powder coatings for decorative and functional purposes [21]. The very short curing time is quite indicative. The decorative coatings have good

Table 3.6 Epoxy powder coating formulations with phenolic resinous crosslinkers for: standard decorative purposes (A and B) and functional pipe coating (C) [21] (Reproduced by permission of Dow Chemical Europe).

Material	Parts by weight		
	A	B	C
DER 663U[a]	39.0	—	—
DER 642U[a]	—	70.0	53.0
DEH 80[b]	—	30.0	—
DEH 81[b]	11.0	—	—
DEM 83[b]	—	—	22.0
TiO_2—rutile	50.0	—	—
Pigment + filler[c]	—	—	25.0
	100.00	100.0	100.0
Bake schedule (min at °C)	6–8/180	8–10/180	1/230

[a] Bisphenol A epoxy resin, trademark of Dow Chemical.
[b] Solid resinous phenolic curing agent of Dow Chemical.
[c] Iron oxides or chrome oxides + $CaCO_3$, baryte or silica.

all-round properties combined with high gloss. In the clear unpigmented coating (B), an epoxy resin with a significantly higher melting range has to be used in order to prevent sticking and sintering during the grinding operation.

Novolac types of epoxy resins can also be used in combination with phenolic hardeners. In this case, an even higher crosslinking density can be obtained, because of the higher functionality of the novolac epoxy resins. A composition suitable for insulation coating for electrical and electronic parts with good moisture resistance, high temperature electrical characteristics and heat-cycle resistance is described in Jap. Pat. 59 226 065 [22] and comprises 24% novolac epoxy resin with an epoxy equivalent weight of 190 and a melting point of 75°C, 16% phenolic resin of novolac type with a melting point of 78°C and an hydroxide equivalent of 220, 58.8% silicon dioxide as a filler, 1% carbon black as pigment and 0.2% imidazole as catalyst.

A distinguishing feature of epoxy powder coatings cured with phenolic resins is an exceptionally high gloss combined with good flow. They have very good mar resistance, chip and impact resistances and good adhesion without an adhesion promoter. With regard to possible contacts with food and skin, they meet the normal toxicity standards. Relatively low baking temperatures can be used for curing, permitting application on heat-sensitive substrates.

Typical applications of the decorative epoxy/phenolic powder coatings include appliances, wire goods, bicycle frames, hot-water radiators, tanks, fire extinguishers, toys and tools, electronic components, housing for electronic instruments, heavy-duty electrical cabinets, control boards, fuse boxes, automotive armatures, etc.

Curing of the epoxy resin can be done with acid anhydrides or polycarboxylic acids. The use of acid anhydrides is almost completely abandoned nowadays, for toxicological reasons. A somewhat lower risk with respect to human health exists when resinous anhydrides described in Section 3.3.5 are used. Polycarboxylic acid anhydrides, described in Section 3.1.2, react with epoxy groups in a complex stepwise manner. The efficiency of curing with acid anhydrides is not very sensitive to the stoichiometry, as is the case with polyacids, since in principle only one hydroxyl group is needed to initiate the reaction. However, this advantage is not big enough to compensate for the heavy irritation of the mucous membranes which is quite common when acid anhydrides are used as hardeners in epoxy powder coatings. On the other hand, acid anhydrides are susceptible to hydrolysis by atmospheric moisture which creates certain problems during storage of the curing agent itself or the powder coating. Therefore, today, the use of anhydrides as curing agents either in the form of low molecular weight compounds or resinous anhydrides of higher molecular weight is very near.

A special feature of polycarboxylic acid anhydride curing agents in epoxy powder coatings is the very rapid curing rate that can be achieved by means of suitable catalysts such as metal salts of organic monoacids. The crosslinked film has excellent hardness, good electrical properties, excellent resistance to solvents and good yellowing resistance on overbake. Curing times of 20–30 seconds can be achieved on preheated metal surfaces at 250–275°C in fluidized bed applications or 10 minutes at 180°C in electrostatic spraying applications. Typical epoxy powder coating formulations using trimellitic anhydride and adduct of trimellitic anhydride are given in Table 3.7 [23].

Polycarboxylic acids are not often used as sole crosslinkers for epoxy powder coatings. From a technological, economical and toxicological point of view, the acid functional polyesters which will be discussed in detail in the following sections have certain advantages. These coatings, known under the name hybrids, hold a dominant position among the epoxy powder coatings. They have almost completely replaced the use of polyacids as curing agents. However, relatively fresh patent applications describing systems based on epoxy resins and polyfunctional acids can be found. For example, matt epoxy powder coatings can be obtained by curing bisphenol A epoxy resin with salts of trimellitic acid and 2-phenylmethyl-2-imidazoline [24]. A patent of Mitsubishi Electric Corp. [25] describes a powder coating obtained by dry blending of acid anhydrides including phthalic, maleic, trimellitic or promellitic, polyethylene or polypropylene glycols with molecular weights of 4000–40 000 and epoxy resins with epoxy equivalent weights of 100–5000.

Epoxy powder coatings with dihydrazides of dicarbonic acids as curing agents which are used for the protection of pipes in pipelines are described in a patent of the 3M company [26]. The binder is bisphenol A type epoxy resin, while the dicarbonic acid dihydrazides are of the following general formula:

$$H_2NNH—CO—R—CO—NHNH_2$$

Table 3.7 Clear epoxy powder coatings with TMA (A) and TMA adduct (B) as curing agents [23] (Reproduced by permission of Shell International Chemical Company Ltd)

Material	Parts by weight	
	A	B
Epikote[a] 1004, 1055	100	100
TMA	10–12.5	—
TMA adduct	—	10–12.5
Catalyst	0.2–0.5	0.2–0.5
Curing schedule at 180°C (min)	10	10

[a] Bisphenol A epoxy resins, trademark of Shell Chemicals.

THERMOSETTING POWDER COATINGS

Although in principle any dicarbonic acid can be used, dihydrazides of aliphatic diacids with more than four atoms between the two carboxyl groups are preferred for providing better flexibility to the cured product. Tertiary amines are suggested as catalysts to speed up the reaction, together with quite high amounts of barium sulphate as filler. An example from this patent is given in Table 3.8 as an illustration. An exceptionally good boiling water and cathodic disbondment

Table 3.8 Epoxy powder coating with dihydrazide curing agent [26]

Material	Parts by weight
DER 663U (bisphenol A epoxy resin—Dow Chemical Company)	70.0
Azelaic dihydrazide	5.4
Barium sulphate	84.0
Phthalocyanine green	0.1
Chromium oxide	0.5
Titanium dioxide	3.0
Flow additive (Modaflow)	0.7
2,4,6-Tris(dimethylaminomethyl) phenol as catalyst	0.7

Table 3.9 Major uses of epoxy resin powder coating systems [27] (Reproduced by permission of Shell International Chemical Company Ltd)

Epoxy resin plus	A	B	C	D	E	F	G	H	I	J
Accelerated DICY	*	*		*	*					*
Substituted DICY	*	*		*	*					
Amine derivatives	*	*	*	*		*				
Polyester resins	*	*	*	*	*	*				
Anhydrides	*	*	*	*	*	*	*	*	*	*
Anhydride adducts	*	*	*	*	*	*	*	*	*	*
Phenolic resins			*	*	*		*	*	*	*

A —domestic equipment.
B —metal furniture.
C —building components and fittings.
D —automotive components.
E —agricultural equipment.
F —radiators, boilers.
G —pipes.
H —containers.
I —concrete reinforcing bars.
J —electrical equipment.

resistance at room and elevated temperatures (60 °C) is claimed as an outstanding achievement. Table 3.9 [27] gives a survey of the major uses of the epoxy powder coatings cured with different curing agents.

Several typical advantages of the epoxy powders over the other coating systems are the excellent chemical and solvent resistances, combined with excellent boiling water resistance. The presence of free hydroxyl groups uniformly distributed along the resin chain provides excellent adhesion on metal substrates. Other outstanding features of the epoxy powder coatings are excellent impact, abrasion and scratch resistances. For use in electrical applications (rotors, stators, electrical housing, etc.), they have the necessary dielectric properties.

Weak points of the epoxies are their inferior yellowing resistance and poor outdoor durability due to the presence of the aromatic ether group which is extremely sensitive to u.v. light. Chalking on outdoor exposure is a phenomenon typical for epoxy powder coatings. Although once it has reached a certain level additional chalking does not occur, the use of epoxy powder coatings has been restricted to functional areas and interior decorative applications.

3.4.2 POLYESTER POWDER COATINGS

Polyester powders dominate the powder coatings market worldwide. The first attempts to make pure polyester powders date from the end of the 1960s. In a patent of Hoechst AG [28] a physical blend of two different types of powdered polyester resins containing carboxyl and hydroxyl functional groups is used as a powder coating with a curing time of 15 minutes at 220°C. In one of the examples of the patent, a melt mixing of both polyesters with titanium dioxide as a pigment is powdered after cooling, resulting in a system with a curing schedule of 15 minutes at 220°C. This process of producing powder coatings closely resembles the powder coating technology of today. The curing reaction involves esterification between the carboxyl and hydroxyl groups from both binders, which unfortunately yields water as a by-product causing surface defect problems. Another disadvantage is a rather long curing time at high temperatures.

Two groups of saturated polyesters containing carboxyl or hydroxyl functional groups represent almost 100% of the resins used for production of thermosetting polyester powder coatings.

In principle all reactions for the preparation of esters can be used for manufacturing polyester resins. Of course, not all of them can be performed on an industrial scale because of technical or price reasons. The industrial preparation of polyesters for powder coatings mainly involves direct polyesterification of polybasic acids and polyfunctional alcohols. Depending on which component is in molar excess, the polyester can be terminated with carboxyl or hydroxyl groups. The chemistry of the process can be represented in the following simplified way:

$$\text{HOOC}-A_2-\text{COOH} + \text{HO}-P_2-\text{OH} + \text{HO}-P_3\!\!\begin{array}{c}\text{OH}\\ \text{OH}\end{array} + \text{HOOC}-A_3\!\!\begin{array}{c}\text{COOH}\\ \text{COOH}\end{array} \longrightarrow$$

$$\text{OH}-P_2-A_2-A_3-A_2-P_2-A_2-P_2-A_2-P_3-A_2-P_2-A_2-P_2-A_2-P_3\!\!\begin{array}{c}\text{OH}\\ \text{OH}\end{array}$$
$$\qquad\qquad\qquad | \qquad\qquad\qquad\qquad\qquad |$$
$$\qquad\qquad\quad P_2 \qquad\qquad\qquad\qquad\qquad A_2$$
$$\qquad\qquad\qquad | \qquad\qquad\qquad\qquad\qquad |$$
$$\text{HO}-P_2-A_2-P_2 \qquad\qquad\qquad P_3-A_2-P_2-A_2-P_2-\text{OH}$$

where P_2, P_3, A_2 and A_3 are residues of the diols, triols, diacids and triacids respectively. The numbers of moles of different constituents in the resin formulation in the case of hydroxyl functional polyesters are interrelated by the following equation [29]:

$$\Sigma N_p = N_{a2} + 2N_{a3} + 1 \qquad (3.128)$$

where ΣN_p is the total number of moles of polyols and N_{a2} and N_{a3} are the number of moles of diacids and triacids.

For carboxyl functional polyesters a similar equation is valid:

$$\Sigma N_a = N_{p2} + 2N_{p3} + 1 \qquad (3.129)$$

where ΣN_a is the total number of moles of the polyacids and N_{p2} and N_{p3} are the number of moles of diols and triols in the resin formulation.

Functionality of the resins can easily be adjusted by incorporation of a certain amount of a branching component, which can be triol, tetraol or triacid. The techniques for calculating formulations for polyester resins are beyond the scope of this book. The interested reader can consult a considerable number of published articles and monographs [30–46].

A big advantage of the polyester resins is the possibility of adjusting the place of functional groups in the polymer chain to a desired position which has an enormous influence upon the mechanical properties of the cured films. This is one of the reasons why polyesters have superior mechanical properties compared to the acrylics, for example, where the functional groups are distributed at random.

Thermosetting polyester resins for powder coatings are produced on the same equipment that is used for production of polyesters for solvent based coatings or high molecular weight polyesters. Although the azeotropic method of processing can be used, it is rather an exception in practice, since the subsequent stripping off of the solvent is an unnecessary complication. The fusion method of cooking followed by removal of the esterification water at the end of the process by means of a vacuum is the most suitable procedure. As in the case of high molecular weight polyesters, the resin is poured onto a cooling belt provided with a rotating crusher which gives the final flake size of the crushed resin.

A standard method of preparation of hydroxyl functional polyester resin for polyurethane powder coating with the reaction details is given in the book of Oldring and Hayward [7, p. 137]. Ingredients 1 to 4 in Table 3.10 are charged together in the reactor and esterified in a temperature interval from 180 to 245°C with removal of the reaction water until an acid value below 2 mg KOH/g is reached. Then components 5 to 7 are charged and esterification continues. In order to facilitate the removal of reaction water, xylene is added to the reaction vessel, and esterification is performed according to the so-called azeotropic method until an acid value below 3 mg KOH/g is obtained. Xylene is removed from the system by vacuum distillation and the resin is discharged onto the cooling belt.

Different patents describe methods for making hydroxyl functional polyester resins for polyurethane powder coatings with minor variations with respect to the acid value, hydroxyl value, functionality of the resin and the choice of raw materials [47–50].

A somewhat different cooking procedure is used with dimethyl terephthalate as a starting raw material in place of terephthalic acid. In such cases the reaction goes in two different steps. In the first step dimethyl terephthalate is transesterified in an excess of glycol with complete removal of methanol as a reaction by-product. In the second step, the polymerization reaction is carried on by means of high vaccum, removing the excess of the glycol from the system, thus shifting the reaction equilibrium in favour of a high molecular compound [51, 52].

An interesting approach using crystalline polyesters to obtain good storage stability of the powder coating and to have at the same time low melt viscosity providing good flow properties to the coating has been developed in the laboratories of Eastman Kodak Co. [53, 54]. In the case of branched polyesters, a

Table 3.10 Hydroxyl functional polyester resin for polyurethane coatings [7, p. 137] (Reproduced by permission of the publishers, Selective Industrial Training Associates Ltd)

Ingredient	Weight
1. Water	15
2. Neopentyl glycol	436
3. Terephthalic acid	552
4. Fascat 4201	1
5. Isophthalic acid	117
6. Sebacic acid	25
7. Trimethylol propane	28
8. Xylene	20

Final characteristics:
  Acid value = 2.2 mg KOH/g
  Hydroxyl value = 35 ± 5 mg KOH/g
  Melting range = 110–115°C

conventional method of preparation is used, starting from dimethyl terephthalate as a source of dicarboxylic acid moieties, which is transesterified and later on polymerized under vacuum with linear glycols (1,6-hexanediol, 1,4-cyclohexanedimethanol or ethylene glycol) and trimethylolpropane or trimellitic anhydride as branching components. The polyesters have a molecular mass between 700 and 3000, OH number from 30 to 160 and melt viscosity at 160°C of 50–3000 cP, preferably between 50 and 100 cP. The curing agent is caprolactam blocked isophorone diisocyanate oligomer. The same method is used in the case of linear polyesters where terephthalic acid and 1,4-cyclohexane dicarboxylic acids in the form of dimethyl esters are transesterified with 1,4-butanediol as a single polyol component. The amount of 1,4-cyclohexane dicarboxylic acid should be at least 50 mol% in order to obtain a melting temperature below 160°C. The main advantage that is claimed is the excellent flow as a result of a very low melt viscosity.

Carboxyl terminated polyesters first developed by DSM Resins BV at the beginning of the 1970s are especially popular binders in Europe [55]. Two main systems are widely accepted by the market: TGIC/polyester for outdoor applications, where good gloss retention and yellowing resistance is a strong request, and so-called hybrids composed of carboxyl terminated polyester and epoxy resin of bisphenol A type as a crosslinker. The hybrids have poor outdoor durability and yellowing resistance, but excellent boiling water resistance and alkali resistance, which makes them very suitable for household equipment applications.

Although the same resins can be used in principle for both systems, there are certain small differences between the polyesters for curing with TGIC and curing with bisphenol A type epoxy resins.

The functionality of the epoxy resins of bisphenol A type is two compared to three in the case of TGIC. Therefore the functionality of the polyester resins for hybrid systems in principle should be higher than the functionality of those intended to be crosslinked with TGIC.

The glass transition temperature of polyester resins for TGIC containing powder coatings should be at least 60°C in order to achieve good storage stability. The storage stability of polyester/epoxy hybrids is not very critical because of a high enough T_g of the bisphenol A epoxy crosslinkers. Therefore the glass transition temperature of polyester resins for hybrids can be approximately 10°C lower compared to TGIC systems. This makes the choice of raw materials for producing polyester resins for hybrids somewhat easier.

With a good approximation, it can be considered that the T_g of the powder coating decreases by 2°C for each weight percentage of TGIC. Since T_g of 40°C is considered to be the lowest value that one can allow for having good storage stability of the coating, the amount of TGIC in the coating formulation in which the polyester resin has a T_g of 60°C usually cannot exceed 10% by weight on the basis of the binder. Consequently, the acid value of the polyesters for TGIC curing is almost never higher than 50 mg KOH/g (which corresponds roughly to 10% TGIC). In the case of polyester resins for hybrids the acid value can go up to 80 mg KOH/g.

Table 3.11 presents typical formulations of polyester resins for TGIC and hybrid systems for which the details of the manufacturing procedure are given by Oldring and Hayward [7, p. 137].

The production of carboxyl functional polyesters can be realized in a one-step cooking procedure having a proper ratio between the reactants, which ensures a certain molecular weight combined with the desired acid value. In many cases, however, the process of production of polyester resins with carboxyl functionality is performed in two steps. The reason is much easier process control, especially in the case where the resin is of a branched character. The first step ends up with a hydroxyl functional polyester resin with a certain molecular mass and hydroxyl value. In the second step a calculated amount of dicarboxylic acid is added, and the esterification is carried on until the desired acid value has been reached [56].

A similar process uses dimethylterephthalate instead of terephthalic acid in the first step, continuing with esterification with dicarboxylic acid in the second step to obtain a carboxyl functional polyester resin suitable for curing with epoxy containing compounds [57].

Addition of acid anhydride to an hydroxyl functional polyester at the last step of the process is also a common method for preparation of carboxyl functional polyesters. Trimellitic anhydride is often used for this purpose [58].

Table 3.11 Polyester resins for TGIC and hybrid systems [7, p. 137] (Reproduced by permission of the publisher, Selective Industrial Training Associates Ltd)

Ingredient	Weight	
	Hybrid	TGIC
Water	28	30
Neopentyl glycol	364	530
Terephthalic acid	423	711
Isophthalic acid	—	88
Adipic acid	41	—
Pelargonic acid	—	58
Trimellitic anhydride	141	43
Fascat 4201	3	5
Final characteristics:		
Acid value (mg KOH/g)	80 ± 5	35 ± 5
Melting range (°C)	110–120	100–110

A somewhat different process is described in the patent of Goodyear Tyre and Rubber Company [59]. In one of the examples of the patent a polyester resin is prepared by transesterification of dimethyl terephthalate with neopentyl glycol in the presence of a small amount of trimethylol propane. Under vacuum in the next step, the excess of neopentyl glycol is removed and hydroxyl terminated polyester is obtained. The latter is endcapped with carboxyl groups by esterification with isophthalic acid in excess of carboxyl groups in respect of the total resin formulation.

Similar techniques starting with terephthalic acid or a blend of terephthalic and isophthalic acid instead of their dimethyl esters are used in many patents [60–65].

A polyester resin suitable for production of powder coatings with a very low amount of TGIC in the coating formulation is described in the patent of DSM Resins BV [66]. The resin is obtained by fusion esterification of neopentyl glycol, 1,4-cyclohexanedimethanol, 1,6-hexane diol, trimethylol propane, terephthalic acid and adipic acid. Products with average molecular masses between 4500 and 12 500 with acid values of 10–26 mg KOH/g and T_g of 40–85°C are obtained which are suitable for making powder coatings containing 1.4–5.3% by weight of TGIC.

A similar patent [67] describes a method for obtaining polyester resins with somewhat higher T_g values (45–85°C), which permits the use of diglycidyl terephthalate as crosslinker in amounts of 2–9% by weight, still giving powders with good storage stability.

A method that differs from the previously described processes involves a depolymerization of high molecular mass polyester resin with polycarboxylic acids [68] resulting in carboxyl functional polyester with a lower molecular mass than the starting material. The polycarboxylic compounds are selected among di-, tri- and tetravalent carboxylic acids. A modification of this method uses diols, triols and tetraols as depolymerizing agents [69] having as the end product an hydroxyl functional polyester resin of lower molecular mass. The molecular weight of the final product and its functionality depend on the amount of depolymerizing agent and the functionality of the same.

An interesting modification of polyester resins leading to polymers which contain epoxy groups, thus being suitable to be crosslinked with carboxyl functional polymers or low molecular polybasic acids, is protected by two patents of Hitachi Chemicals [70, 71]. The process involves a reaction of epoxy resin of bisphenol A, hydrogenated bisphenol A type or TGIC with monoalkyl ester of terephthalic acid (typically monomethyl terephthalate) in the first step. In the second step, this product is transesterified with an hydroxyl functional polyester resin having an hydroxyl value between 20 and 200 mg KOH/g and an acid value lower than 20 mg KOH/g, stripping of the methanol. The resulting polyester contains epoxy groups and can be used as a binder or crosslinker in a powder coating formulation together with carboxyl functional materials.

Polyester resins terminated with epoxy groups can be produced by reaction of carboxyl terminated polyester resins in an excess of low molecular weight diepoxide. The patent literature [72] describes such a process using vinyl cyclohexane dioxide of formula given below and linear carboxyl terminated polyester composed of neopentyl glycol, phthalic anhydride and hexahydrophthalic anhydride, with a molecular weight of 1800 and an acid number of 62.4 mg KOH/g. Of this polyester 900 g were reacted with 140 g of vinyl cyclohexane dioxide to obtain a polyester resin with a melting range between 90 and 90°C, an acid value of 3.8 and an epoxy equivalent weight of 1035.

$$\underset{O}{CH_2 - CH} - \underset{O}{\bigcirc}$$

A general weak point of polyesters is their low resistance to saponification. By the proper choice of raw materials used in building up the backbone of the polyester resin, this property can be improved to a considerable extent.

Under normal conditions of hydrolysis, resistance to ester cleavage increases with increasing alkyl substituent at the alpha and beta positions on both alcohol and acid. Glycols like neopentyl glycol, trimethyl pentanediol, 2,2-diethyl-1,3-propanediol, 2-methyl-2-butyl-1,3-propanediol, Ester diol 204 (an ester of hydroxypivalic acid and neopentyl glycol) and cyclohexyl dimethylol are the raw materials that can be used for this purpose.

It is interesting that a patent describing a process of making polyester powder coatings in which bisphenol A epoxy resin is used as crosslinker had already been filed in 1961 [73]. The coating in one of the examples of the patent intended to be used for electrical insulation purposes is a physical blend of powdered carboxyl functional polyester resin and powdered epoxy resin. It is applied in a fluidized bed on preheated aluminium wire and cured for 2 hours at 205°C, which is certainly an unacceptable curing schedule for practical reasons.

A period of 10 years was necessary to realize the advantages that are offered by this system over the pure epoxy powder coatings. In 1970, Huneke [74] reported that the gloss retention of powder coatings based on blends of epoxy and polyester resins is considerably improved in comparison with pure epoxy powders. However, nothing was said about what kind of polyester resins were used and what curing reaction takes place.

A patent of Unilever NV from 1971 [75] refers to the powder coating composition manufactured by melt extrusion of carboxyl functional polyester resin and epoxy resin of bisphenol A type. The coating is applied by the electrostatic spraying method on a metal surface and is cured for 10 minutes at 200°C. Later it was realized that the gloss retention and yellowing resistance are not the only advantages that the polyester powder coatings can offer. Excellent mechanical properties, attractive visual appearance, easy tailoring of processing, application and end properties and an acceptable price/performance ratio are some of the attributes that contributed to these systems becoming the most popular among a wide range of powder coatings offered on the market.

The sensitivity of the bisphenol A epoxy resins to u.v. light restricted their use only in polyester powder coatings for indoor applications. The first attempts to make polyester powder coatings suitable for exterior use were naturally concentrated on the well-known curing chemistry from the solvent borne coatings which employ melamine resins as curing agents. More than 20 years of practical experience which had already been built up with the melamine solvent borne coatings was a guarantee for good outdoor performance of these materials. The use of hexamethoxymethyl melamine (HMMM) as crosslinker for hydroxyl functional polyesters in powder coating formulations is described in a few patents from the early 1970s [76, 77]. Several problems caused this system to disappear very soon from the market. The liquid nature of the melamine resins restricts the production of the polyester melamine powder coatings to the Z-blade kneaders. When this system was developed, extruders were already dominating production equipment for powder coatings with all of the advantages over the Z-blade kneaders which will be discussed in Chapter 5. The second problem facing powder manufacturers and users was surface defects in the form of blisters, due to liberation of methanol during curing. In order to avoid it, the system should not be catalysed, which resulted in a slow curing rate requiring a curing schedule of 30 minutes at 200°C.

Two different curing agents of a similar type to the melamine based crosslinkers available in a solid form were also developed in the 1970s: tetramethoxymethyl benzoguanimine and tetramethoxymethyl glycoluryl [78]. Although the processing problems were overcome in this way and normal extrusion equipment could be used for compounding, the surface defects and high prices prevented successful commercialization of these powder coatings.

In the meantime, the development in the direction of polyester resins curable with epoxy compounds and blocked isocyanates was so fast and successful that it swept polyester/melamine systems out of the market when they had only just appeared. It was probably the main reason why further research on the amino/hydroxyl systems has ceased, although the biggest part of solvent borne coatings for industrial application employs this curing chemistry.

In the last 15 years, three different types of polyester powder coatings established quite strong positions in the market. Polyester/epoxy powder coatings, well known by the name hybrids, came as a result of attempts to improve the yellowing resistance of the pure epoxy powder coatings. Polyester/TGIC systems dominating the market for exterior durable coatings were developed almost at the same time. Polyester/isocyanate or polyurethane powder coatings were the natural consequence of the successful introduction of the solvent borne polyurethane coatings. These three types of polyester powder coatings are discussed in further detail in the following sections.

3.4.2.1 Interior Polyester Powder Coatings

Hybrids is the usual name for powder coatings with polyester resin as a binder and bisphenol A epoxy resin as crosslinker. They are the main representatives of powder coatings used for interior applications. It has always been a problem as to how to classify the hybrid systems. The word crosslinker usually refers to the component of the coating which is used in minor amounts to produce a network, reacting with the functional groups of the main binder. In the case of hybrids, the polyester and epoxy resin are used in comparable amounts. In the so-called 50/50 systems their quantities in the powder coating formulation are equal. Therefore it is difficult to distinguish between the resin and hardener in this case. Because of better yellowing resistance and for price reasons, the contemporary tendency is towards coatings that employ less epoxy resin than polyester. Let us use this argument as a justification for classifying the hybrids in the group of polyester powder coatings.

Since epoxy resins as curing agents are in most cases two functional, the main binder must have a higher functionality than two. Deviations from the stoichiometric ratio can lead to a marked deterioration of the film properties, such as solvent resistance, boiling water resistance and impact resistance. Polyester resins for powder coatings are usually characterized by a certain acid number which can be used to calculate the necessary amount of epoxy partner in the coating formulation, knowing its epoxy equivalent value. The following formula can be used for this purpose:

$$EC = \frac{R \times AV \times EEW}{56\,100} \quad (3.130)$$

where

EC = amount of epoxy resin (g)
R = amount of polyester resin (g)
AV = acid value of the polyester resin (mg KOH/g)
EEW = epoxy equivalent weight of the epoxy resin (g)

The reaction speed between the epoxy groups coming from bisphenol A epoxy resin and the carboxyl groups of the polyester is rather slow. Normally, the hybrid powder coating should be catalysed in order to achieve an acceptable curing time at baking temperatures between 160 and 200°C which are mostly used in practice. It is a rule that the polyester resin manufacturers incorporate the catalyst in the resin during its production, providing in this way the most uniform distribution of the catalyst in the powder coating. This is very important, because it is rather difficult to obtain a uniform distribution of very small amounts of catalyst during the extension of the powder coating. When additional amounts of catalyst have to be added, a master batch technique should normally be used. Catalysts normally used to speed up the reaction rate are tertiary amines or quaternary ammonium salts.

In order to improve the levelling properties of the coating and to reduce the surface defects, suitable additives, whose roles will be discussed later, are usually incorporated. The levelling additives are almost never used as such. Normally, a master batch containing 10% of the additive in the main binder is used in order to obtain uniform distribution of the additive. However, certain additives like benzoin or polyvinyl butyral resins do not require master batching. The same holds for different types of colloidal silicas which are used as thixotropic agents.

The ratio between polyester and epoxy resins depends on the acid value of the polyester resin and epoxy equivalent weight of the crosslinker. The ratio between the resins is often used to name these systems, such as 50/50, 70/30, etc., the first number indicating the amount of polyester resin and the second the amount of epoxy resin in the powder formulation.

Table 3.12 [79] contains different formulations of hybrid polyester powder coating as an illustration. The films after curing are characterized with König pendulum hardness above 200 seconds, slow Erichsen penetration over 8 mm, reverse impact of 160 inch-lb and crosshatch adhesion of G_{t0}. Formulation for

Table 3.12 Polyester hybrid powder coatings [79] (Reproduced by permission of DSM Resins BV)

System	50/50	60/40	70/30	80/20
Uralac 2127[a]	330	—	—	—
Uralac 3649[b]	—	360	—	—
Uralac 3560[c]	—	—	700	—
Uralac 3850[d]	—	—	—	800
Araldit GT 7004[e]	330	240	300	200
Kronos CL310[f]	330	300	500	500
Benzoin	7	—	7	7
Modaflow[g]	7	6	10	10
Choline chloride	—	3	—	—
Stoving cycles (min/°C)	15/180 10/200	15/180 10/200	15/180 10/200	15/180 10/200

[a] Saturated polyester—acid value of 70–85 (DSM Resins BV).
[b] Saturated polyester—acid value of 45–55 (DSM Resins BV).
[c] Saturated polyester—acid value of 30–40 (DSM Resins BV).
[d] Saturated polyester—acid value of 18–22 (DSM Resins BV).
[e] Epoxy resin—epoxy equivalent weight of 715–835 (Ciba-Geigy).
[f] Titanium dioxide—rutile type (Kronos Titan GmbH).
[g] Flow control agent (Monsanto).

60/40 systems contain choline chloride as catalyst, which is FDA approved so can be used to cover surfaces that come in direct contact with food.

Higher amounts of epoxy resin in general improve chemical and boiling water resistance of powder coatings, but on the other hand have a negative effect on the yellowing resistance and weatherability. The price should also be considered, since epoxy resins are in general more expensive than the polyesters. Therefore, for price reasons, systems with a low epoxy content have been developed. A patent describing a hybrid that uses only 15% bisphenol. A epoxy resin in hybrid formulation [80] claims excellent yellowing resistance of the cured film combined with excellent flow and impact resistance.

At present, the most important end uses of polyester/epoxy hybrids cover the following: household equipment such as refrigerators, deep-freezers, washing machines, cookers and other kitchen equipment, metal furniture, ceiling panels for the building industry, shower cabinets, automotive components, agricultural equipment and machinery, engineering and electrical parts.

3.4.2.2 Exterior Polyester Powder Coatings

Powder coatings for exterior application have to fulfil the typical request addressed to all coatings being exposed to water in the form of rain and snow, frost, ultraviolet radiation from sunlight, aggressive industrial atmosphere containing sulphur dioxide, etc.

Water and humidity resistance of polyester powder coatings is determined mainly by the nature of the main binder containing ester groups prone to hydrolysis. The new generation of polyester resins built up of monomers, which provide improved hydrolytic resistance, in most cases satisfies the normal requirements with respect to water and humidity resistance. On the other hand, they are characterized by considerable resistance to u.v. radiation.

The right choice of crosslinker plays an important role in formulating polyester powder coatings with good outdoor performances. The crosslinker should also be resistant to u.v. light, which excludes the bisphenol A epoxy resins containing ether oxygen. Unfortunately, aliphatic and cycloaliphatic epoxy compounds with good u.v. light resistance are liquids at room temperature. This makes them unsuitable to be easily compounded into powder coatings using the present melt extrusion technique. Only a few of them are solid at room temperature, but only two have a commercial use: triglycidyl isocyanurate (TGIC) and diglycidyl terephthalate (DGT), the former being by far the more important.

The first patent protecting the use of TGIC as crosslinker for powder coatings was filed in 1969 [81]. According to the invention, 77–95% hydroxyl functional powdered polyester resin with a particle size below 300 μm is dry blended with 5–10% powdered TGIC and 6.5–13% powdered acid anhydride. The powder blend is applied by electrostatic spraying or by the fluidized bed technique on a metal object which is then cured at 270°C for 4 minutes or only 4 seconds at 350°C. The curing reaction goes in two steps. In the first step, the anhydride reacts with the hydroxyl groups of the polyester resin giving as a result carboxyl functional polyester. In the second step, TGIC reacts with the formed carboxyl groups, crosslinking the systems and building up a network. Although it is claimed that the obtained films have a good appearance, an obvious disadvantage of this system is the inhomogeneity of the coating on a molecular scale, since the constituents are only blended in a dry form.

One year later, a process of melt extrusion of carboxyl functional polyester resins with TGIC as crosslinker was disclosed in a patent application from 1970 [82]. Very soon in 1972, a polyester/TGIC powder coating for protection of aluminium extrusions and claddings in outdoor architecture was used in Switzerland for the first time [55]. More than fifteen years of experience in manufacture and application of TGIC powder coatings for outdoor application have confirmed the reliability of this system for providing long term protection. At this moment, polyester/TGIC systems undoubtedly dominate the exterior market for powder coatings.

TGIC as a low molecular weight compound dramatically decreases the glass transition temperature of the coating. It is a rule that for every weight percentage of TGIC, the glass transition temperature decreases by 2°C. The amount of TGIC normally used in a standard powder coating formulation is 7%, although for economical reasons there is a tendency to reduce the TGIC content to lower values. A patent of DSM Resins [83] describes a system containing only 4% TGIC. The resins that are used have low acid values between 10 and 26 mg KOH/g and molecular weights between 4500 and 12 000. The high molecular weight of the polyester, and consequently the high shear modulus, decrease the grinding efficiency of the powder coating which could be considered as a small disadvantage.

Hoppe [84] refers to a system containing 5% TGIC which still has a sufficient crosslinking density, excellent mechanical properties and good flow.

The amount of TGIC in the powder coating depends on the acid value of the polyester resin. The best results are obtained when there is a stoichiometric amount of epoxy groups coming from TGIC with respect to the carboxyl groups of the polyester. The following formula can be used to calculate the stoichiometric amount of TGIC knowing the acid value of the polyester resin:

$$\text{TGIC} = \frac{R \times \text{AV} \times 107}{56100} \qquad (3.131)$$

where

TGIC = amount of TGIC (g)
R = amount of polyester resin (g)
AV = acid value of the polyester resin (mg KOH/g)

Table 3.13 illustrates three formulations of powder coatings with different TGIC contents [79].

Table 3.13 Typical polyester/TGIC powder coating formulations [79] (Reproduced by permission of DSM Resins BV)

System	90/10	93/7	96/4
Uralac P 2200[a]	900	—	—
Uralac P 3500[b]	—	558	—
Uralac P 3800[c]	—	—	576
TGIC[d]	100	42	24
Kronos 2160[e]	500	300	300
Flow agent[f]	15	9	9
Benzoin	—	4	—
Catalyst[g]	—	18	—
Curing schedule (min/°C)	15/180 10/200	15/160	15/180 10/200
Pendulum hardness (König), s	200–220	190–210	—
Slow penetration (Erichsen)	> 8 mm	> 8 mm	> 8 mm
Reverse impact (inch lb)	160	160	160
Cross-hatch adhesion	G_{t0}	G_{t0}	G_{t0}

[a] Polyester resin DSM Resins BV, acid value 48–55 mg KOH/g.
[b] Polyester resin DSM Resins BV, acid value 33–38 mg KOH/g.
[c] Polyester resin DSM Resins BV, acid value 19–23 mg KOH/g.
[d] Araldit PT 810, Ciba-Geigy.
[e] Titanium dioxide—rutile, Kronos Titan GmbH.
[f] Modaflow powder III, Monsanto.
[g] XB 3126, Ciba-Geigy.

The normal curing conditions for polyester/TGIC systems existing on the market at present are 10 minutes at 200°C object temperature. Usually it is not necessary to catalyse the system, since the reactivity of TGIC itself is high enough to obtain the desired curing cycle. If faster curing schedules are required or lower curing temperatures are requested, an incorporation of suitable catalyst, as in the formulation 93/7 (Table 3.13), is necessary. This is especially the case when aluminium extrusions with plastic inserts have to be coated. Then the temperature of the object should not exceed 180°C.

The yellowing resistance of TGIC powder coatings at overbake is extremely good; 30 minutes exposure at 220°C usually does not give a colour change (delta E) of more than 0.2–0.3 NBS.

The first polyester/TGIC powder coatings offered a dramatic improvement in outdoor durability compared to that of the hybrids. While the hybrids show a 50% reduction of the initial gloss after 50–200 hours in the QUV tests dependent on the amount of the epoxy resin in the coating formulation, TGIC systems exhibit 50% retention of the initial gloss over a period of 500–600 hours, and the results with new experimental resins show an extension of this time to 800 hours [55] (see Figure 3.23).

Properly formulated and cured polyester/TGIC powder coating gives film

THERMOSETTING POWDER COATINGS

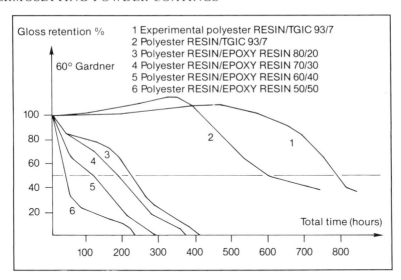

Figure 3.23 The QUV test results for several polyester powder coatings [55] (Reproduced by permission of Products Finishing Magazine, Cincinnati, Ohio)

with excellent hardness, flexibility, impact resistance, abrasion resistance and scratch resistance. The adhesion properties of TGIC systems are in general good, and it is not necessary to use an adhesion promoting primer.

Weak points of polyester/TGIC systems are the flow and levelling of the coatings. The orange peel effect is more emphasized than with other powder coating systems. Considerable work has been done to improve the flow properties of the TGIC powder coatings resulting in new generation resins with lower viscosity and slower viscosity build-up during curing, leading to improved flow and a less emphasized orange peel effect [84].

Polyester/TGIC powder coatings are mainly used for protection of outdoor architectural elements. The surface is in most cases aluminium, but it can also be galvanized steel. In general, the adhesion of the coating on a galvanized steel surface is somewhat inferior in comparison with aluminium. The usual film thickness deposited in one layer is 60–80 μm.

Another market segment where TGIC powder coatings are used is for metal furniture for outdoor use (garden furniture) and automotive trim parts such as wheel rims, fenders, bumpers, window frames, windscreen wipers, bicycle frames, motor-cycles, etc.

Polyester powder coatings using crosslinkers other than TGIC are also described in the patent literature. A patent of DSM Resins [85] gives a powder coating composition using carboxyl functional polyester resin with an acid value between 10 and 30 mg KOH/g as a main binder and 2–9% diglycidyl terephthalate (DGT) as a crosslinker. The use of diglycidyl hexahydrophthalate is also described in the same patent. The outdoor durability of DGT/polyester powder coating is comparable with that of the TGIC systems. However, a disadvantage of DGT is its higher epoxy equivalent weight than TGIC. Therefore, higher amounts of DGT by weight have to be used in order to maintain a stoichiometric ratio between the epoxy and carboxyl groups. Consequently, the decrease of T_g of the powder coating is stronger and therefore polyester resins with higher T_g's have to be used in combination with DGT as crosslinker. This narrows the choice of raw materials that can be used for making the polyester resin.

A number of other solid aliphatic or cycloaliphatic polyepoxy compounds which have not been commercialized are described in the patent literature.

The use of hydantoin epoxides containing two to four epoxy groups as crosslinkers for carboxyl functional polymers in powder coatings is described in a patent assigned to Du Pont [86]. It is claimed that the outdoor durability, especially in the case where carboxyl acrylics are used as main binders, should meet the requirements of automotive clear top coats.

In a patent of Bayer AG [87] the use of triglycidyltriazolidine-3,5-dione as a crosslinker for powder coatings for outdoor use is disclosed. Comparable or slightly better gloss retention on outdoor exposure is claimed compared to the powder coating compositions where TGIC is used as a crosslinker.

Polyester powder coatings cured with acid anhydrides exhibit good outdoor durability. Walz and Kraft [88] compared the gloss retention of powder coating based on saturated polyester cured with modified ethylene glycol bistrimellitate (MEGBM) with that of epoxy/DICY and polyester/epoxy hydrid powder coatings. Results presented in Figure 3.24 clearly indicate the potentials of the polyester/anhydride system for exterior use.

An advantage of polyester/anhydride powder coatings is the fast curing rate which allows stoving schedules of 10 minutes at 180°C to 30 minutes at 140°C. Powder coatings in which the polyester resin is based on isophthalic acid need higher stoving temperatures or longer stoving times to bring about complete curing. Despite the short curing cycles and high reactivity, storage and processing stability of polyester/anhydride powder coatings are good enough from a practical standpoint. In Table 3.14 a typical formulation of a polyester/anhydride powder coating is presented [88].

Polyester/anhydride systems have not gained market shares compared to their performances. The main reason is the toxicity of the hardeners which are heavy respiratory and mucous membrane irritants. In some countries their use is even prohibited. Although the so-called resinous anhydrides or bistrimellitates are less problematic from a toxicological point of view, they did not contribute very much to a greater acceptance of these powder coating systems by the market.

The full success of polyurethane solvent borne coatings was naturally followed by trials to use the same curing chemistry in the powder field. To prevent the

THERMOSETTING POWDER COATINGS

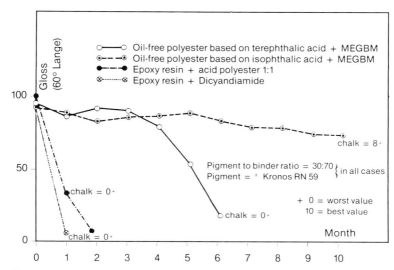

Figure 3.24 Results of exposure tests in South Florida for different powder coating systems [88] (Reproduced by permission of International Conference in Organic Coatings, Athens)

Table 3.14 Powder coating composition based on saturated polyester resin and ethylene glycol bistrimellitate hardener (EGBM) [88] (Reproduced by permission of International Conference in Organic Coatings, Athens)

Component	Parts by weight
Oil-free polyester	51.7
Modified EGBM	17.2
Titanium dioxide	29.6
Flow agent	1.5
Stoving schedule (min/°C)	10/200
Gloss (60°)	94
Impact resistance (inch-pound)	160
Processing stability expressed as viscosity increase during 10 min processing at 120°C (μ_{10}) compared to initial (μ_0) (μ_{10}/μ_0)	4

premature reaction of the isocyanate groups with the functional groups of the polyester resin it is necessary to use blocked isocyanates as crosslinkers. Blocked IPDI and TDI derivatives with caprolactam as the blocking agent are the most commonly used crosslinkers for polyurethane powder coatings. The toxicity and very poor weatherability and yellowing resistance of TDI restricts its use to indoor cheap polyurethane powders. IPDI derivatives which are described in detail in Section 3.3.2 are for the time being the most important commercial crosslinkers. Caprolactam blocked IPDI oligomers release the blocking agent during curing. Therefore, for thick layer application these crosslinkers are not the best choice because of the possible surface defects in the form of blisters. On the other hand, the blocking agent contributes very much to the flow improvement of the polyurethane powder coatings; namely the curing reaction does not start until deblocking takes place. This results in a permanent viscosity decrease during the increase of the temperature at the stoving period of the coating. This is not compensated by a viscosity increase as a result of a network formation and an increase of the molecular weight as is the case with epoxy or polyester/TGIC systems. Moreover, once the blocking agent is released, it only remains for a short time in the film before it evaporates, acting as a solvent, contributing to a further decrease of the melt viscosity of the coating. Therefore polyurethane powder coatings have the best flow properties among the thermosetting powders. In many cases, the flow and levelling properties of the polyurethane powder coatings can even be compared with those of the solvent borne coatings. The presence and release of caprolactam, on the other hand, is a nuisance from an ecological point of view, slightly spoiling the image of the powder coatings as environmentally friendly materials. However, the released amount of caprolactam per square metre of coated surface is much less than the amount of solvents emitted by the liquid coatings.

Polyurethane powder coatings using uretidione oligomers of IPDI can be considered as coatings containing internally blocked isocyanate. As a result there is no release of blocking agent during curing. This is an advantage from an ecological point of view, but it must be admitted that the flow properties and gloss of these coatings are somewhat inferior compared to caprolactam blocked systems.

IPDI derivatives like all aliphatic or cycloaliphatic isocyanates have a slow reaction rate. A proper catalyst is a prerequisite for adjusting the curing schedule of the system to the most common requests in practice. Tertiary amines can be used for this purpose. However, metal catalysts are by far the most effective for this system. It has been found that compounds containing tin, bismuth, lead, zinc, titanium, iron, antimony, zirconium, cobalt, nickel, aluminium and many other metals can in principle be used as catalysts for NCO/OH curing. However, catalysts of real commercial importance are the organotin compounds such as dibutyltin dilaurate (DBTDL), dibutyltin oxide (DBTO) and stannous octoate (SnOct). It is common practice for the resin producer to incorporate the catalysts in the resin during the production process in order to ensure the most homogeneous distribution of the catalyst in the system.

Polyesters used for polyurethane powder coatings contain hydroxyl and carboxyl functional groups. The ratio between them depends on the stoichiometry between the reactive monomers in the formulation of the polyester resin, and on the degree of conversion of the carboxyl groups during esterification.

THERMOSETTING POWDER COATINGS

Both hydroxyl and carboxyl groups can react with the polyisocyanate crosslinker. In an excellent article McLafferty, Figlioti and Camilleri discuss the influence of the hydroxyl/carboxyl ratio in the polyester resin upon the curing characteristics of IPDI and TDI based polyurethane coatings [89].

Typical formulations of polyurethane powder coatings using caprolactam blocked IPDI oligomer and internally blocked uretidione derivatives of IPDI are presented in Table 3.15 [79].

A double curing mechanism as a means of producing matt polyurethane coatings employs carboxyl/epoxy curing together with an hydroxyl/isocyanate curing reaction [90]. A special hardener containing carboxyl and blocked isocyanates as functional groups is made, reacting 2 moles of isophorone diisocyanate with 2 moles of caprolactam and 2 moles of 2,2-bis-(hydroxymethyl)-propionic acid. In combination with TGIC as co-crosslinker, hydroxyl functional polyester, pigments and suitable flow additive, a powder coating can be obtained having gloss below 30% after stoving. It is essential to keep the ratio between the NCO content and the acid number in the hardener between 0.1 and 0.3 in order to obtain matt films. The control powder coating formulation in which the ratio mentioned above is 0.8 has a gloss of 77%.

Table 3.16 gives a comparison between three different powder coating systems for outdoor application [55]. It is interesting that the acrylic coatings which will be discussed in the following section do not range higher than the polyester/TGIC systems with respect to weathering resistance, although in

Table 3.15 Polyurethane powder coatings for outdoor application [79] (Reproduced by permission of DSM Resins BV)

Material	Parts by weight		
	A	B	C
Uralac P 2115[a]	800	—	—
Uralac P 2504[a]	—	800	800
IPDI-B 1530[b]	200	200	—
IPDI-BF 1540[c]	—	—	200
Kronos CL 310[d]	500	500	500
Tin octoate	—	10	10
Modaflow powder III[e]	10	10	10
Stoving schedule (min/°C)	20/200 15/190	8/200 13/190	8/200 13/190
Stoving loss (%)	2.5	2.5	0

[a]Hydroxyl functional polyester resins (DSM Resins BV).
[b]Caprolactam blocked IPDI derivative (Chemische Werke Hüls AG).
[c]Uretidione derivative of IPDI (Chemische Werke Hüls AG).
[d]TiO$_2$ rutile (Kronos Titan GmbH).
[e]Flow promoting agent (Monsanto).

Table 3.16 Comparison of outdoor resistant powder coatings [55] (Reproduced by permission of Products Finishing Magazine, Cincinnati, Ohio)

Property	Acrylic	Polyurethane	Polyester
Weathering resistance	+	+	+ +
Mechanical properties	−	+	+ +
Corrosion resistance	O	+	+
Storage stability	O	+ +	+ +
Decomposition products	+	−	+ +
Levelling	+ +	+ +	+(1) + +(2)
Edge covering	O	O −	+ +
Resistance to overcuring	+	O +	+ +

+ + — very good O — average. (1) — high gloss.
+ — good. − − bad. (2) — semi-gloss or matt.

the liquid systems acrylics are by far the best ones for outdoor use. This corresponds with the results reported by Gulpen and van de Werff who performed weathering tests on different powder coatings and compared them with the conventional wet systems [17]. Although the comparison in Table 3.16 is qualitative only, and some objections can always be addressed, it gives a good indication of expectations from various outdoor powder coating systems.

3.4.3 ACRYLIC POWDER COATINGS

It is interesting that, despite the fact that acrylic solvent borne coatings are the best thermosetting materials for outdoor application, in the powder field they do not have the importance which might normally be expected. Recent market analysis shows that even in Japan, which has always been the biggest market for acrylic powder coatings, their market share is declining.

Several reasons can be considered as relevant for this position of the acrylic powders. The mechanical properties of acrylics, which are generally inferior compared to the other coating systems, are even worse in the case of acrylic powder coatings. Surprisingly, the weathering resistance, which is the strongest point of acrylics, was not as good as with the solvent borne acrylic coatings. The latest developments indicate that the weathering properties of pure acrylic powder coatings have improved very much [91] and it might be expected that in the near future much more attention will be paid to this system. This could especially be the case if improved flow properties, combined with environmental compliances, contribute to more extensive use of powder coatings in the automotive field.

Polymethyl acrylate and polymethyl methacrylate produced by free radical polymerization were the first acrylic polymers manufactured on an industrial scale. The products under the names of Plexigum and Perspex were introduced in 1927 by Rohm and Haas AG in Germany and Imperial Chemical Industries in England [92]. Although acrylic ester polymers can be obtained by anionic polymerization, thermal polymerization in the absence of catalyst, photopolymerization and radiation-induced polymerization, free radical polymerization is the predominant synthetic route used in industrial manufacture.

Since powder coating resins are 100% solid materials, bulk polymerization of acrylic esters should be the most attractive production route. However, technical problems with respect to the removal of the exothermic heat during polymerization restricts the use of this method almost exclusively to manufacturing articles by casting. Although the bulk polymerization process using vinyl versatate as a comonomer is a known process, there is no evidence in the literature for practical use of this production method in manufacturing powder coating resins.

Polymerization in solution is carried out by adding the monomers to an organic solvent in which both the monomer and the polymer are soluble. Typical catalysts are organic peroxides, hydroperoxides and azo catalysts. The molecular weight of the polymer is controlled by selection and concentration of the initiator, the type of solvent, monomer concentration and temperature. When the polymerization is accomplished, the solvent is removed either by vacuum distillation or by a spray drying technique.

Emulsion polymerization is widely used for the manufacture of acrylic dispersions for water borne coatings. The monomer blend is dispersed in the water phase by means of agitation and suitable surfactant. Polymerization takes place in the water phase initiated by water soluble catalyst. Emulsions with a particle size between 0.01 and 5 μm are easily obtained. The precipitation and drying of such small particles is rather a difficult process. Therefore, the suspension polymerization process, by which particles with a diameter of a few millimetres are obtained, is preferred.

Suspension polymerization is carried out in a similar manner to emulsion polymerization. The monomer added to water is dispersed by agitation in droplets of a few millimetres. Unlike emulsion polymerization, reaction is initiated by a catalyst soluble in the monomer blend. As a result, the polymerization takes place entirely in the monomer phase. To overcome sticking of the droplets at certain critical moments when polymerization is partially complete, a small amount of protective colloids such as polyvinyl alcohol is added to the batch. Removal of the particles by filtration, their purification by washing with water and drying are the following steps after polymerization has been completed. The product is in the form of small beads which are easy to handle and process.

Typical monomers used for the production of acrylic resins are acrylic and methacrylic esters of C1–C8 alcohols. Different functional groups can easily be introduced by copolymerization with monomers such as acrylic and methacrylic acid, hydroxyl ethyl acrylate and methacrylate, hydroxyl propyl acrylate and methacrylate, glycidyl methacrylate, acryl amide, methylol acryl amide, butoxymethyl acryl amide, etc. In addition, acrylic and methacrylic esters copolymerize readily with practically all types of vinyl monomers. This versatility provides a possibility for development resins with a broad variety of properties.

The importance of the glass transition temperature as an essential parameter determining the stability of powder coatings is discussed in Chapter 4. Glass transition temperatures of acrylic homopolymers vary in a broad range, as presented in Table 3.17. By copolymerizing acrylic monomers, polymers with intermediate glass transition temperatures can be prepared. An empirical equation proposed by Fox [94] can be used to predict the glass transition temperature of the copolymer from the known values of the homopolymers:

$$\frac{1}{T_g} + \frac{W_1}{T_{g1}} + \frac{W_2}{T_{g2}} + \frac{W_3}{T_{g3}} \cdots + \frac{W_n}{T_{gn}}$$

where $W_1, W_2, \cdots, W_n$ are the weight fractions of each acrylic monomer in the copolymer and T_{g1}, and $T_{g2}, \cdots, T_{gn}$ are the glass transition temperatures of the respective homopolymers.

Reactivity of acrylic monomers differs considerably. Consequently, wide differences in the reactivity ratios of many comonomer mixtures can be expected. This leads to heterogenicity in the composition of the chains and in the distribution of the monomer units along the chain. Although methods for controlling heterogenicity of the copolymer based on programmed addition of the comonomers have been developed, it is believed that this is one of the main reasons for rather poor mechanical properties of acrylic powder coatings.

In polycondensation types of polymers, such as polyesters, the functional groups can be located at the chain ends and the distance between the branching points in the chain can easily be regulated by choosing the correct ratio between the reacting monomers. In the case of acrylic resins, it is impossible to choose the position of the functional groups. They are distributed at random not only along the chain but also among the macromolecules, thus having chains with very low

Table 3.17 Glass transition temperatures of various acrylic homopolymers [93] (Reproduced by permission of John Wiley & Sons, Inc.)

Polymer	T_g (°C)	Polymer	T_g (°C)
Polyacrylates		Polymethacrylates	
Methyl	8	Methyl	105
Ethyl	−22	Ethyl	65
n-Propyl	−51	n-Propyl	33
n-Butyl	−54	n-Butyl	20
2-Ethylhexyl	−55	n-Octyl	−20

functionality or no functionality at all, and chains with very high functionality. Since the part of the chain with the so-called dead end having no functional groups does not contribute much to the impact resistance (chains with low or no functionality) and, on the other hand, the impact resistance is inversely proportional to the crosslinking density for the same molecular weight of the crosslinked polymer (chains with high functionality), the overall result is crosslinked film with low flexibility and poor impact resistance.

Acrylic powder coatings, which are used in practice, are almost exclusively based on epoxy functional acrylic resins cured with long-chain dicarboxylic acids as hardeners. The long aliphatic chain of the crosslinker provides flexibility and impact resistance to the cured film, but still far below the values which are usually obtainable with the other powder coating systems. In most cases the impact resistance of the acrylic powder coatings does not exceed 30 inch-pounds. This is still several times lower compared to the values of polyester and polyurethane powder coatings.

A typical acrylic powder coating with glycidyl functionality and dicarboxylic aliphatic acid as hardener is described in the patent of the Glidden Company [95]. Acrylic resin prepared by suspension polymerization of ethylene unsaturated monomers, including glycidyl monomers, is formulated into a powder coating together with a stoichiometrical amount of 1,10-decane dicarboxylic acid (with respect to the epoxy/carboxyl ratio), flow agent and titanium dioxide in the case of pigmented coatings, or without pigments to produce a clear coating. The coatings are prepared by extrusion of the components at 80°C and pulverization in the presence of liquid nitrogen to keep the temperature below 20°C during the milling operation.

A similar powder composition [96] containing glycidyl acrylic copolymer and aliphatic dibasic acid as hardener is obtained by melt mixing of both components, cooling and pulverizing to give a coating with good antiblocking properties and a smooth weather resistant surface after curing.

A patent of Du Pont [97] discloses a system in which the glycidyl functional acrylic resin contains 12–20% glycidyl methacrylate as the comonomer, the main crosslinking agent being aliphatic dicarboxylic acid with 10–20 carbon atoms, blocked aliphatic diisocyanate, hexamethoxymethyl melamine or tetramethoxymethyl glycoluryl as co-crosslinkers in minor amounts and u.v. stabilizer of the benzotriazole type. The unpigmented composition can be used as an automotive clear top coat characterized with high gloss, smooth surface and excellent weathering resistance.

Better flexibility of the cured films is claimed in a disclosure of Labana and Chang [98] where glycidyl functional acrylic resins are cured with carboxyl terminated polyethers obtained by termination of hydroxyl functional polyalkylene glycols with diacid anhydride.

Other systems having acrylic resin with carboxyl functional groups as the main binder and epoxy functional low molecular weight crosslinkers are also described in the patent literature. In a patent of Bayer [99], carboxyl groups bearing acrylic resin are used in combination with triglycidyl isocyanurate to produce clear or pigmented powder coatings which are claimed to have good mechanical properties, gloss and weather resistance.

The use of hydantoin epoxides containing two to four epoxy groups as crosslinkers for carboxyl functional polymers in powder coatings is described in a patent assigned to Du Pont [100]. It is claimed that the outdoor durability, especially in the case when carboxyl acrylics are used as the main binders, meets the requirements of automotive clear top coats.

Self-crosslinking acrylic powder coating can be produced using acrylic resin containing hydroxylalkyl acrylic comonomers that have been reacted before or after copolymerization with other unsaturated monomers with partially caprolactam blocked isocyanates [101].

Blocked isocyanates are also used as crosslinkers for acrylic powder coatings. An example is a powder coating composition containing hydroxyl functional acrylic resin as the binder and caprolactam blocked adduct of hexanediol and isophorone diisocyanate [102]. After curing at 180°C for 20 minutes, films with good flexibility, gloss and chemical and solvent resistances are obtained.

A powder coating composition containing hydroxyl functional acrylic resin and blocked isocyanate prepolymer is described by Yousuf [103] and compared with two polyester based polyurethane powder coatings. Obvious advantages of the acrylic system are better solvent resistance and gloss retention during the accelerated weathering tests. However, the direct and reverse impact resistances of the acrylic coating are much worse compared to the polyester powders.

Several attempts have been made to benefit from the good mechanical properties of the polyester resins and good weathering resistance of the acrylics, combining them into one system. A patent of Dainippon Ink and Chemicals, Inc. [104] refers to a system containing 70–90% carboxyl functional polyester resin and 10–30% glycidyl functional acrylics. The average number of carboxyl groups per molecule of polyester resin is very low, largely below 2, in order to obtain good flow properties. Acrylic resin itself has a rather low molecular weight between 500 and 3000, having a relatively high content of glycidyl methacrylate of at least 20% by weight. It is claimed that after curing at 160°C for 20 minutes, smooth high glossy films with good impact resistance and flexibility are obtained.

A combination of polyester and acrylic resin is described in a patent of Nippon Paint [105] and comprises 65–95% (by weight) of polyester resin with an hydroxyl value of 200–100 and an acid value of 1–20, 4–30 wt% blocked isocyanate and 2–30 wt% acrylic resin containing hydroxyl and glycidyl groups with an hydroxyl value of 5–150, a glycidyl group content of 0.0035 equivalents/g and molecular mass of 3000–50000 with $T_g = 40$–$85°C$. It is claimed that the coating has an improved resistance to staining and solvents.

Another system developed by S. C. Johnson & Son, Inc., is based on a blend of hydroxyl functional acrylic resins and hydroxyl functional polyesters with

THERMOSETTING POWDER COATINGS

blocked isocyanates as crosslinkers. An improved outdoor durability is claimed as a result of the positive influence of the acrylic cobinder in the coating composition [106].

In both systems, the amount of acrylic resin is rather low. In the case of Dainippon Ink, the low amount of acrylic resin comes as a result of the low epoxy equivalent weight. Higher epoxy equivalent weights will lead to glycidyl acrylates with higher molecular weights, resulting in limited compatibility with the polyester. Although in both systems there is an improvement in the weathering resistance, the effect is not as high as would be expected if higher amounts of acrylic resin were used in the coating formulation. In the absence of a real solution to the problems related to the inferior mechanical properties of the acrylic systems, even when acrylic polyester blends are used, the so-called dual coat dry-on-dry powder coating system has been proposed offering excellent physical film performance properties [107]. The first coat is a thin application of a polyester based powder coating. The top coat, applied directly over the polyester basecoat, dry-on-dry without previous curing, is based on acrylic resin. In this way the advantages of both acrylic and polyester coatings are combined into one system.

The dry-on-dry system can certainly be considered as an improvement of the performances of the acrylic powder coatings. However, the performances of the coating and the reproducibility of the results depend to a large extent on the skill of the operator. One coat system will not suffer from such a disadvantage. A solution of the compatibility problems, leading to acrylic rich polyester/acrylic hybrids, could be a way to substantially improve the weathering resistance of powder coatings for outdoor use.

3.4.4 UNSATURATED POLYESTERS POWDER COATINGS

Unsaturated polyester resins have found very limited application as binders for in-mould powder coatings for sheet moulding compounds (SMC). The principle of curing is a free radical polymerization of the double bonds of the resin initiated with peroxides or other standard initiator/accelerator systems widely used in the field of unsaturated polyester resins for construction purposes. The problem of oxygen inhibition which is typical for this type of curing reaction restricts their use for coating surfaces that are in direct contact with air during curing. In the case of in-mould powder coatings this problem is avoided simply because the crosslinking of the powder coating takes place on the surface that is in direct contact with the mould wall. So far this system has never found a single application outside the SMC area.

Polyester resins for in-mould powder coatings are not different in principle from the unsaturated polyester resins for glass reinforcement, and concern the resin backbone. Since the coating has to be in a powder form, the use of styrene and other liquid monomers which serve as reactive diluents is excluded, and only solid sources of unsaturation should be considered.

The production of unsaturated polyester resins proceeds by polyesterification of unsaturated diacids and glycols. Although unsaturated glycols and saturated diacids can also be used, because of the availability and price of the former, this is not the case in practice. Most widely used unsaturated diacids are maleic anhydride and its *trans* isomer fumaric acid. Among the diols, ethylene glycol and 1,2-propylene glycol are the most important. The degree of unsaturation can be easily adjusted by replacing part of the unsaturated diacid with saturated diacid. For this purpose phthalic anhydride has been widely accepted. Isophthalic acid, the *meta* isomer of phthalic acid, has found wide use as the diacid imparting excellent hydrolytic stability of the cured resins. Adipic acid can be used for flexibility improvement. For the same reasons diethylene glycol can also be used, sacrificing to a certain extent the hydrolytic stability of the cured material [108]. In that respect 1,6-hexanediol is a better choice, while the other commercially available higher glycol, 1,4-butane diol, has to be esterified carefully because it forms tetrahydrofuran during esterification [109]. Neopentyl glycol, due to the presence of the two methyl groups in the α position to the hydroxyl groups, provides excellent hydrolytic stability to the polyesters.

Although the production process of saturated and unsaturated polyester resins is the same, there are certain specific differences.

During the polyesterification of the maleic acid, there is always a certain equilibrium with the *trans* isomer, fumaric acid [110–112]:

$$\begin{array}{c} \text{HC—COOH} \\ \| \\ \text{HC—COOH} \end{array} \longleftrightarrow \begin{array}{c} \text{HC—COOH} \\ \| \\ \text{HOOC—CH} \end{array}$$

Maleic acid   Fumaric acid

The degree of isomerization depends on the reaction conditions and the catalysts used during esterification. A high esterification temperature promotes the isomerization. Even the carboxylic groups from the maleic acid are good isomerization catalysts. Therefore a longer esterification time due to a slow reaction rate leads to a higher degree of isomerization.

Fumaric acid is more reactive during free radical polymerization at the crosslinking stage, shortening considerably the curing time of the unsaturated polyester.

An example of the preparation of unsaturated polyester for an in-mould powder coating will be presented as an illustration of the process described by DSM Resins [113]. According to this process, 2672 parts by weight of 1,6-hexane diol, 692 parts of terephthalic acid and 0.45 parts of diphenylol propane are esterified at 230°C in an inert atmosphere until an acid number of 7.5 mg KOH/g is reached. The reaction mass is cooled down to 170°C and 1864 parts by weight of fumaric acid are added. The esterification continues at 210°C until an acid number of 9.3 mg KOH/g is obtained. Before cooling down and discharging, the polyester resin is inhibited with 0.45 parts of hydroquinone. The resulting resin has an unsaturation number of 3.6 mol/kg, a melting point of 99°C, an acid value of 8 and an hydroxyl value of 52 mg KOH/g.

Crystalline unsaturated polyesters have an advantage over the amorphous types, giving powder coatings that have improved antiblocking properties during storage and better flow after melting before curing takes place [114]. On the other hand, the impact resistance of powder coatings based on amorphous unsaturated polyester is better, and the cured films exhibit a higher flexibility and less brittleness. Isocyanate modified unsaturated crystalline polyesters, however, have an improved impact resistance, without sacrificing the stability of the system during storage [113].

Next to the unsaturated binder, in-mould powder coatings contain high boiling point copolymerizable monomers with a functionality higher than two, pigments and additives such as initiators, release agents, flow promoters, stabilizers, etc.

The boiling point of the monomers used in combination with the unsaturated polyesters must be in principle 25°C higher than the temperature at which the powder coating is applied and cured, in order to avoid possible losses during curing or the appearance of surface defects. The main role of the polyfunctional monomers in the powder coating composition is to increase the crosslinking density of the cured material. Suitable monomers used in practice are triallylcyanurate, triallylisocyanurate, trimethylolpropane triacrylate and triallyltrimellitate.

Catalysts which are mostly used for the curing of unsaturated compounds belong to the group of peroxides, such as hydroperoxides, ketoneperoxides or peresters. They are present in amounts between 0.5 and 1% by weight with respect to the unsaturated binder. Conventional accelerators such as cobalt salts or tertiary amines which are very often used in curing conventional unsaturated resins for construction purposes can be incorporated in the powder coating composition in order to speed up the curing rate. A patent of Morton Thiocol, Inc. describes an unsaturated polyester powder coating composition of which the curing rate can be easily altered by a combination of a slow and a fast initiator [115].

Although the powder coating can be prepared by physical mixing of finely ground constituents, the best results are of course obtained by melt mixing of all components, cooling and grinding of the obtained homogeneous blend [116, 117]. In order to avoid prepolymerization during melt compounding, a certain amount of a suitable polymerization inhibitor (quinone or catechol compounds) can be added. As a rule, the addition of the catalysts takes place just before cooling. An example of an in-mould powder coating described by DSM Resins [113] is given in Table 3.18.

In-mould powder coatings are usually applied by an electrostatic spraying technique on preheated walls of the mould at a temperature between 100 and 160°C. Immediately after application the powder particles melt down and form a

Table 3.18 Unsaturated polyester in-mould powder coating [113]

Material	Parts by weight
Unsaturated polyester resin	100
Acrylate flow agent	1
Titanium dioxide	10
Triallylisocyanurate	3
p-Benzoquinone	50 ppm
t-Butylperbenzoate	2
Curing schedule (min/°C)	20/120

continuous layer. In order to ensure that the coating layer can withstand some mechanical load, the coating is allowed to cure partially, before the construction material from which the article is made enters the mould. A complete curing in this stage can cause interadhesion problems between the coating layer and the article that is coated. Unsaturated polyester in-mould powder coatings are mainly used for articles made from reinforced unsaturated polyester resins at elevated temperature and pressure via the SMC or BMC moulding techniques, or by injection moulding.

REFERENCES

1. Greenspan, F. P., in *Chemical Reactions of Polymers*, Ed. E. M. Fettes, Interscience, New York, 1964, p. 152.
2. Paul, S., in *Surface Coatings: Science and Technology*, John Wiley and Sons, 1985, p. 219.
3. Wise, J. K. *J. Paint Technol.*, **42** (540), 29, 1970.
4. Ranney, M. W., in *Epoxy Resins and Production: Recent Advances*, Noyes Data Corp., 1977.
5. Brushwell, W., *Am. Paint and Coat. J.*, 2 November, 65, 1981.
6. Brushwell, W., *Farbe und Lack*, **87** (3), 201, 1981.
7. Oldring, P., and Hayward, G., in *Resins for Surface Coatings*, Vol. II, SITA Technology, London, 1987, p. 4.
8. Asahi Chemical Ind., Jap. Pat. 58 187 464, 1982.
9. Sumitomo Durez, Jap. Pats. 59 226 066 and 59 226 067, 1983.
10. Greensitt, E. A. *Paint Technol.*, **28** (9), 16, 1964.
11. Kimmo Peltonen, *Thermal Degradation of Epoxy Powder Paints*, Institute of Occupational Health, Helsinki, 1986.
12. Peltonen, K., *Analyst*, **111**, 819, 1986.
13. 'Dow epoxy resins in powder coatings', Dow Chemicals Brochure, 1981 ed, Nos. 1451 and 1495.
14. Hitachi Chemical KK, Jap. Pat. 58 087 123, 1981.
15. Shell Oil Company, US Pat. 3 824 035, 1974.

16. Dow Chemical Europe, Eur. Pat. 0 072 371, 1981.
17. Gulpen, N. J. H., and Werff A. J. van de, *J. Paint Technol.*, **47** (608), 81, 1975.
18. Nitto Electric Industry, Jap. Pat. 57 042 760, 1980.
19. Yuka Shell Epoxy, Jap. Pat. 58 118 818, 1982.
20. 'Shell resins, amine-cured Epikote resin powders', Technical Bulletin EP 2.10.7, 1980.
21. 'Dow epoxy resins in powder coatings, Dow Chemicals Borchure', 1981 ed., Nos. 1003, 974 and 9.1.C.
22. Sumitomo Durez KK, Jap. Pat. 59 226 065, 1983.
23. 'Shell resins, Epikote resin powder coatings', Technical Bulletin EP 2.10, 1983, p. 11.
24. Chemische Werke Hüls AG, Eur. Pat. 0 044 030, 1980.
25. Mitsubishi Electric Corp., Jap. Pat. 57 018 765, 1980.
26. Minnesota Mining and Manufacturing Company, US Pat. 3 876 606, 1975.
27. 'Shell resins, Epikote resin powder coatings', Technical Bulletin EP 2. 10, 1983, p. 12.
28. Hoechst AG, Ger. Pat. 1 913 923, 1969.
29. Misev, T., *XIX FATIPEC Congress, Proceedings*, Vol. III, Aachen, 19–23 September 1988, p. 99.
30. Patton, T. C., *Off. Dig.*, **32** (430), 1544, 1960.
31. Lynas-Gray, J. I., *Paint Technol.*, **11** (124), 129, 1946.
32. Lynas-Gray, J. I., *Paint Technol.*, **12** (133), 7, 1947.
33. Burrel, H., *Paint Oil and Chem. Rev.*, **110**, 19, 1947.
34. Wangsness, I. L., Jerabek, R. D., Murphin, A. T., and Naponen, G. E., *Off. Dig.*, **26**, 1062, 1954.
35. Seaborne, L. R., *Paint Technol.*, **19** (208), 6, 1955.
36. Wiederhorn, N. M., *Am. Paint J.*, **41** (2), 106, 1956.
37. Kraft, W. M., *Off. Dig.*, **29**, 780, 1957.
38. Vaughan, C. L. P., and Schmitt Jr, F. E. *Off. Dig.*, **30** (405), 1131, 1958.
39. Glaser, D. W., *Off. Dig.*, **33**, 642, 1961.
40. Patton, T. C., *Alkyd Resins Technology*, Interscience Publishers, John Wiley, New York, 1962.
41. Misev, T., Ban, N., and Bravar, M., *Hemijska Industrija*, **33** (5), 177, 1979.
42. Misev, T., *Hemijska Industrija*, **34** (6), 164, 1980.
43. Misev, T., Ban, N., and Bravar, M., *Hemijska Industrija* **34** (7), 179, 1980.
44. Mleziva, J., in *Polyestery*, SNTL, Praha, 1964.
45. Tysall, L. A., in *Calculation Technique in the Formulation Alkyds and Related Resins*, Paint Research Association, May 1982 ed.
46. Misev, T., *J. Coat. Technol.*, **61** (772), 49, 1989.
47. Cargil Inc., US Pat. 4 275 189, 1980.
48. Kodak Ltd, WP 8 300 328, 1981.
49. Hüls AG, Eur. Pat. 317 741, 1987.
50. Mitsui Toatsu Chem. Inc., Eur. Pat. 314 447, 1987.
51. Chemische Werke Hüls AG, Ger. Pat. 3 004 876, 1980.
52. Chemische Werke Hüls AG, Ger. Pat. 3 322 719, 1983.
53. Eastman Kodak Co., Eur. Pat. 0 070 118, 1981.
54. Eastman Kodak Co., US Pat. 4 352 924, 1981.
55. Bodnar, E., *Product Finishing*, August 1988, p. 22.
56. UCB, Belg. Pat. 0 841 213, 1975.
57. UCB, Belg. Pat. 0 841 681, 1975.
58. Unilever NV, US Pat 4 147 737, 1979.
59. Goodyear Tire and Rubber Co., US Pat. 4 379 895, 1982.
60. DSM Resins BV, NL Pat. 8 204 206, 1982.
61. DSM Resins BV, Belg. Pat. 0 898 100, 1982.

62. Du Pont, US Pat. 4 242 253, 1979.
63. Dainippon Ink Chem., Jap. Pat. 58 034 869, 1981.
64. Hitachi Chemical, Jap. Pat. 57 055 920, 1980.
65. Nippon Ester, Jap. Pat. 59 011 375, 1982.
66. DSM Resins BV, Belg. Pat. 898 099, 1982.
67. DSM Resins BV, NL Pat. 8 204 206, 1982.
68. Nippon Ester, Jap. Pat. 57 135 829, 1981.
69. Nippon Ester, Jap. Pat. 57 126 822, 1981.
70. Hitachi Chemicals, Jap. Pat. 57 055 958, 1980.
71. Hitachi Chemicals, Jap. Pat. 57 055 923, 1980.
72. Takeda Chemical Ind., Jap. Pat. 59 230 069, 1983.
73. Minnesota Mining and Manufacturing Company, US Pat. 3 340 212, 1961.
74. Huneke, H., von, *Metalloberflache*, **24** (9), 315, 1970.
75. Unilever NV, Ger. Pat. 2 163 962, 1971.
76. Scado BV, US Pat. 3 624 232, 1970.
77. UCB SA, Ger. Pat. 2 352 467, 1972.
78. SCM Corporation, Eur. Pat. 0 021 770, 1979, and US Pat. 4 271 277, 1979.
79. DSM Resins BV, 'Powder coating resins', Brochure, July 1988.
80. DSM Resins BV, Belg. Pat. 0 898 100, 1982.
81. Metallgesellschaft AG, Ger. Pat. 1 905 825, 1969.
82. Unilever NV, GB Pat. Appl. 61107-70, 1970.
83. DSM Resins BV, Eur. Pat. 0 107 888, 1982.
84. Hoppe, M., *J. Coat. Technol.*, **60** (763), 53, 1988.
85. DSM Resins BV, NL Pat. 8 204 206, 1982.
86. Du Pont De Nemours Co., US Pat. 4 402 983, 1980.
87. Bayer AG, Eur. Pat. 0 024 680, 1979.
88. Walz, G., and Kraft, K., *Proceedings of 4th International Conference on Organic Coatings, Science and Technology*, Athens, 1978, p. 56.
89. McLafferty, J. J., Figlioti, P. A., and Camilleri, L. T., *J. Coat. Technol.*, **58** (733), 23, 1986.
90. Bayer AG, Ger. Pat. 3 232 463, 1982.
91. R. M. Gallas, *Acrylic Modified Polyester System for Powder Coatings*, Johnson Wax Brochure, 1988.
92. Horn, M. B., *Acrylic Resins*, Reinhold Publishing Corp., New York, 1960, p. 15.
93. Adapted from Luskin, L. S., and Myers, R. J. in *Encyclopedia of Polymer Science and Technology*, Ed. N. M. Bikales, Vol. I, Interscience Publishers, New York, 1964, p. 299.
94. Fox, T. G., *Bull. Am. Phys. Soc.*, **1** (3), 123, 1956.
95. The Glidden Company, Eur. Pat. 0 256 369, 1986.
96. Kansai Paint Co. Ltd, US Pat. 3 876 578, 1975.
97. Du Pont De Nemours Co., Eur. Pat. 0 045 040, 1980.
98. Ford Motor Company, US Pat. 3 880 946, 1975.
99. Bayer AG, US Pat. 3 836 604, 1974.
100. Du Pont De Nemours Co., US Pat. 4 402 983, 1980.
101. BASF Farben and Faser, Ger. Pat. 3 310 545, 1981.
102. Deutsche Gold und Silber Scheideanstalt, US Pat. 3 867 347, 1975.
103. Yousuf, M. K., *Modern Paints and Coatings*, June 1989, p. 48.
104. Dainippon Ink and Chemicals, Inc. Eur. Pat. 0 038 635, 1980.
105. Nippon Paint, Jap. Pat. 115 516, 1986.
106. R. M. Gallas, *Acrylic Modified Polyester System for Powder Coatings*, Internal Publication, S. C. Johnson & Son, Inc., 1988.

107. Yousuf, M. K., and Pfeffer, W. G., *Dual Coat, Dry-on-Dry, Powder Coating Systems*, Internal Publication, S. C. Johnson & Son, Inc., July 1988.
108. Williams, H., *Plast. Inst. Trans.*, **24**, 37, 1956.
109. Bayer AG, Ger. Pat. 1 105 610, 1961.
110. Feuer, S. S., Bockstahler, T. E., Brown, C. A., and Rosenthal, I., *Ind. Eng. Chem.*, **46**, 1643, 1954.
111. Hayes, B. T., Read, W. J., and Vaughan, V. H., *Chem. Ind.*, **1957**, 1165.
112. Mleziva, J., and Vladyka, J., *Farbe und Lack*, **68**, 144, 1962.
113. DSM Resins BV, Eur. Pat. 0 188 846, 1984.
114. DSM Resins BV, Eur. Pat. 0 106 399, 1983.
115. Morton Thiocol, Eur. Pat. 309 088, 1987.
116. Stamicarbon BV, US Pat. 4 228 113, 1980.
117. Stamicarbon BV, US Pat. 4 287 310, 1980.

4

Parameters Influencing Powder Coating Properties

4.1	Molecular Weight of the Binder	175
4.2	Functionality of the Coating Composition	177
4.3	Glass Transition Temperature	181
	4.3.1 T_g and Powder Stability	183
	4.3.2 T_g and the Melt Viscosity	183
	4.3.3 T_g and the Thermal Stress Development	185
	4.3.4 Molecular Weight and T_g	186
	4.3.5 Chemical Structure and T_g	188
	4.3.6 Glass Transition Temperature of Polymer Blends	189
4.4	Viscosity	191
	4.4.1 Viscosity and Processing Performances	192
	4.4.2 Viscosity and Film Forming Properties	193
4.5	Resin/Crosslinker Ratio	196
4.6	Catalyst Level	199
4.7	Surface Tension	204
	4.7.1 Wetting Properties and Surface Tension	205
	4.7.2 Surface Tension and Cratering	206
	4.7.3 Surface Tension and Film Levelling	208
4.8	Pigment Volume Concentration and Pigment Dispersion	211
4.9	Particle Size	215
4.10	Stoving Temperature Profile	217
	References	220

Powder coatings have a specific place in the paint industry, being different in many aspects from the liquid coating systems. Compared to the conventional organic coatings similarities can be drawn with respect to the ultimate exploitation properties of cured films, where the same general rules are valid for both systems. However, all similarities end there. The analysis of the differences between powder and conventional solvent borne coatings should also include a comparison between the production technologies, ways of application and the process of film formation. With respect to the manufacturing process, powder coatings should be classified in the group of plastics rather than paint technologies. The application techniques for powder coatings are unique and not applicable to the other organic surface coatings used commonly by the industry.

PARAMETERS INFLUENCING POWDER COATING PROPERTIES 175

Finally, despite the same crosslinking chemistry, the physics of the film formation of powder coatings differs completely from that of the solvent borne paints.

The problems with powder coatings begin when the processing and application characteristics have to be combined with the final coating performances. In most cases the requirements attributed to these properties are contradictory. The application characteristics of the solvent borne systems can be adjusted by choice of a proper solvent or solvent blend with little influence on the ultimate properties of the cured film. In the case of powder coatings this possibility does not exist, simply because the solvent is absent. Although it seems to be a simple system considering the number of constituents in the coating formulation, powder coatings are rather complex with regard to the combination of processing, storage, film forming and final service properties of the coating.

The absence of solvent, which is desirable from an ecological point of view, decreases the degree of freedom of the coating formulator. Most of the properties of a powder coating are determined exclusively by the binder. Therefore a good understanding of the resin parameters influencing the coating properties is a prerequisite for the resin and paint chemist in order to produce powder coatings with well-balanced overall properties.

A basic understanding of the correlations between different resin and paint parameters and their influence upon the final coating properties can help the paint chemist more than the most extensive list of powder coating formulations. Therefore this chapter addresses the important resin parameters such as molecular weight, functionality, glass transition temperature and viscosity, with the aim of correlating them with the coating properties. The description of the elements of the paint formulation, such as the binder/crosslinker ratio, pigment volume concentration, catalyst level, surface tension of the molten powder paint follows the discussion of the resin parameters. The last two sections consider the influence of powder particle geometry and the characteristics of the stoving line to the application and film forming properties of the powder coating during the curing stage.

4.1 MOLECULAR WEIGHT OF THE BINDER

Like all polymeric materials, the resin used in powder coatings are blends of molecules of different molecular weights. Therefore averaging of the molecular weights is necessary. From the several different ways of averaging the molecular weights two are most important for the properties of the powder coatings: the number average molecular weight (M_n) and the weight average molecular weight (M_w), expressed by the well-known equations:

$$M_n = \frac{\Sigma M_i N_i}{\Sigma N_i} \tag{4.1}$$

$$M_w = \frac{\Sigma N_i M_i^2}{\Sigma N_i M_i} \qquad (4.2)$$

Mechanical properties of the powder coatings such as tensile strength and impact resistance are mainly dependent on the number average molecular weight, while the weight average molecular weight mainly governs the melt viscosity of the resin.

For polymers, mechanical properties can often be expressed by the following formula [1]:

$$X = X_\infty - \frac{A}{M_n} \qquad (4.3)$$

where X is the property considered, X_∞ is its value at a very high molecular weight and A is an empirical constant.

Based on a wide range of collected data it has been established that molecular weight should be between 20 000 and 200 000 for a commercial polymer in order to ensure that it exhibits a good impact resistance and tensile strength [2]. Let us consider this fact and try to implement it in the powder coatings.

Suppose a linear carboxyl terminated polyester resin with a number average molecular weight of M_{np} is to be cured with a linear type of bisphenol A based epoxy resin with a number average molecular weight M_{ne}. If there is complete consumption of the epoxy groups during curing, in the presence of a slight excess of carboxyl groups coming from the polyester resin, the number average molecular weight M_n of the cured coating can be easily calculated by the following formula:

$$M_n = (x+1) M_{np} + x M_{ne} \qquad (4.4)$$

where x is the degree of polymerization of the block copolymer composed of blocks of polyester (P) and epoxy (E) nature:

$$P - (E - P)x$$

Obviously, the molar ratio between the functional groups can be expressed by

$$z = \frac{x+1}{x} \qquad (4.5)$$

It is clear that $x = \infty$ only for an ideal stoichiometric ratio between the carboxyl end epoxy groups, i.e. for $z = 1$. Let us assume that the lowest value for M_n to achieve good mechanical properties after curing has to be 20 000. Using the equations (4.4) and (4.5) and supposing an equivalent weight of the crosslinker of 725 (Araldite GT 7004 from Ciba-Geigy, for example), the values of z for obtaining this molecular weight can be calculated. These figures are presented in Table 4.1 for polyesters with molecular weights ranging between 1000 and 6000. The results show that the lower the molecular weight of the polyester the closer the

Table 4.1 Mole ratio between carboxyl and epoxy groups for obtaining a molecular mass of 20 000

M_n polyester	z
1000	1.13
2000	1.19
3000	1.26
4000	1.34
5000	1.43
6000	1.53

polyester/epoxy ratios to the stoichiometric ratio must be. In other words, the system is more sensitive to the deviations that are quite common—i.e. in weighing raw materials during the coating manufacture, determination of acid, hydroxyl or epoxy equivalent values, degree of hydrolysis of the oxyrane rings during the storage of the coating or the epoxy crosslinker itself and the extent of the reaction during the curing cycle.

The possible manufacture of polyesters with higher molecular masses has the unfortunate consequence of increasing the viscosity that causes problems concerning processability and flow of the coating.

4.2 FUNCTIONALITY OF THE COATING COMPOSITION

The problem of sensitivity of the coating formulation to the correct ratio between the functional groups can be overcome by increasing the functionality of the crosslinker or the functionality of the resin. In this way, the system will be less sensitive to the necessary stoichiometry for obtaining an infinite network. Using Gordon's theory of branching processes with cascade substitution [3–6], calculations have been made for a system based on carboxyl functional polyester with a number average molecular weight of 3800 and functionality between 2 and 3.25 and bisphenol A epoxy resin with a number average molecular weight of 1500 and epoxy functionality of 2. Figure 4.1 represents the dependence of the number average molecular weight of the system at different degrees of conversion of the epoxy groups during the curing process. As expected, the build-up of the molecular weight is much faster with increasing functionality. A number average molecular weight of 20 000 is obtained at a conversion of 86% in the case of a functionality of 2, and 62, 24 and 8% when the functionality of the polyester resin is 2.5, 3 and 3.25 respectively.

As usual, the formulator is once again faced with several difficulties when taking this option as a solution to the problem. High functionality will have as a consequence a fast increase of viscosity, thus shortening the time available for

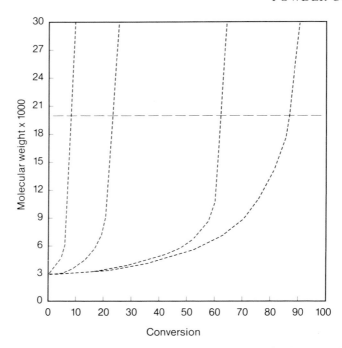

Figure 4.1 M_n as a function of conversion during the curing of polyesters with different functionalities with bisphenol A epoxy resins (Reproduced by permission of the Oil and Colour Chemist's Association) [27]

flowing of the coating during film formation. The result will be poor flow with an emphasized orange peel effect. Scholtens, van der Linde and Tiemersma-Thoone [7], using thermoviscoelastic analysis (t.v.a.), have determined the influence of the variations of the molecular mass and the functionality of the polyester resins on the rheological behaviour of a molten polyester/epoxy unpigmented hybrid powder coating at 120°C and the change of the dynamic modulus of the system in a dynamic run from 120 to 200°C. The results obtained are presented in Figures 4.2 and 4.3 where in all cases the epoxy resin is Araldite GT 7004 with an epoxy equivalent weight of 725 (Ciba-Geigy) and the polyester resins differ in functionality (ϕ_n) and number average molecular mass (M_n). The lowest viscosity and almost Newtonian behaviour ($\delta(w) = 90°$) is reached at the highest frequency by composition 4E, which contains a linear polyester with the lowest molecular weight (Figure 4.2). The same formulation is characterized with the lowest dynamic modulus. This is reflected in the best flow of the coating. The other two compositions with a functionality of 2.5 have a higher viscosity and modulus. Composition 5E with a higher molecular weight reaches the Newtonian behaviour at a lower frequency compared to composition 6E. It has also a higher

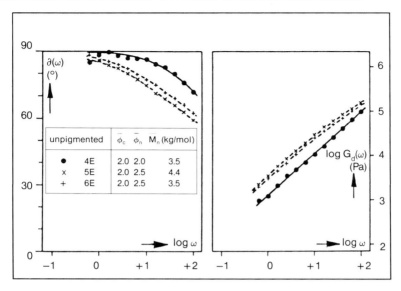

Figure 4.2 The t.v.a. isotherms at 120°C for three unpigmented polyester/epoxy systems with variations in the polyester structure [7] (Reproduced by permission of DSM Resins BV)

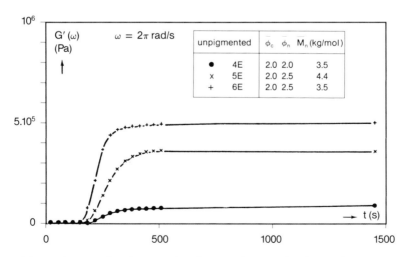

Figure 4.3 Isofrequent heating runs for three unpigmented polyester/epoxy systems at $T = 120–200°C$ at a heating rate of 20°C/min. G' is plotted on a linear cale [7] (Reproduced by permission of DSM Resins BV)

dynamic modulus. Both effects result in the least flow of the coating. This behaviour is explained by the longer relaxation times of the branched structures compared to the linear polymers and to the species with a lower molecular weight. The changes in the systems during curing are presented in Figure 4.3. An increase in functionality increases the crosslinking density (comparison between 6E and 4E), whereas an increase of molecular mass decreases the crosslinking density (comparison between 5E and 6E).

Functionality (F_n), the number average molecular weight (M_n) and the acid value (AV) or the hydroxyl value are not independent variables of the polyester resins. In the case of acid terminated polyester resins which are to be crosslinked with epoxies, this dependence can be expressed by the following equation:

$$F_n = \frac{AV \times M_n}{56\,100} \quad (4.6)$$

This means that for the same acid value of the resin (in order to keep constant the weight ratio of polyester/epoxy) if one wants to increase the functionality a higher molecular weight has to be chosen. The other option is to keep the same molecular weight but then to increase the acid value, which has a direct influence on the desired polyester/epoxy ratio.

The increase in the functionality also leads to a higher mole fraction of branching components in the resin formulation. In the case of a random esterification process the number average molecular weight increases linearly with the functionality at a constant acid value, but the increase in the weight average molecular weight is very rapid.

Using the Flory–Stockmayer theory for molecular weight distribution in non-linear polymerization [8,9] a calculation has been made for a series of resins based on neopentyl glycol, isophthalic acid and trimethylol propane as a branching component with an acid value of 35 and functionality between 2 and 4. The results are presented in Table 4.2. At a functionality of 3.5 the resin is already beyond the gel point. Because of the assumptions that have been made in deriving the Flory–Stockmayer theory (especially neglecting the intramolecular reactions that lead to cyclization), there is a certain discrepancy between the theoretical

Table 4.2 Dependence between F_n, M_n and M_m in the case of non-linear polyesterification

F_n	M_n	M_m
2	3205	6 550
2.5	4000	12 830
3	4800	384 000
3.5	5600	Infinite
4	6400	Infinite

predictions and the practical results [10]. Therefore it is possible to make a resin with a functionality of 4, but the viscosity will be too high for practical application. In addition, the gel point distance defined as the ratio between the extent of the reaction at the gel point and the extent of the reaction necessary to obtain a resin with certain characteristics will be too small to produce a safe production process [11].

4.3 GLASS TRANSITION TEMPERATURE

Although a wide range of crystalline polymers are used in the production of powder coatings, their use is almost exclusively restricted to the thermoplastic coatings. Thermosetting powder coatings have become more important because of the reasons discussed in Chapter 3.

While the crystalline melting temperature is one of the decisive factors in choosing a suitable thermoplast as binder for a powder coating, in the case of thermosetting powder coatings that use mainly amorphous polymers the glass transition temperature of the coating composition is a parameter to which the resin and paint chemist must pay great attention. It influences directly or indirectly the physical and chemical stability of the coating composition during storage, the rheological behaviour during the production and film formation, and finally the development of internal stress in the cured film during its service life.

Although the glass transition temperature (T_g) as an indication of a phase change resembling a thermodynamic second-order transition is exhibited by many low molecular weight compounds, it is a parameter that is considered typical for polymers. Very often, those approaching polymer science for the first time are not familiar with this term [12]. In the case of a second-order transition, the plot of a primary quantity (specific volume, for example) shows a change in slope, while the plot of a secondary quantity (such as the expansion coefficient) shows a sudden jump. In contrast, in the case of a first-order transition the primary quantities (specific volume, for example) show an abrupt change in their values.

Thermodynamically speaking, the T_g is not in fact a true second-order transition, since it is kinetically controlled. The value of the glass transition temperature is dependent on the rate of cooling or heating. Although from an exact thermodynamic point of view the classification of T_g as a second-order transition cannot be fully defended, it is very suitable for the paint chemist. It is directly related to the specific and the free volume of the polymers helping to explain many phenomena in the paint field. At least, the definition of the T_g as the temperature above which there is an increase in the temperature coefficient of expansion of the polymer, is much better than that which defines T_g as the temperature below which the polymers are brittle and above which they are flexible [13]. It is not necessary to go far into the powder coating world to find evidence that the latter definition is wrong. As will be seen, all powder coatings in cured states have T_g much higher than their exploitation temperature, but they still exhibit a certain flexibility. On the other hand, the definition of the T_g by means of a brittle and flexible state is short and simple, and possibly because of this it is widely used.

There are several different theories of the glass transition temperature that can be classified into three main groups [14]: the free volume theory, the kinetic theory and the thermodynamic theory. The concept of the free volume and the free volume theory of the glass transition is very convenient for elucidating many phenomena in the paint world, and is probably the most acceptable for the paint chemist.

The term free volume, defined as the volume not occupied by the molecules making up the material, was introduced in the 1930s by Eyring [15] who assumed that the molecular motion in the bulk state depends on the presence of holes into which the molecules can jump. Moving into these holes, the molecules and the holes exchange their places which results in a flow of the material. This concept was applied to the polymers in the 1950s by Fox and Flory [16, 17]. They found that the value of the specific free volume above T_g, in the case of polystyrene, is proportional to the temperature only for a polymer with an infinite molecular weight. However, below T_g, this relation is valid independently of the molecular weight. A very obvious conclusion is that below T_g the configurational arrangement of the polymer segments is the factor determining the free volume and that it is not dependent on the molecular weight and temperature. In other words, below T_g the free volume reaches a constant value and does not decrease as the material is cooled further. The contribution of the chain ends to the creation of a free volume differs from that of the other polymer segments. When the molecular weight is lowered, the contribution of the chain ends to the free volume is more emphasized, and, in fact, the dependence of the free volume on molecular weight is valid only for high molecular weight polymers.

Polymers exibit two types of motion: a localized mobility of the segments called segmental motion and total mobility of the molecule called molecular motion. In a glassy state the motions in the polymeric materials are restricted to vibrations on a short range and rotational movements. Below T_g only one to four chain atoms are involved in the motions. Reaching T_g with an input of thermal energy, vibration of segments on a long range in a coordinated manner begins. From a plot of T_g versus the molecular weight between the crosslinking points (M_c) it has been calculated that this segmental motion in the polymers involves 10 to 50 chain atoms [18–20]. Above T_g the vibrations of the segments are strong enough to free the neighbouring segments, thus creating new empty holes, which contributes to an increase of the free volume of the polymer. The overall result is a change in the slope of the curve representing the dependence of the specific volume of the polymer on temperature. This is the so-called rubbery or viscoelastic state. If the system is not crosslinkned, the molecules jump into the holes in a random Brownian motion exhibiting mobility on a molecular scale. When stress is applied, the jumps are in preferred directions relieving the stress, and the polymeric material flows.

4.3.1 T_g AND POWDER STABILITY

Assume now a layer of powder coating particles exposed to a certain pressure derived from the weight of the powder particles above. If the T_g of the powder is higher than the storage temperature, because of the absence of segmental mobility there is no diffusion of the material on a segmental or molecular scale between different particles. In the case of having T_g lower than the storage temperature, the segmental mobility is high enough to produce, in the long term, a considerable degree of interpenetration of the molecular chains between different powder particles, resulting in blocking of the same. This phenomenon is known as physical instability of the powder coating. Therefore a high T_g value of the powder coating is a prerequisite for good physical stability. It is difficult to judge what should be the minimum T_g that guarantees good powder stability. It is dependent, of course, on the storage temperature of the coating. For practical reasons it is widely accepted that the T_g of the powder coatings should not be lower than 40°C.

In the case of thermosetting powder coatings, the glass transition temperature also influences the chemical stability of the system. Since both reaction partners, the resin and the crosslinker, are present in the coating, it could be expected that a certain reaction will take place during storage. From a thermodynamic point of view, a reaction takes place only when the reactants that meet each other possess enough energy, at least equal to the activation energy. When the powder coating has a T_g higher than the storage temperature, the probability that two functional groups belonging to the resin and the crosslinker meet each other is very low because of the restricted mobility of the polymer segments. Paints, which must be formulated as two-component systems because of limited storage stability in the case of liquid coatings, can exist as one-component systems in the form of a powder. Good examples are polyester/epoxy powder coatings based on bisphenol A epoxy resins and polyesters with carboxyl functional groups. While they can be formulated into powder coatings with practically unlimited storage stability, the solvent based counterparts have a storage stability of only a few days at room temperature or a few months when not catalysed [21, 22].

4.3.2 T_g AND THE MELT VISCOSITY

According to the free volume theory, the glass transition temperature is related to the unoccupied volume by the polymer segments or molecules. The latter is directly proportional to the difference between the glass transition temperature and the temperature at which the viscosity of the polymer is monitored. The unoccupied volume facilitates the flow of the material or, with viscosity of the polymer defined as a resistance to flow, should be directly dependent on T_g. Although many aspects of the rheological behaviour of polymers in melts still have to be elucidated, the dependence of the melt viscosity on T_g is a fact that was recognized, and was the reason for many attempts to find expressions giving a quantitative description of this dependence. Among several empirical, semiempirical and theoretical equations, the relation suggested by Williams, Landel and Ferry, or the well-known WLF equation [23], best fits the experimental data.

$$\log \mu_T = \log \mu_{T_g} - \frac{C_1(T - T_g)}{C_2 + (T - T_g)} \tag{4.7}$$

where

μ_T = viscosity of the polymer with the glass transition temperature T_g at temperature T

μ_{T_g} = viscosity of the polymer at a temperature equal to its glass transition temperature

C_1, C_2 = constants

Although it has been suggested that C_1 and C_2 have universal values, they differ in fact for different polymers. When data are not available, with good approximation, values for $C_1 = 17.44$ and $C_2 = 51.6$ can be used.

The validity of the WLF equation is restricted to the temperature interval $(T - T_g) < 100°C$ [24, 25]. It has been presumed that for $(T - T_g) > 100°C$ the melt viscosity is governed more by the thermal activation of the bond rotation and secondary segment interactions than by the free volume availability, which makes an Arrhenius type of equation more suitable for such a case [26]:

$$\log \mu = \log K + \frac{E_a}{RT} \tag{4.8}$$

where E_a is the activation energy of viscous flow and K is the constant.

The WLF equation cannot be directly used to calculate the absolute values for the viscosity of the powder coatings, because of the complicated interdependence of the elements influencing the rheology of the melt, but it can be very useful for comparing the expected values for resins with different glass transition temperatures.

For example, a comparison of viscosities of two resins with $T_g = 60°C$ and $T_g = 30°C$ obtained by using the WLF equation for a temperature of 150°C gives the following results [27]:

$$\mu_{T_g = 60} = \mu_{T_g} \times 10^{-11} \tag{4.9}$$

$$\mu_{T_g = 30} = \mu_{T_g} \times 10^{-12} \tag{4.10}$$

Since the viscosities of different polymers at T_g are of the same order (1×10^{13}

PARAMETERS INFLUENCING POWDER COATING PROPERTIES

poises) the comparison shows that the resin with a T_g of 30°C will have at 150°C a ten times lower viscosity than the other one.

The T_g of the powder coating resins are much higher than the T_g of the resins for solvent borne coatings, for reasons related to the physical and chemical stability of the coatings, as discussed in the previous section. Consequently, the melt viscosities of the powder coatings are several hundred times higher than those of the solvent borne coating at the curing temperature. The fact that there is no solvent present to reduce the viscosity during the film formation emphasizes this difference even more. This has an adverse effect on the flow and the levelling characteristics of powder coatings.

4.3.3 T_g AND THE THERMAL STRESS DEVELOPMENT

The internal stress development in a cured coating is not a characteristic typical only of powder coatings. This phenomenon exists in almost all types of coatings, but it is certain that powder coatings, together with the radiation curing materials, suffer the most.

The reasons for internal stress development in the coatings during curing and its determination are described in many papers [28–34]. The dimensional changes in the coating during film formation are the main reason for the stress formation. The shrinkage of the coating during crosslinking is very noticeable in the case of the radiation curing coatings, where the stress can be a reason for development of microcracks in the cured film or in extreme cases can cause even a detachment of the coating from the surface. Shrinkage of the other coating materials, except the high solids coatings, is not of such a magnitude.

Powder coatings, especially the carbon/epoxy curing types, do not undergo major shrinkage due to crosslinking. The ring opening reaction of crosslinking in some cases is even followed by expansion [35, 36]. Stress in this case is mainly of a thermal character, and is related to the high curing temperature combined with a high glass transition temperature of the cured films.

The expansion coefficients of the coating and the substrate are in general different. Therefore, during cooling of the thermally cured coatings the dimensional changes in the substrate and the coating are different. As a result of this difference an internal stress will develop as the film cools. When the exploitation temperature of the coating is above T_g, the internal stress is completely relieved due to the fast relaxation processes. Below T_g the relaxation processes are very slow, and the coating layer cannot follow the dimensional changes of the substrate. Therefore a certain stress will build up, proportional to the difference between the T_g of the cured film and the temperature to which the coating layer is cooled down after the curing; this is given by the following equation [34]:

$$S_T = \frac{M(\alpha_f^T - \alpha_s^T)(T_g - T)}{1 - v} \quad (4.11)$$

Table 4.3 Stress values at 21°C and 0% relative humidity

Type of coating	T_g (°C)	S (MPa)
Polyurethane thermoplast	<0	0.24
Air drying alkyd	<0	0.60
Conventional polyester/melamine	25	0.80
High solids polyester/melamine	33	2.10
Powder coating	62	3.60

where

S_T = thermal stress at temperature T
M = Young's modulus of the coating
α_f^T = thermal expansion coefficient of the coating at temperature T
α_s^T = thermal expansion coefficient of the substrate at temperature T
v = Poisson's ratio of the coating

Values of internal stresses for different types of coatings are listed in Table 4.3 [37]. It is clear that powder coating has the highest value for the stress, but is still comparable with one of the high solids coatings.

Although the stress development in principle is not a typical problem related to powder coatings, as it is with radiation curing coatings, for example, it should be considered especially in cases where problems concerning adhesion of the coating to substrates are encountered.

4.3.4 MOLECULAR WEIGHT AND T_g

For a given molecular structure, the T_g of the resins depends up to a certain value on the molecular weight. The influence of molecular weight can be expressed by the equation of Fox and Loshaek [38] as

$$\frac{1}{T_g} = \frac{1}{T_{g\infty}} - \frac{a}{x} \qquad (4.12)$$

where a is a constant and x is the degree of polymerization, or more conveniently for practical use by

$$T_g = T_{g\infty} - \frac{K}{M_n} \qquad (4.13)$$

where $T_{g\infty}$ is the glass transition temperature for a polymer of infinite molecular weight and K is a constant for a polymer with a given molecular structure [39, 40]. The original form of this equation comes as the result of the theoretical

PARAMETERS INFLUENCING POWDER COATING PROPERTIES   187

analysis of Fox and Flory [16]. It shows that the glass transition temperature rises with an increase of the molecular weight, levelling off asymptotically to a certain value typical for a specific polymer. This can be qualitatively explained by the free volume theory of glass transition, assuming that the chain ends contribute more free volume to the system than similar segments along the chains.

Since the resins used for manufacture of thermosetting powder coatings are of relatively low molecular weight, in most cases around 3000, they are in principle in the region where the slope of the curve representing the T_g–molecular weight dependence is steep. This means that a small increase or decrease in molecular weight could have a considerable influence upon the T_g of the resin.

The molecular weight distribution will influence the T_g of the resins. This is to be expected and is easily explained. The low molecular fractions will decrease considerably the T_g of the resin acting as plasticizers. Kumler, Keinath and Boyer [41] found for polystyrenes of different polydispersities that the T_g values are a function of M_n rather than M_w. For example, polystyrene with M_n of 2740 and M_w of 25 100 has a T_g of 43°C, while the blend with M_w of 650 and M_w of 23 300 has a T_g of -25°C. Holda [42] considers this phenomenon as an explanation for the lower T_g of the polyesters for powder coatings based on phthalic anhydride, in comparison with those where terephthalic and isophthalic acids are used as diacids.

The glass transition temperature in principle should be influenced by the chain entanglements which restrict the mobility of the system. The molecular weights of most polymers for thermosetting powder coatings fall below the critical value at which there is a change in slope of the curve representing the viscosity versus log M_w. This change in slope is an indication of the contribution of the entanglements to the rheological behaviour of the polymer. Therefore, for most thermosetting powder coating resins, the existence of some critical molecular weight, concerning molecular mass–T_g dependence, that may be attributed to entanglements should not be expected. In fact, there is no extensive evidence of such behaviour of the polymers, although hints of this character have been made in the literature [43, 44].

According to the free volume theory, branching of the polyesters should contribute to a higher free volume and decrease the glass transition temperature. There is no broad evidence for this phenomenon in the literature. In a paper dealing with saturated polyesters for powder coatings, Holda [42] gives a comparison between different polyesters with a different degree of branching, concluding that the branched types have a lower T_g than the linear ones. The presented results are easily discussed, because the polyesters in question differ in molecular weights, and moreover their molecular masses are in the region of 2000, when small differences between them result in large differences in T_g. It is likely that the influence of the branching upon the T_g in the case of powder coating resins is too small to play an important role.

4.3.5 CHEMICAL STRUCTURE AND T_g

The influence of the chemical structure on T_g is not fully elucidated, although some general rules are well established. Since the T_g is related to the cooperative motion of the groups or atoms making up the polymer chain, any molecular parameter affecting the chain mobility will influence the glass transition temperature. More precisely, any group or structure that reduces the chain mobility will increase T_g.

Polymers with perfectly repeating units prone to crystallization usually show a higher T_g than should be expected. This is the result of limited chain mobility in the crystalline state. Some of these polymers crystallize very fast and can not be quenched easily in an amorphous form. Very often their T_g is a matter of disagreement. For example, in the case of polyethylene, the published values for T_g vary from -30 to $-108°C$ [45]. Polymers with carbon–carbon bonds in the main chain, without side groups, have a very low T_g, as mentioned for polyethylene. The rigid bulky substituents in the chain decrease the flexibility and affect the steric hindrance to rotation, increasing T_g. Thus, for example, in contrast to polyethylene, polypropylene, polystyrene and poly(α-methylstyrene) have T_g values of -12, 100 and 180°C respectively.

The nature of the side chain also plays quite a large role. Long side chains in the case of acrylic polymers, which act as internal plasticizers, have little effect on the freedom for rotation, since they are nearly flexible. On the other hand, they keep the main chains further apart, contributing more free volume. According to the free volume theory the result will be a decrease of the glass transition temperature in the series of different polyacrylic or polymethacrylic esters, which is the case in practice. Polymethyl methacrylate, for example, has a T_g of 105°C, and the higher polymethacrylic esters like polyethyl, polybutyl and polyhexyl methacrylate have T_g values of 65, 22 and $-5°C$ respectively.

The chain stiffness and the nature of the constituents in the heterochain polymers, like polyesters for example, play an enormous role concerning the T_g of the polymer. Polytetramethylene oxalate has a T_g of $-5°C$, while the corresponding polytetramethylene terephthalate has a 50°C higher glass transition temperature. Again, the presence of side bulky groups has the same influence as in the case of homochain polymers. Polyhexamethylene terephthalate ($T_g = 8°C$) is a good example to be compared with polyneopentyl glycol terephthalate ($T_g = 63°C$).

Intermolecular attractions introduced by the presence of polar groups or hydrogen bonding increase the T_g of polymers, as one would expect. Atactic polystyrene with a T_g of $-19°C$ should be compared with polyacrylonitrile with a T_g of 104°C. It has been suggested that the influence of the polarity of the polymers (in terms of their cohesive energy density, CED) to the T_g should be expressed as

$$T_g = K_1 (CED)^{0.5} \tag{4.14}$$

where K_1 depends on the structure of the polymer [46].

PARAMETERS INFLUENCING POWDER COATING PROPERTIES   189

The hydrogen bonding will have the same effect as the polarity, increasing the T_g. Polyhexamethylene adipate has a glass transition temperature of $-63°C$, while polyhexamethylene adipamide has a T_g of $80°C$. The effect of the polar groups on the T_g plays an important role, since in most cases the thermosetting powder coating resins contain functional groups of a polar nature, or groups introducing hydrogen bonding forces.

In systems without solvents, like the powder coatings, one could expect that the intermolecular attraction forces between the binders and the pigments and fillers should increase the T_g. In an extensive review on the effect of the fillers on T_g, Toussaint [47] noted that it is difficult to find a predictable dependence. The fillers can increase or decrease the glass transition temperature, or have no remarkable effect. The increasing effect on T_g resulting from the buildup of reversible internal structures between the pigments and fillers on one side and the resin on the other, can be compensated by looser packing, which is an explanation for the noticed decrease of density of the filled polymers [48–50]. In general, the influence of the pigments and fillers on the T_g of the powder coatings, in concentrations in which they are used, is too low to be considered from a practical point of view.

4.3.6 GLASS TRANSITION TEMPERATURE OF POLYMER BLENDS

The phenomenon of decreasing the glass transition temperature of powder coatings by incorporating crosslinkers is well known by coating chemists. Very often it is the reason for having unstable systems during storage. The possibility of predicting the T_g of a blend resin/crosslinker should be very helpful, especially in developing new powder coating formulations or examining new crosslinkers.

The hardeners that are used for powder coatings are solid materials at room temperature. Many of them are low molecular weight compounds, also having a low glass transition temperature. The fact that they are solids at room temperature is related to the molecular symmetry which allows crystallization. In a blend with the resin after melting in the extruder and cooling down to the storage temperature (which is always lower than the T_g of the coating), the crosslinker does not recrystallize, because of the restricted mobility in the system below its glass transition temperature. Since there is no phase separation, the resin and the crosslinker in the powder coating are in a form of homogeneous blend exhibiting only one T_g. Because the T_g of the crosslinker is generally low, the resulting glass transition temperature of the blend is always lower than that of the resin.

Many attempts have been made to relate the T_g of the blends to the blend composition. The results have shown that, in principle, equations that are used for predicting the T_g of the amorphous random copolymers are valid for polymer blends or blends of polymers with low molecular compounds.

A well-known formula for predicting the glass transition temperatures of polymer blends or amorphous random copolymers is derived by Gordon and Taylor:

$$T_g = \frac{w_1 T_{g1} + K w_2 T_{g2}}{w_1 + K w_2} \quad (4.15)$$

where T_{g1} and T_{g2} are the glass transition temperatures of the homopolymers present in weight fractions w_1 and w_2 [51]. The constant K is related to the thermal expansion coefficients of the homopolymers in the rubbery α_r and glassy α_g states and is given by

$$K = \frac{\alpha_{r2} - \alpha_{g2}}{\alpha_{r1} - \alpha_{g1}} \quad (4.16)$$

A very simple and useful equation is that of Fox [52] involving only the glass transition temperatures of the blend constituents and their weight fractions:

$$\frac{1}{T_g} = \frac{w_1}{T_{g1}} + \frac{w_2}{T_{g2}} \quad (4.17)$$

An empirical equation describing the glass transition temperature of one phase system involves the parameter K, which, once determined for one ratio of the components in the blend, can be used for the whole concentration range [53]:

$$T_g = w_1 T_{g1} + (1 - w_1) T_{g2} + w_1 (1 - w_1) K \quad (4.18)$$

The value of K can also be determined from the initial slope:

$$\frac{dT_{g(0)}}{dw_1} = T_{g1} - T_{g2} + K \quad (4.19)$$

Couchman [54–57] has developed, on the basis of the thermodynamic theory, an equation for predicting the glass transition temperature of a blend, knowing the heat capacities of the components at constant pressure C_{p1} and C_{p2}:

$$T_g = \frac{w_1 C_{p1} T_{g1} + w_2 C_{p2} T_{g2}}{w_1 C_{p1} + w_2 C_{p2}} \quad (4.20)$$

Supposing that $C_{pi} T_{gi} = $ constant, the equation of Couchman takes the form of the previously given Fox equation.

If the heat capacities of the blend constituents are equal, or close to each other, then the following formula, which is also used semiempirically, results from the Couchman equation:

$$T_g = w_1 T_{g1} + w_2 T_{g2} \quad (4.21)$$

Another relatively simple expression gives the following relation between the T_g of the blend and the blend composition [58]:

$$\ln T_g = w_1 \ln T_{g1} + w_2 \ln T_{g2} \quad (4.22)$$

This equation can also be treated as a special case of that of Couchman, assuming that $C_{p1} = C_{p2}$ and avoiding the use of the expansion $\ln(T_g/T_{g1}) = T_{g1}/T_{g2}$ and $\ln(T_{g2}/T_{g1}) = T_{g2}/T_{g1}$, which is involved in the derivation of the Couchman equation.

4.4 VISCOSITY

The discussion about molecular weight and glass transition temperature of the binder, related to the desired properties of the powder coating, leads to the conclusion that both parameters should have as high values as possible. This is in favour with better powder stability and a powder coating formulation with mechanical properties which are not too sensitive to the possible deviations from the best coating composition. However, high molecular weight combined with high glass transition temperature will lead to an increase of the melt viscosity of the polymer, adversely affecting the processing performances of the powder coating during its manufacture and the flow properties during film formation.

The viscosity dependence on T_g was discussed in the previous chapter, and can be quantitatively described by the WLF equation.

Melt viscosity of the polymers will increase with increasing molecular weight, as could be expected. The plot of the logarithm of the melt viscosity versus the logarithm of the molecular weight in arbitrary units (Figure 4.4) exhibits two distinct regions: region A in which the melt viscosity in terms of molecular weight

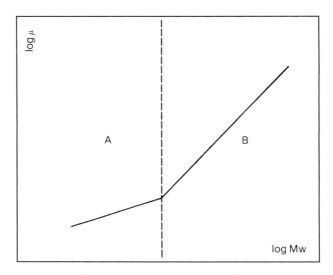

Figure 4.4 Melt viscosity as a function of molecular mass

can be expressed as

$$\log \mu = A \log M_w + K_1 \qquad (4.23)$$

and region B for which the following expression is valid:

$$\log \mu = B \log M_w + K_h \qquad (4.24)$$

where M_w is the weight average molecular weight of the polymer, μ is viscosity and K_1 and K_h are constants for low and high degrees of polymerization.

Fox [59] has shown that below a certain critical value of the weight average molecular weight, melt viscosity is proportional to M_w ($A = 1$). Above this value, melt viscosity of the high molecular weight polymers depends on M_w to the 3.4 power ($B = 3.4$).

The critical value for M_w, above which, due to the chain entanglements, a viscosity increase follows equation (4.24), differs for different polymers. However, it has been found that, in general, the critical weight average molecular weight corresponds to a weight average degree of polymerization of 600 or a number average molecular weight of 30 000 [14, p. 389]. It is interesting that this number average molecular weight is close to the low end of molecular weights at which most of the polymers exhibit good mechanical properties, as mentioned in Section 4.1.

The number average molecular weights of powder coating resins for thermosetting powder coatings are far from this region. Therefore, equation (4.23) can be considered as the correct way to describe melt viscosity dependence on molecular weight.

In contrast to equation (4.3), which indicates that the mechanical properties of the polymers level off after a certain molecular mass has been reached, equation (4.23) indicates that with increasing molecular mass the viscosity increases exponentially to infinity.

In an article dealing with the melt viscosity of saturated polyester resins for powder coatings, Holda [42] has proposed that the viscosity–molecular weight relationship obeys the following equation:

$$\log \mu = A \log M_n + B \qquad (4.25)$$

Holda relates the failure of the general equation (4.23) involving M_w to the choice of technique used to determine the M_w in the case of the powder coating polyesters in question. He also points out the contribution that polarity and hydrogen bonding make to the melt viscosity of the resins with relatively low molecular weights.

4.4.1 VISCOSITY AND PROCESSING PERFORMANCES

Viscosity of the resins has a direct influence on the processing characteristics of the powder coatings during hot melt compounding. The probability that a

PARAMETERS INFLUENCING POWDER COATING PROPERTIES 193

pigment cluster will be deagglomerated during extrusion is directly proportional to the energy density present to overcome the forces holding the primary particles together [60]. The energy supplied by the motor in the case of smearing types of machines like the extruder is proportional to the viscosity of the media in which the pigments are to be dispersed. Also in these types of machines, where the pigments are dispersed in a gradient of shear, the shear stress will be the determining quantity with respect to the efficiency of the deagglomeration process. The equation that describes the mutual dependence between the shear stress and the viscosity is as follows:

$$\tau = \mu D \qquad (4.26)$$

where

τ = shear stress
μ = dynamic viscosity
D = shear rate

In other words, high resin viscosity is favourable considering the efficiency of the extruder as a dispersing machine. On the other hand, the wetting of the pigment, which is in fact displacement of the air or other contaminants (water, for example) adsorbed on the surface of the pigment particles by the resin, is controlled by the velocity U with which the resin penetrates the pores on the pigment surface given by [61]

$$U = \frac{Kr}{\mu} \qquad (4.27)$$

where r is the pore radius, μ is viscosity of the resin and K is a constant that comprehends the surface tension coefficient of the resin. It is quite obvious that the viscous resin will facilitate the wetting of the pigments. From these two equations it can be concluded that there is an optimum viscosity at a given temperature (which is again restricted by the reactivity of the system) that provides the best dispersing efficiency of the extruder. This optimum depends on the type of extruder, the coating formulation and the types of pigments used.

4.4.2 VISCOSITY AND FILM FORMING PROPERTIES

The layer of powder coating, when applied on the surface to be coated, cannot be regarded as a continuous system. This is, among others, one of the principal differences from the solvent borne coatings, where the film formation comes as the result of evaporation of the solvent from a continuous low viscous layer of the applied coating, followed by subsequent crosslinking in the case of thermosetting coatings.

Conversely, film formation of powder coatings results from melting powder particles and their coalescence to form a continuous film. This is followed by flow of the formed film, to obtain a smooth surface from the irregular one produced after the melting.

Different workers considering the coalescence of particles during film formation have derived equations which describe this phenomenon with considerable success [62–66]. The derivations of these equations are beyond the scope of this book. Only two of them will be presented to illustrate the influence of viscosity on the speed of sintering during film formation of powder coatings. In both equations the gravitational forces are neglected, and it is assumed that for films thinner than 100 μm the single driving force for coalescence is the surface tension of the coating [67].

One of the first equations derived by Frenkel [62] relates the time (t) necessary for two spherical particles to coalescence, radius of the particle (R), radius of the contact area between the particles (x), viscosity of the material (μ) and its surface tension (τ):

$$\left(\frac{x}{R}\right)^2 = \frac{3\tau t}{2R\mu} \qquad (4.28)$$

Nix and Dodge [63] have proposed something different, assuming that the surface tension is the only reason for coalesce and flow and viscosity of the molten particle the only resistance to flow. In such a case they propose that the pressure causing two particles to flow together will be proportional to the surface tension divided by the average radius of the curvature (R_c). The latter is in a first approximation equal to the mean particle radius. According to them, the time necessary for coalescence will be given by the following equation:

$$t = f\left(\frac{\mu R_c}{\tau}\right) \qquad (4.29)$$

Using an instrumental technique to compare the surface profiles of coatings with different viscosities, they have confirmed in principle the validity of the above equation, admitting that some other factors concerning the pigment concentration and dispersion have to be included for a better quantitative description of the sintering process.

Although these equations refer to a simplified model of a single spherical particle, which is not the exact description of a real powder coating composed of irregular particles of different size, they both show that the higher the melt viscosity, the longer the time necessary for particles to coalesce.

Melt viscosity of the powder coating also influences the possibility of the coating flowing during the second stage of the film forming process, after the sintering has been accomplished. The equation of Orchard [68], which was originally derived for liquid coating systems applied by brush, can in principle be used here to illustrate the influence of the viscosity on the flow. This equation connects the time (t) necessary for a coating layer of a thickness (h), melt viscosity (μ), surface tension (τ) and a wavelength of striation (λ), to decrease the initial

PARAMETERS INFLUENCING POWDER COATING PROPERTIES

amplitude of striation ($a°$) to a value (a_t), as presented in Figure 4.5:

$$\ln \frac{a_t}{a°} = -\frac{16\pi^4 h^3}{3\lambda^4} \int_0^t \frac{\tau}{\mu} dt \qquad (4.30)$$

Supposing that the viscosity of the powder coating does not change with time at a given temperature, which is the case with the thermoplastic powder coatings, then the integrated form of Orchard's equation will be

$$t = \frac{3\lambda^4 \mu \ln a°/a_t}{16\pi^4 h^3 \tau} \qquad (4.31)$$

The integrated form of the term dt/μ in the Orchard equation known as integrated fluidity, often designated by F, is

$$F = \int_0^t \frac{dt}{\mu} \qquad (4.32)$$

Spitz [69] has shown in the case of thermoplastic powder coatings that the experimental results fit Orchard's equation. On a doubly logarithmic plot against $h^n t/\mu$, where $n = 3$, the wavelength increased linearly with a slope of 0.25.

Hannon, Rhum and Wissbrum [70] confirmed the validity of the equation of Orchard for thermosetting powder coatings performing rheological measurements during curing using a thermomechanical spectrometer and the obtained data for calculation of the integrated fluidity F. They have developed a model which takes into consideration the viscosity change (dx) in non-isothermal curing of a powder coating due to the change of the temperature $(\delta X/\delta T)_t$ and the network formation during curing $(\delta X/\delta t)_T$:

$$dX = \left[\frac{\delta X}{\delta T}\right]_t dT + \left[\frac{\delta X}{\delta t}\right]_T dt \qquad (4.33)$$

In their model they have assumed that the activation energy for viscous flow and the cure rate depend only on temperature. These assumptions allow numerical

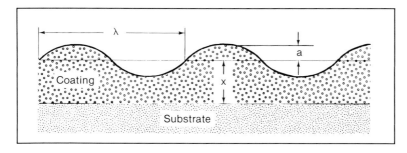

Figure 4.5 Illustrative explanation of the terms involved in the equation of Orchard

integration of the above equation. Knowing $\mu(t)$ from equation (4.33) the integrated fluidity can be calculated by numerical integration and replaced in the Orchard equation. The model was successfully tested on a commercial white TiO_2-pigmented epoxy powder coating, using a thermomechanical spectrometer to measure the time and temperature dependence of the viscosity, showing good agreement with experimental results.

An improved model describing the non-isothermal viscosity changes in unpigmented acrylic systems has been developed by Kuwano, as reported by Nakamichi [71]. He has taken into consideration the influence of an increase in molecular weight during curing on the activation energy of viscous flow. This effect was disregarded in the model of Hannon.

The equation of Orchard shows that for good levelling properties of the powder coating, a low melt viscosity is desirable, as was also the case for fast coalescence of the powder particles in the first stage of the film formation process. However, too low a viscosity leads to bad edge coverage and even sagging on vertical surfaces. Walz and Kraft [72] have suggested that a good compromise should be within a viscosity range of 6000 and 10 000 mPa s.

The flow of powder coatings cannot be completely described by integrated fluidity only, because it is a more complex phenomenon. However, for comparison, integrated fluidity can be quite a useful parameter. Thermoviscoelastic analysis (t.v.a.), described in Section 3.2.3, is a very suitable experimental method for determination of integrated fluidity of powder coatings. The temperature profile can easily be adjusted to simulate the normal stoving conditions. The area under the reciprocal viscosity–time curve gives the value for the integrated fluidity.

Nakamichi [71] and others [73–75] have used the rolling ball technique to determine the viscosity changes during the non-isothermal curing of acrylic resins with different functionalities and melt flow index as an indication of the viscosity. Although Nakamichi addresses some remarks to the reproducibility of the technique, it seems to be a very simple and practical method for determination of the non-isothermal viscosity change during the curing process, with even better simulation of the real stoving conditions than with thermoviscoelastic analysis.

4.5 RESIN/CROSSLINKER RATIO

The ratio between the resin and the crosslinker has already been discussed in the section dealing with the molecular mass of the resin. It was stated that the proper ratio between the resin and the hardener is very important for obtaining a certain molecular weight of the cured material, and indirectly for obtaining good mechanical properties of the coating. Since the mechanical properties are directly proportional to the molecular weight up to a certain point, in principle the higher the molecular weight of the cured coating, the better. The highest molecular weights of the cured coating can be obtained when the resin and the crosslinker are in stoichiometric proportion. Because powder coating resins and crosslinkers are usually of a much lower functionality compared to the solvent based systems, it is very important to have a proper amount of crosslinker in the coating composition. The self-condensation reactions which are present in the polyester/melamine or acrylic melamine systems leading to harder films do not proceed in the powder coating systems based on epoxy or polyisocyanate crosslinkers. Therefore more hardener in a powder coating does not mean a harder film. The result can even be the opposite—a softer film without flexibility and impact resistance.

The resin producer usually gives the parameters of the resin important for determination of the stoichiometric amount of the crosslinker in the coating formulation. These parameters are the acid value of the resin in the case of carboxyl functional binders and the hydroxyl value in the case of resins that are to be crosslinked by polyisocyanates. The following formulas can be useful for calculating the amount of crosslinker in the coating formulation:

$$EC = \frac{R \times AV \times EEW}{56\,100} \qquad (4.34)$$

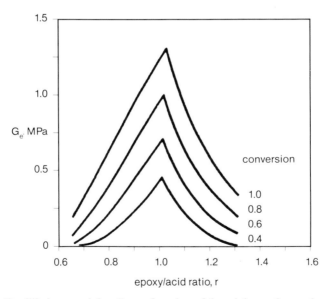

Figure 4.6 Equilibrium modulus G_e as a function of the mixing ratio r and conversion for polyester with functionality of 2 and TGIC [76]

$$\mathrm{IC} = \frac{0.0749 \times R \times \mathrm{HV}}{\mathrm{NCO}} \quad (4.35)$$

where

EC = amount of epoxy crosslinker (g)
IC = amount of isocyanate crosslinker (g)
R = amount of carboxyl or hydroxyl functional resin (g)
AV = acid value of the resin (mg KOH/g)
HV = hydroxyl value of the resin (mg KOH/g)
EEW = epoxy equivalent weight of epoxy crosslinker (g)
NCO = content of isocyanate groups in the isocyanate crosslinker (%)

Powder coatings cured by epoxy/carboxyl curing reactions, like TGIC powders, for example, are extremely sensitive to the ratio between the reactive groups. Figure 4.6 [76] represents the variation of the equilibrium modulus of the polyester/TGIC system as a function of the reactant stoichiometry (r) and degree of conversion during curing. It is obvious that a combination of incomplete

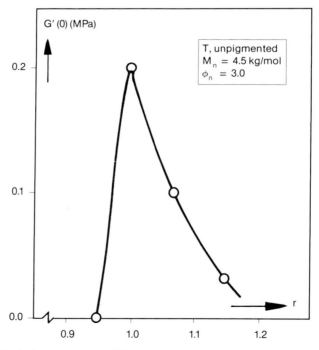

Figure 4.7 Variation of the equilibirum modulus with the acid/epoxy ratio at full conversion for polyester with functionality of 2 and TGIC [77] (Reproduced by permission of DSM Resins BV)

curing and non-stoichiometrical ratio between the reactive groups in the coating composition can be disastrous for mechanical properties. Even when the curing reaction proceeds completely, the ratio between the reactants is of extreme importance. In Figure 4.7 [77] it can be seen that, in the case of polyester/TGIC powder coatings, a shortage of TGIC in the coating composition has a more dramatic effect on the mechanical properties than an excess of the same. Therefore, a small excess of TGIC in polyester powder coatings for outdoor application is desirable. Greidanus [76] explains this phenomenon by a higher consumption of the epoxy groups than that calculated with respect to the amount of available carboxyl groups, assuming that part of the oxirane rings undergoes esterification with the formed hydroxyl groups during curing. Although this could be a reason for such a behaviour, the difference in crosslinking density between compositions with an excess and shortage of three functional crosslinkers like TGIC should also be considered.

4.6 CATALYST LEVEL

The kinetics of the curing process in thermosetting powder coatings is greatly influenced by the choice and level of the catalyst. The normal tendency towards lower curing temperatures and shorter curing times in order to reduce the energy consumption during curing stimulates the paint chemist to search for systems which offer a combination of high curing rate and acceptable storage stability. On the other hand, in the case of powder coatings, premature curing during the processing in the extruder should be minimized in order to avoid undesirable viscosity increase leading to inferior flow and levelling properties. The influence of the catalyst on the integrated fluidity as an indication of the flow properties of certain coating compositions has already been mentioned (Section 4.4.2). The graph of a dynamic DSC run of thermosetting powder coating exhibiting good flow (Chapter 3) is characterized by a horizontal plateau between the temperature onset flow (T_{of}) and temperature onset cure (T_{oc}). This means that the ideal catalyst should not be active at room and processing temperatures, and will manifest its activity at the stoving temperature, permitting shorter cure times.

In contrast to the solvent borne coatings where the activity of the catalyst at room temperature can cause considerable problems with respect to the storage stability, chemical stability of powder coatings during storage is significantly less influenced by the reaction kinetics. The storage temperature of powder coatings is always lower than the glass transitions of the coating composition in order to avoid blocking of the powder particles. In such conditions, the mobility of the system is restricted to vibrations on short range and rotational movements only, which are not favourable conditions for reaction. Therefore, curing systems such as epoxy/carboxyl, which cannot be used in one-component solvent borne coatings because of storage instability, can be employed without incurring large problems in powder coatings. However, catalysis of the curing has a considerable influence on the processing stability of the powder coating and its flow and levelling during film formation.

Since in the early stage of the reaction at relatively low viscosities the curing rate is determined predominantly by the reaction kinetics, a logarithmic form of the following well-known Arrehenius equation, which describes the temperature dependence of the rate constant, can be quite helpful for providing general guidelines in choosing the correct catalyst:

$$\ln K = \ln A - \frac{E_a}{RT} \tag{4.36}$$

In the diagram of $1/T$, $\ln K$ in equation (4.36) represents a straight line with an intercept $\ln A$ and slope $-E_a/R$ (Figure 4.8). Let us suppose a powder coating

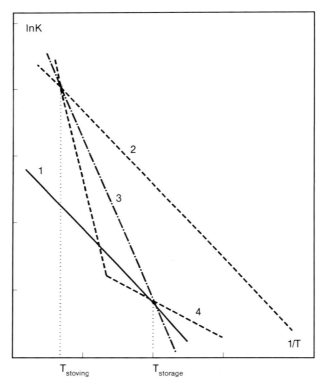

Figure 4.8 Arrhenius plot of curing reactions catalysed with different concentrations of the same catalyst (1 and 2), a catalyst with a higher activation energy and preexponential parameter (3) and a latent catalyst (4) (arbitrary units) (Reproduced by permission of DSM Resins BV)

PARAMETERS INFLUENCING POWDER COATING PROPERTIES

with reaction kinetics described by line 1 in Figure 4.8 that has, at the stoving temperature T_s, a rate constant K_{s1} and an unacceptably long curing time. The curing process could be accelerated by increasing the concentration of the catalyst, reaching the rate constant $K_{s2} > K_{s1}$. As a consequence it will have an increasing rate constant at the onset flow temperature to a value K_{of2} (line 2). In other words, more prereaction will take place during the extrusion of the powder coating, and viscosity build-up will be faster in the period until the stoving temperature is reached. This will worsen the flow properties of the powder coating composition, resulting in a film with a more emphasized orange peel effect.

The kinetics of the curing process of powder coatings has been studied by means of differential scanning calorimetry [78–80]. It has been found that the overall curing reaction could be considered to be of an order between 1 and 2 [78, 80].

Supposing for simplicity that the curing reaction is of second order; then the rate constant can easily be calculated by the following equation, which is derived from basic kinetics principles:

$$K = \frac{1/C_a - 1/C_{a0}}{t} \quad (4.37)$$

where C_{a0} and C_a are the concentrations of the reactive groups at the beginning of the reaction and after time t.

For an analysis of the system let us choose a polyester/epoxy powder coating with 70% polyester resin and 30% epoxy crosslinker, or the so-called 70/30 hybrid. Polyester resin with an acid value of 37 mg KOH/g and epoxy resin with an epoxy equivalent value of ca. 700 is a typical binder–crosslinker combination for such a system and will correspond to an initial concentration of the carboxyl groups of 0.46 equivalents/litre. In most cases powder coatings are formulated in such a way that 90% conversion during curing should be enough to obtain good mechanical properties. In order to reach 90% conversion at a typical stoving schedule of 10 min at 200 °C, the rate constant calculated from equation (4.37) should be 3.24×10^{-2} L/eqs. Five different rate constants at extrusion conditions at 100 °C are calculated supposing that 0.1, 0.5, 1.0, 5.0 and 10.0% conversion has taken place within the residence time in the extruder of 1 minute, the first one being the most favourable and the last one the worst concerning the flow and stability of the system. Using a pair of values for K_e and K_s (rate constants at extrusion and stoving temperatures), by means of the equation of Arrhenius the values for the activation energy E_a and preexponential parameter A are calculated. The results are presented in Table 4.4 and shown graphically in Figure 4.9.

Experimental results in the same range as those calculated and presented in Table 4.4 are reported by Franiau for the catalysed polyester/hybrid system [80].

Table 4.4 Rate constants, activation energies and preexponential parameters for polyester/epoxy powder coating at extrusion and stoving temperatures supposing different degrees of prereaction during the processing time

Pre reaction (%)	K_e (L/eq s)	K_s (L/eq s)	E_a (kJ/mol)	A (L/eq s)
0.1	2.78×10^{-8}	0.0324	103.6	8.8×10^9
0.5	1.81×10^{-4}	0.0324	76.1	8.1×10^6
1.0	3.65×10^{-4}	0.0324	65.8	5.9×10^5
5.0	1.90×10^{-3}	0.0324	41.6	1.3×10^3
10.0	4.01×10^{-2}	0.0324	30.6	0.8×10^2

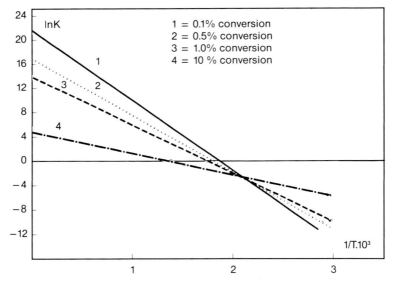

Figure 4.9 Arrhenius plots for powder coatings having different degrees of prereaction during the processing time

He has found by d.s.c measurements a value for E_a of 104 kJ/mol and for the preexponential parameter of 3.6×10^9 L/eq s.

It is obvious that in searching for a good catalyst in a proper concentration at a given stoving schedule, one should look for a combination that gives an increase in the activation energy of the system and the preexponential parameter, resulting in a series of Arrhenius plots passing the same point and representing the necessary rate constant at the stoving temperature, as presented in Figure 4.9.

The same approach of choosing a catalyst by which the reaction kinetics will be characterized by a large Arrhenius activation energy (E_a) and large pre-exponential parameter (A) to compensate for the small values of K (line 3 in Figure 4.8) has been suggested by Pappas and Hill for one-component liquid coatings [81]. For an optimal combination of a desired curing rate and good storage stability in the case of first-order curing reactions, A values of $10^{12}\,\text{s}^{-1}$ are recommended [76]. Whereas bimolecular reactions have in general A values not exceeding 10^{11} L/mol s, unimolecular first-order reactions are characterized by preexponential parameters of up to $10^{16}\,\text{s}^{-1}$. Most of the thermosetting coatings employ bimolecular curing processes. For good storage and, in the case of powder coatings, processing and levelling characteristics, these curing reactions have to be preceded by rate-controlling unimolecular reactions, utilizing latent catalysts or latent reactants. A typical example of latent reactants are blocked isocyanates, where the first-order deblocking reaction precedes the formation of the urethane linkage by a second-order condensation between the isocyanate and hydroxyl groups. An Arrhenius plot of reactions involving blocked catalyst or blocked reactants is presented in Figure 4.8 (line 4).

The type and concentration of the catalyst depends on the binder/crosslinker system, the stoving schedule and the requested flow and levelling of the film. Even for the same type of curing reaction the effect of the catalyst on the flow properties of the system can be very different. Figures 4.10 and 4.11 present results obtained

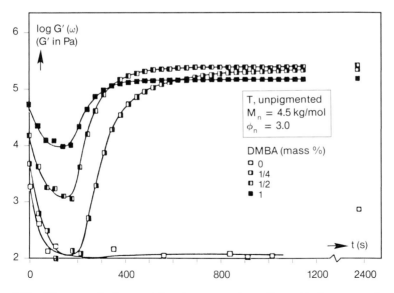

Figure 4.10 The t.v.a. isofrequent heating runs at $w = 2\pi$ rad/s for an unpigmented polyester/epoxy hybrid system at four concentrations of DMBA catalyst at $T = 110-200°\text{C}$ (heating rate 20°C per minute) [77] (Reproduced by permission of DSM Resins BV)

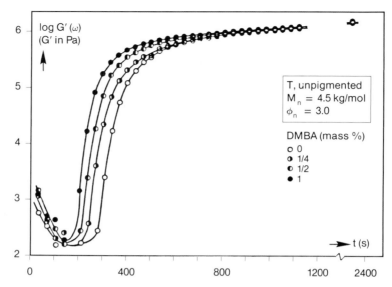

Figure 4.11 The t.v.a. isofrequent heating runs at $w = 2\pi$ rad/s for an unpigmented polyester/TGIC at four concentrations of DMBA catalyst at $T = 110-200°C$ (heating rate 20°C per minute) [77] (Reproduced by permission of DSM Resins BV)

by thermoviscoelastic measurements on mixtures of polyester resins with TGIC and bisphenol A epoxy resin respectively, using N,N-dimethyl benzyl amine as catalyst [77]. The storage modulus is plotted semilogarithmically versus time. As the heating proceeds at a rate of 20°C/min there is an initial decrease of the storage modulus (and viscosity), and after 200 seconds an abrupt increase can be observed which is caused by the crosslinking process. The minimum modulus is almost independent of the catalyst concentration in the TGIC systems, but it varies strongly in the epoxy systems. The subsequent increase in the modulus due to network formation begins at an earlier stage with increasing catalyst concentration. An additional observation is that the reaction in the uncatalysed epoxy formulation is extremely slow, in contrast to the TGIC formulation.

4.7 SURFACE TENSION

Equation (4.31), describing the necessary time for two powder particles to coalesce as a function of the relevant parameters, and the Orchard equation (4.30), referring to the flow behaviour of the powder coating after melting, indicate that high surface tension will have a positive influence on the flow properties. On the other hand, wetting of the pigment particles and of the substrate is dependent also on the surface tension of the resin and the coating. It will be shown later that

PARAMETERS INFLUENCING POWDER COATING PROPERTIES 205

good wetting of the pigment particles is essential for a low yield value and melt viscosity of the powder coating, both having a considerable influence on the levelling of the film. Good wetting of the substrate is a prerequisite if surface defects are to be avoided, a phenomenon known as cratering of powder coatings. All of this shows that the influence of surface tension on the film forming properties of powder coatings is very complex. It is the intention in this section to elucidate the most important aspects of surface tension relevant to the levelling, wetting and cratering of powder coatings.

4.7.1 WETTING PROPERTIES AND SURFACE TENSION

A molecule present in a drop of liquid in contact with a solid surface, depending on its position, can undergo attraction from the liquid phase only (position A), the liquid and vapour phase (position B), the liquid and solid phase (position C) or the liquid, solid and vapour phase (position D), as depicted in Figure 4.12.

In position A the molecule is in an equilibrium in a uniform field. However, in positions B, C and D the molecule undergoes an attraction that is not uniform in all directions, since the cohesive and adhesive forces at the liquid/vapour, liquid/solid and solid/vapour interfaces are of different magnitudes. In order to reach an equilibrium, the system will tend to minimize its free energy per unit area, which results in a contraction of the surface to a minimum, thus obtaining a typical form of drop. According to the well-known Young equation, the three existing interfacial forces in the equilibrium state are connected by the following dependence:

$$\gamma_{sv} = \gamma_{lv} \cos \theta + \gamma_{sl} \qquad (4.38)$$

where

γ_{sv} = surface tension of the substrate in contact with the vapour from the powder coating
γ_{lv} = surface tension of the powder coating in contact with its vapour
γ_{sl} = interfacial tension between the powder coating and the substrate
θ = contact angle, representing the angle between the tangent of the liquid surface drawn at the three phase boundaries and the plane of the substrate surface

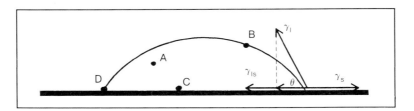

Figure 4.12 Drop of liquid in contact with the surface

A spontaneous spreading will occur only when $\gamma_{sv} > \gamma_{lv} \cos\theta + \gamma_{sl}$, or the so-called coefficient of spreading S defined by the following equation has a positive value:

$$S = \gamma_{sv} - \gamma_{lv} - \gamma_{sl} \tag{4.39}$$

The value of the spreading coefficient S can be calculated from the three γ values. Since it is very difficult to measure γ_{sv}, equation (4.39) is of limited practical use.

On the other hand, the contact angle θ can be easily measured in a relatively simple way by a contact angle goniometer. Since it is related to the partial wetting of the substrate, it is a useful parameter for estimating the wetting properties of the powder coatings.

From Figure 4.12 the contact angle θ can be defined by the following equation:

$$\cos\theta = \frac{\gamma_{sv} - \gamma_{sl}}{\gamma_{lv}} \tag{4.40}$$

or the equation for the spreading coefficient after proper substitution has the following form:

$$S = \gamma_{lv}(\cos\theta - 1) \tag{4.41}$$

Since $\cos\theta$ can never be greater than 1, from equation (4.41) it follows that no spontaneous spreading can occur; it can happen only in an extreme case when $\theta = 0$ and the value for the spreading coefficient is also zero. On the other hand, a lower contact angle means a less negative value for S and promotes better wetting. Equation (4.40) shows that for lower values of θ the surface tension of the powder coating γ_{lv} should be as low as possible and that the surface tension of the substrate should be maximized, having no contaminants with low surface tension.

4.7.2 SURFACE TENSION AND CRATERING

The cratering phenomenon of the coatings is a very common problem facing the paint chemist and has been a matter of interest to many scientists [82–90]. In solvent borne coatings, the viscosity of the film forming material is relatively low and the cratering is caused mainly by surface tension gradients, which are responsible for the flow of the material from one point to another (from low surface tension areas to areas with higher surface tension). This is the so-called Maragony effect. Mathematically it can be described by the equation of Fink-Jensen [82] which gives the dependence of the flow of the material (q) as a function of the film thickness (h), viscosity (μ) and the surface tension gradient across the surface (γ):

$$q = \frac{h^2 \gamma}{2\mu} \tag{4.42}$$

It seems that, for good control of the cratering, thinner films with higher viscosities and equalized surface tension gradients across the surface should be applied.

Scholtens, Linde and Tiemersma-Thoone [77] believe that the Maragony effects do not play a role in cratering of the powder coatings and that the crater formation is caused by an insufficient wetting of the substrate by the molten coating.

Feyt and Bauwin [83] have found that the surface tension of the substrate decreases linearly with an increase in temperature from 10 to 185 °C. The same dependence was found by Gabriel [86] for bisphenol A type epoxy resins. It is interesting that the slopes of the straight lines representing the values of the surface tension against temperature are almost the same in both cases. This means that the contribution of the temperature in an eventual decrease of the contact angle due to the change of surface tension of the powder coating and the substrate is almost negligible. However, Gabriel has shown that the wetting of the substrate is improved by increasing the temperature, and has attributed this to the reduction of the interfacial tension between the coating and the substrate (τ_{sl}) which according to equation (4.41) will increase the value of the spreading coefficient.

The results obtained by Lange [67], in experiments where the surface tension of an epoxy powder and polyester powder coatings was varied by addition of acrylic additive, support at first sight the idea that the cratering is caused by bad wetting of the substrate; namely, in both cases, the powder coatings without acrylic additive having a contact angle with the substrate of 60° (epoxy powder) and 35° (polyester powder) showed obvious cratering. The addition of 0.25% of acrylic additive decreases the contact angles and reduces the cratering tendency to a certain extent, but does not prevent the cratering completely. The cratering disappears completely at a concentration of 0.5% of acrylic additive. Further addition has a positive effect on the flow, reducing the orange peel effect.

Figure 4.13 [67] represents the dependence of the contact angles of the polyester and epoxy powder coating with the substrate as a function of the concentration of the acrylic additive. In both cases there is an obvious minimum at a concentration of 0.25%, after which the contact angles increase again and level off at concentration higher than 0.5%. Very similar results were obtained by Feyt and Bauwin for epoxy powder coating and linear acrylic polymer additive of medium molecular weight bearing unbranched side chains. The presence of craters at the minimum contact angle and their absence at higher concentrations when the contact angle is increased does not support the hypothesis that the cratering is caused only by poor wetting of the substrate. If this is true, then the best results concerning the cratering should be obtained with 0.25% of acrylic additive. Obviously the cratering is a more complex phenomenon.

Feyt and Bauwin, and de Lange give an identical explanation of the action of the acrylic additive. They suppose that the lowering of the interfacial tension between the substrate and the coating, manifested by the lower contact angle, has

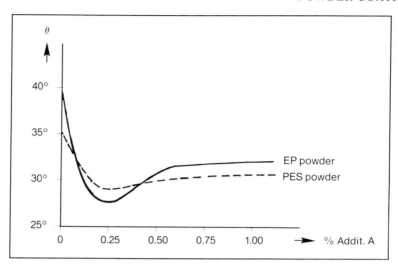

Figure 4.13 Effect of percentage of acrylic additive on contact angle for unpigmented epoxy and polyester powder coatings [67] (Reproduced by permission of Federation of Societies for Coatings Technology)

a positive effect on the wetting of the substrate, but it is not the only element necessary to prevent the crater formation. The acrylic additive, which is linear acrylic polymer with unbranched side chains, has a lower surface tension than the main resin in the coating formulation. Consequently, it will migrate to the surface with the polar backbone dissolved in the polymer, while the alkyl chains will tend to escape into the surrounding environment, arriving at the surface and thus equalizing its chemical potential [91]. A certain minimum concentration of the linear side chains of the acrylic additive per unit surface area is necessary for equalization of the surface tension across the film surface, thus minimizing the surface tension gradient. In this way, the flow of the material on the surface is reduced (Fink-Jensen equation), resulting in a decreased cratering tendency. The further addition of acrylic additive has no effect upon the cratering.

4.7.3 SURFACE TENSION AND FILM LEVELLING

According to the equation of Nix and Dodge (equation 4.29), which gives a quantitative description of the coalescence of powder particles during their fusion in a continuous layer, and the equation of Orchard (equation 4.30), describing the levelling of the film in the next stage, high surface tension promotes better flow of the powder coatings during the film forming process. On the other hand, it was shown that low surface tension facilitates the wetting properties. This lowers the yield value and viscosity of the coating, thus improving the flow of the material and preventing the formation of craters during curing. It is obvious that in order to obtain a powder coating with good flow and without surface defects, a compromise must be found to combine these two contradictory requirements.

Gabriel [86] and de Lange [67] have suggested that a good balance between the two opposite influences of the surface tension can be obtained by a combination of two different types of additives—one decreasing and the other increasing the surface tension of the coating. It has been found that additives being copolymers of acrylic esters (ethyl acrylate/2-ethyl hexyl acrylate, for example) with a number average molecular weight of ca. 8000 reduce the contact angle, thus improving the wetting. The other types of additives which increase the contact angle are epoxidized soybean fatty acids or abiethyl alcohol. These should improve the flow by raising the surface tension [67]. In this way it seems that no advantage can be obtained because one additive nullifies the effect of the other, unless a higher freedom is expected when only the surface tension is adjusted. In fact, this is only the case for certain fixed temperatures. The curing of the powder coating is not a completely isothermal process. The object temperature varies between the normal ambient, when the article enters the stove, and the stoving temperature at the end of the curing process. The surface tension of the coating decreases linearly with increasing temperature. The addition of an acrylic based additive will improve the wetting of the pigments during the dispersion and the wetting of the surface, thus avoiding crater formation. However, since the surface tension, once lowered, undergoes additional decrease during curing due to the temperature increase, the levelling properties of the coating are badly affected. An ideal powder coating should have a low enough surface tension to facilitate pigment dispersion and substrate wetting, which will remain constant during the curing, thus not affecting the levelling of the film.

Gabriel [86] has found that the slopes of the straight lines representing the change of the surface tension with temperature for pure bisphenol A epoxy resin, additive which decreases the surface tension (additive O), additive increasing the surface tension (additive P), epoxy resin with additive O and epoxy resin with additive P are not the same. Figure 4.14 represents the plots of the surface tension for the pure resin and the additives against temperature, as well as blends of the resin with 0.4% additives.

It can be seen that neither the resin alone, nor the resin with additive O only, or additive P only behave in a desirable way. The resin without additives has a high surface tension affecting the pigment and surface wetting, which decreases rapidly with increasing temperature. The addition of additive O improves the wetting properties, but affects the levelling at a late stage of curing. Addition of additive P only has as a consequence an unacceptably high surface tension over the whole temperature range, promoting cratering problems. Fortunately, the combination of both additives in equal concentrations gives a good compromise, lowering the slope of the straight line and bringing it closer to an ideal horizontal position.

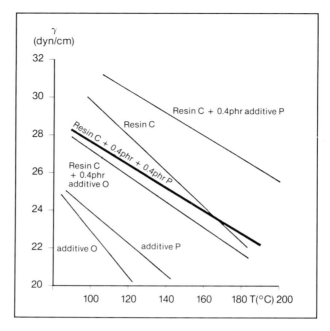

Figure 4.14 Surface tension versus temperature for bisphenol A epoxy resin, the same resin with 0.4% additive O, 0.4% additive P, 0.4% additive O + 0.4% additive P and the pure additives [86] (Reproduced by permission of the Oil and Colour Chemist's Association)

De Lange comes to the same conclusions, summarizing them in the following four statements:

1. Epoxy resin without additives leads to cratering and the orange peel effect.
2. Epoxy resin + type P additive (increasing surface tension) leads to cratering but no orange peel effect.
3. Epoxy resin + type O additive (decreasing surface tension) leads to the orange peel effect but no cratering.
4. Epoxy resin + both types of additives leads to good flow without craters and the orange peel effect.

He suggests that the term 'flow control agent', which is commonly used in practice, is misleading and should be replaced by 'flow promoters' for additives increasing surface tension and 'wetting promoters' for those decreasing surface tension.

The effect of different additives on flow and levelling properties of powder coatings depends very much on the type of binder. The results obtained with epoxy resins should not be automatically transferred to other systems. In the case

PARAMETERS INFLUENCING POWDER COATING PROPERTIES

of polyester powder coatings, the addition of type P additives has completely the opposite effect, promoting even more orange peel effect. Therefore, in this coating system it is common practice to use acrylic types of additives only in order to prevent cratering.

4.8 PIGMENT VOLUME CONCENTRATION AND PIGMENT DISPERSION

The ideal liquid has a constant viscosity for a given temperature for any shear rate. This is the definition of a Newtonian liquid or Newtonian flow behaviour. A linear plot of the shear stress against the shear rate is a straight line passing through the origin with the slope representing viscosity.

Coatings usually exhibit a different behaviour than Newtonian liquids. For many of them, and especially the pigmented coatings, a certain minimum shear stress must be exceeded before the material begins to flow. This non-Newtonian flow behaviour is known as plastic flow, and the liquids exhibiting it are commonly called Bingham liquids.

A linear plot of the shear stress against shear rate for plastic flow results in a straight line which does not pass through the origin. The slope of the line represents the plastic viscosity, and the intercept on the Y axis (Figure 4.15) is the minimum shear stress that has to be exceeded to initiate flow. This minimum shear stress (Y) is called the yield value and is a typical characteristic for powder coatings.

The reason for the non-Newtonian behaviour of powder coatings exhibiting

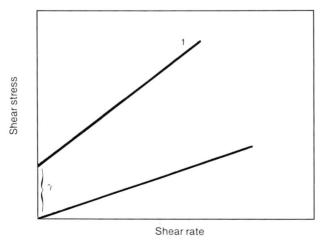

Figure 4.15 Shear stress/shear rate graphs of plastic (1) and Newtonian flow

plastic flow is the interaction forces between the pigment particles. The yield value results from the work necessary to break the reversible physical network structure built up by the interparticle pigment interaction. This interaction increases with increasing pigment volume concentration, resulting in higher yield values and higher viscosities. Consequently, it will affect the flow and levelling properties of the powder coatings.

The role of the yield value in determining the extent of flow of pigmented epoxy powder coatings has been examined by Stachowiak [92], who has found that for excellent flow the yield value of the powder coatings should not exceed 3 Pa. A rough classification made by him could serve as a guideline for the expected flow behaviour of the powder coatings dependent on the yield values:

Yield value	Flow properties
0–3 Pa	Very good to excellent
3–7 Pa	Moderate
7–50 Pa	Poor
> 50 Pa	Very poor

Figures 4.16 and 4.17 represent results reported by Lange [67], who measured the yield values and viscosities for different powder coatings as a function of the pigment volume concentration (titanium dioxide) at various temperatures, for three different powder coatings: epoxy/DICY, epoxy/polyester (50/50) and polyester/TGIC (93/7). In both figures, a dramatic increase of viscosity and yield value is quite obvious at PVC levels of approximately 20%.

It has already been mentioned that powder coatings with excellent flow should not have yield values that exceed 3 Pa. Since the usual PVC levels in powder coatings are about 15%, from the intercept between the curves representing the yield values as a function of PVC and the line $Y = 3$ Pa (Figure 4.16), Lange draws the conclusion that for good flow epoxy powders should be cured at 160°C, polyester/epoxy hybrids at 180°C or higher and TGIC/polyester powder coatings at a minimum temperature of 200°C [67].

The interaction between the pigment particles and the binder in powder coatings has been extensively studied by Ghiljadow [93] and others. Three acrylic resins with carboxyl functionality (RCOOH), hydroxyl functionality (ROH) and without functional groups (RH) were used in combination with a rutile type of titanium dioxide treated with aluminium, zinc and silicon salts. The polymer/pigment interaction was estimated by i.r. spectroscopy.

The spectrum of the rutile pigment taken after mixing in melt with RCOOH acrylic resin shows an absorption band at 1550–1580 cm^{-1} typical for the COO— ion, indicative of the formation of a strong adsorption band (chemisorption) between the polymer and the pigment surface. This absorption band does not exist in the spectrum of the pigment itself, but it is still present, and even more pronounced, after separating the pigments from the binder by washing

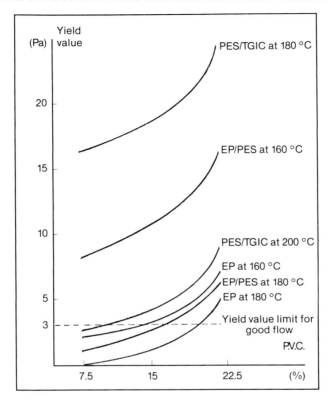

Figure 4.16 Influence of the pigment volume concentration and stoving temperature on yield values for three powder coatings [67] (Reproduced by permission of Federation of Societies for Coatings Technology)

the pigment particles with a solvent in a Soxhlet apparatus. The same results were obtained with a blend containing 75% non-functional acrylic resin and 25% carboxyl functional acrylic resin. This indicates the preferential interaction between the pigment particles and the carboxyl functional polymer.

The analogous experiments with ROH acrylic resin also showed that the resin cannot be washed out from the pigment surface with a solvent, due to the strong hydrogen bond formation between the hydroxyl groups from the polymer and the surface hydroxyl groups of the pigment. Although the absorption band typical for the C=O group remains after washing with a solvent in the case of the ROH acrylic resin, it almost disappears when the RH non-functional acrylic resin is treated in the same way.

These results show that a considerable part of the polymers containing functional groups capable of interaction with the pigment forming salt-type

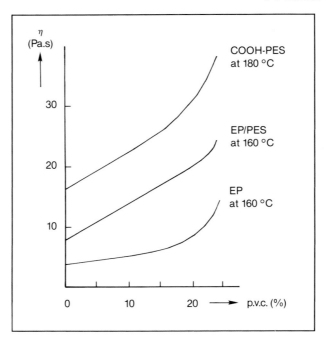

Figure 4.17 Effect of pigment volume concentration on melt viscosity for three different powder coatings [67] (Reproduced by permission of Federation of Societies for Coatings Technology)

compounds or hydrogen bonds is irreversibly adsorbed on the pigment surface.

The strong pigment/polymer interaction will result in the formation of a thin polymer layer around the particles, reducing in this way the interparticle interaction. This should lead to a reduced melt viscosity of the powder. The values for the activation energy of viscous flow presented in Table 4.5 [93] for different pigment concentrations (by weight) in the RCOOH polymer, the RH polymer and a powder coating based on the RH acrylic resin in which the RCOOH polymer was added as modifier confirm this.

The influence of the addition of the RCOOH polymer modifier is quite remarkable. The activation energy of viscous flow is reduced even below the original value for the unpigmented non-functional acrylic resin, due to the formation of an adsorption layer on the pigment. This indicates the importance of good wetting of the particles during dispersion of the pigments in the extruder. Nix and Dodge [63] have shown that the degree of dispersion of the pigment in an acrylic powder coating has a considerable influence on the orange peel effect typical for these coatings. The worse flow of the coating with poorly dispersed pigment is accompanied by much higher viscosities, which in extreme cases for

Table 4.5 Activation energy values of viscous flow of RCOOH and RH acrylic resins and their blends with TiO_2—rutile [93]

System	Polymer concentration (%)	Pigment concentration (%)	Activation energy (kcal/deg mol)
RCOOH	100	—	26.5
RCOOH/TiO_2	85	15	27.6
RCOOH/TiO_2	70	30	28.0
RCOOH/TiO_2	50	50	29.0
RH	100	—	32.0
RH/TiO_2	85	15	37.0
RH/TiO_2	70	30	38.4
RH/TiO_2	50	50	42.5
RH/TiO_2/RCOOH	50	50	30.0

the same coating formulation can vary by a factor of 10. This could be expected, since in the case when the pigments are not completely wetted by the binder, the intermolecular attraction forces between the pigment particles will build up reversible internal structures immobilizing the system.

4.9 PARTICLE SIZE

Particle size is an important parameter of the powder coatings and is often expressed by two different terms: micrometres and mesh screen number. On the other hand, the measures for the film thickness are reported in micrometres, Hegman gauge and mils. For example, 1 mil film corresponds to a thickness of 25.4 μm and 150 mesh screen allows spherical particles of up to 100 μm to pass during the sieving operation. Figure 4.18 [94] gives the relationship between these different units commonly used in the powder paint industry.

Because of the specific nature of the film forming process, the particle size of powder coatings plays an important role in determining the levelling properties of the coating. It is difficult to expect good levelling of a coating consisting of coarse large particles which have to be sintered in a uniform continuous film during stoving. Higher particle packing densities will promote the formation of films with less voids, pinholes, orange peel and shrinkage during curing. Supposing spherical particles to be of uniform size, the voids amount to 48% when particles are packed in a cubical array or 26% when the package has a tetrahedral or pyramidal structure. Even when it is possible to freely choose the best possible packing of particles with a particular particle size, the voids will still amount to 15% [94].

During the application by electrostatic spraying, smaller particles will be charged higher with respect to the unit weight than larger ones. Because of the

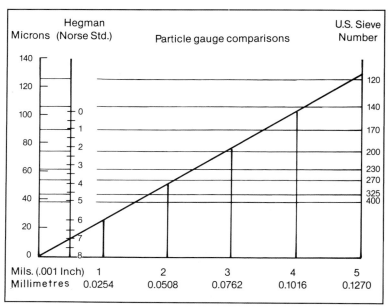

Figure 4.18 Relationship of the film thickness to particle size and the sieve number [94] (Reproduced by permission of Federation of Societies for Coatings Technology)

Faraday cage effect which is specially emphasized in corona spraying guns, the smaller particles will penetrate less into the openings on the article to be coated, thus causing non-uniform distribution of the particles across the coated surface. On the other hand, due to the larger mass and lower velocity, the larger particles are more likely to fall out of the spray cloud. In the longer term, in continuous operation with recirculation of the powder coating this means a permanent change of the particle size distribution with time.

Crowley *et al.* [94] report that a probe introduced into freshly applied powder causes particle movement over a hundred particles distant from the point of the probe, concluding that the charged particles behave as individual dipoles attracted to each other in chains, circles and loose agglomerates as presented in Figure 4.19.

Two stages can be distinguished during film formation of the powder coating. In the first stage a continuous film that flows is formed through coalescence of the molten particles, transforming the irregular film surface to a smooth one in the second stage. Below a thickness of 100 μm the gravitational forces play no significant role in the film formation [67]. It has already been pointed out that the major driving force that causes flow is the surface tension. The time required for two particles to coalesce can be expressed by the already mentioned equation of Nix and Dodge [63]:

PARAMETERS INFLUENCING POWDER COATING PROPERTIES

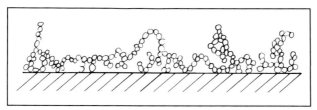

Figure 4.19 Observed configurations of loose powder formed on freshly coated metal surface [94] (Reproduced by permission of Federation of Societies for Coatings Technology)

$$t = \frac{k\mu R}{\gamma} \qquad (4.43)$$

where

- t = time necessary for two particles to coalesce
- k = constant
- μ = viscosity of the powder coating
- R = radius of the powder particle
- γ = surface tension of the coating

Equation (4.43) indicates that finer powder particles will improve the coalescence speed, which means more time is available for flow of the formed continuous film in the second stage. The end result will be a coating with improved flow properties.

To obtain a powder coating with good flow and gloss when applied at low film thicknesses down to 25 μm, the particles with a radius larger than 50 μm should be removed by proper particle classification.

4.10 STOVING TEMPERATURE PROFILE

Melt viscosity of the resin as a function of temperature can be expressed very well by the WLF equation. This equation is valid only for thermoplasts. In the case of thermoreactive systems, which most powders are, the situation is more complicated. Immediately after melting the coating viscosity decreases to a certain value with an increase in temperature, but at the same time the network formation that follows causes a dramatic increase in the viscosity.

It is very difficult to derive a mathematical expression that will quantitatively describe the viscosity profile during curing. Viscosity build-up depends on the kinetics of the curing reaction. The Arrhenius equation can be used to calculate the degree of conversion with time in a dynamic temperature profile of the coated surface, but it will be very difficult to correlate these results with their implication to the viscosity of the system. The situation is even more complicated because the viscosity decreases with an increase in temperature. The viscosity profile for a certain temperature pattern during curing will be the result of several factors combined in a rather complicated manner for mathematical interpretation.

The approach of Hannon, Rhum and Wissbrun [70] (Section 4.4.2) which takes into consideration the viscosity change in non-isothermal curing of the powder coating due to the change of temperature with time $(\delta X/\delta T)_t$ and the network formation during curing $(\delta X/\delta t)_T$ can be used to simulate the melt viscosity change with time, performing numerical integration with a computer.

The experimental results obtained by Ghijsels and Klaren [95] referring to the viscosity profile of epoxy powder coatings cured with different heating rates showed that they attained lower minimum viscosities and consequently better flow by the use of higher heating rates. Viscosity profiles obtained at three different heating rates are presented in Figure 4.20. In Figure 4.21 the minimum

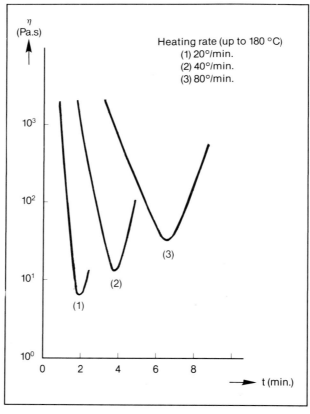

Figure 4.20 Viscosity profiles of epoxy powder coatings cured at different heating rates [67] (Reproduced by permission of Federation of Societies for Coatings Technology)

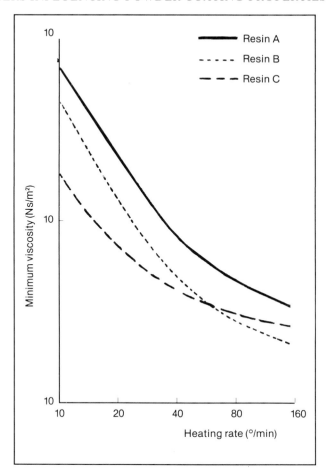

Figure 4.21 Minimum viscosity levels of epoxy powder coatings cured under different heating rates [71] (Reproduced by permission of Elsevier Sequoia SA.)

viscosity levels of three different epoxy resins cured with the same curing agent are presented as a function of the heating rate. This is in agreement with the fact, already known by powder coating users, that better flow is obtained when powder coatings are applied on preheated substrates or on non-massive articles which have a low heat capacity.

Figure 4.22 shows the d.s.c. curing exotherms at four different heating rates (2.5, 5.0, 7.5 and 10.0°C/min) for a composition containing stoichiometrical amounts of digylicdyl ether of bisphenol A (DGEBA) and diamino diphenyl sulphone (DDS) reported by Galy, Sabra and Pascault [96]. It can be seen that the peak of the exotherms shifts to higher temperatures with an increase in the

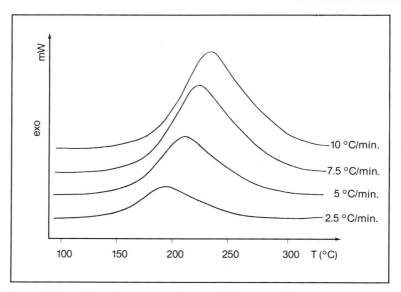

Figure 4.22 Effect of heating rate on the cure behaviour of the DGEBA/DDS system [96] (Reproduced by permission of the Society of Plastics Engineers)

heating rate. This results in lower viscosity of the coating obtained during the stoving process.

Walz and Kraft [72] have established that for good flow the time necessary to reach a viscosity of 50 000 mPa s at a certain curing temperature should be longer than 100 s. The graphs in Figure 4.20 show that even at heating rates higher than those normally used in practice (10–30°C/min), this time of 100 seconds is not enough to reach the minimum viscosity during curing. In other words, high heating rates are favourable for obtaining good levelling of the powder coating during film formation.

REFERENCES

1. Krevelen, D.W. van, in *Properties of Polymers*, Elsevier Sci. Publ. Co., Amsterdam, 1976, p. 19.
2. Govarkier, V. R., in *Polymer Science*, John Wiley & Sons, New York, 1976, p. 98.
3. Gordon, M., *Proc. Roy. Soc. London*, *A268*, 240, 1962.
4. Gordon, M., and Malcolm, G. N., *Proc. Roy. Soc. London*, *A295*, 29, 1966.
5. Dusek, K., *Adv. Polym. Sci.*, **78**, 1, 1986.
6. Tiemersma-Thoone, G. P. J. M., Scholtens, B. J. R., and Dusek, K., *1st International Conference on Industrial and Applied Mathematics (ICIAM '87)*, *CWI track 36*, 1987, p. 295.

7. Scholtens, B. J. R., Linde, R. van der, and Tiemersma-Thoone, P. J. M. *DSM Resins—Powder Coating Resins Symposium,* Nordwijk, May 1988.
8. Flory, P. J., in *Principles of Polymer Chemistry,* Cornell University Press, Ithaca, N.Y., 1953.
9. Stockmayer, W. H., *J. Polym. Sci.,* **IX** (1), 69, 1952.
10. Munari, A., *La Chimica e L'Industria,* **V67** (N12), 694, 1985.
11. Misev, T., and Belder, E., *DSM Resins—Powder Coating Resins Symposium,* Nordwijk, May 1988.
12. Collins, A. E., Bares, J., and Billmeyer Jr, F. W., in *Experiments in Polymer Science,* John Wiley & Sons, New York, 1973, p. 216.
13. Wicks Jr, Z. W., *J. Coat. Technol.,* **58**, 743, 23, 1986.
14. Sperling, L. H., in *Introduction to Physical Polymer Science,* John Wiley & Sons, New York, 1986, p. 254.
15. Eyring, H., *J. Chem. Phys.,* **4**, 283, 1936.
16. Fox, T. G., and Flory, P. J., *J. Appl. Phys.,* **21**, 581, 1950.
17. Fox, T. G., and Flory, P. J., *J. Polym. Sci.,* **14**, 315, 1954.
18. Boyer, R. F., in *Encyclopedia of Polymer Science and Technology,* Suppl., Vol. 2, Ed. N. M. Bikales, Interscience, New York, 1977, p. 822.
19. Katz, D., and Zervi, I. G., *J. Polym. Sci.,* **46C**, 139, 1974.
20. Brandruys, J., and Immergut, E. G. (Eds.), *Polymer Handbook,* Vol. III, 2nd ed., John Wiley & Sons, New York, 1975, p. 139.
21. Union Carbide Corp., US Pat. 3 397 254, 1968.
22. Kurka, A. K., and Herbert, M. B., US Pat. 3 027 279, 1959
23. Williams, M. L.., Landel, L. F., and Ferry, J. D., *J. Am. Chem. Soc.,* **77**, 3701, 1952.
24. Ferry, J. D., in *Viscoelastic Properties of Polymers,* 3rd Ed., John Wiley, New York, 1980, pp. 280–8.
25. Nielsen, L. E., in *Polymer Rheology,* Marsel Dekker, New York, 1977, p. 33.
26. Paul, S., in *Surface Coatings—Science and Technology,* John Wiley & Sons, Chichester, 1985, p. 711.
27. Misev, T., and Belder, E., *J. Oil Colour Chem. Assoc.,* **72** (9), 363, 1989.
28. Corcoran, E. M., *J. Paint Technol.,* **41** (538), 635, 1969.
29. Shimbo, M., Ochi, M., and Arai, K., *J. Coat. Technol.,* **56** (713), 45, 1984.
30. Shimbo, M., Ochi, M., and Arai, K., *J. Coat. Technol.,* **57** (728), 93, 1985.
31. Perera, D., and Eynde, D. V., *J. Coat. Technol.,* **53** (677), 39, 1981.
32. Aronson, P. D., *J. Oil and Colour Chemists' Assoc.,* **57**, 66, 1974.
33. Sato, K., *Prog. Org. Coat.,* **8**, 143, 1980.
34. Perera, D., and Eynde, D. V., *J. Coat. Technol.,* **59** (748), 55, 1987.
35. Tagoshi, H., and Endo, T., *J. Polym. Sci., Polym. Lett.,* **C26**, 77, 1988.
36. Bailey, W. J., Sun, R. L., Katsuki, H., Endo, T., Iwama, H., Tsutshima, R., Saigou, K., and Bitritto, M. M., *ACS, Symp. Ser.,* **59**, 38, 1977.
37. Perera, D., Greidanus, P., and Misev, T., unpublished results.
38. Fox, G. T., and Loshaek, S., *J. Polym. Sci.,* **15**, 371, 1955.
39. Cowie, J. M. G., *Eur. Poly. J.,* **11**, 297, 1975.
40. Boyer, R. F., *Macromolecules,* **7**, 142, 1974.
41. Kumler, P. L., Keinath, S. E., and Boyer, R. F., *J. Macromol. Sci., Phys.,* **B13**, 631, 1977.
42. Holda, E. M., *J. Paint Technol.,* **44** (570), 75, 1972.
43. Bueche, F., *Physical Properties of Polymers,* Interscience Publishers, New York, 1962.
44. Boyer, R. F., *Rubber Chem. Technol.,* **36**, 303, 1963.
45. Davis, G. T., and Eby, R. K., *J. Appl. Phys.,* **44**, 4274, 1973.
46. Schalaby, W. S., in *Thermal Characterisation of Polymeric Materials,* Ed. E. A. Turi, Academic Press, New York, 1981, p. 261.

47. Toussaint, A., *Prog. Org. Coat.*, **2**, 273, 1973–4.
48. Kwei, T. K., *J. Polym. Sci.*, **A3**, 3299, 1964.
49. Kumins, C. A., Roteman, J., and Ralle, C. J., *J. Polym. Sci.*, **A1**, 541, 1963.
50. Lipatov, Yu, S., and Geller, T. E., *Vysokomol. Soedin.*, **A9**, 222, 1967.
51. Gordon, M., and Taylor, J. S., *J. Appl. Chem.*, **2**, 493, 1952.
52. Fox, T. G., *Bull. Am. Phys. Soc.*, **1**, 123, 1956.
53. Jenckel, E., and Heusch, R., *Kolloid-Z*, **130**, 89, 1958.
54. Couchman, P. R., *Macromolecules*, **11**, 1156, 1978.
55. Couchman, P. R., *Polym. Eng. Sci.*, **21**, 377, 1981.
56. Couchman, P. R., *J. Mater. Sci.*, **15**, 1680, 1980.
57. Couchman, P. R., and Karasz, F. E., *Macromolecules*, **11**, 117, 1978.
58. Pochan, J. M., Beatty, C. L., and Hinman, D. F., *Macromolecules*, **11**, 1156, 1977.
59. Fox, T. G., *J. Polym. Sci.*, **C9**, 35, 1965.
60. Winkler, J., Klinke, E., and Dulog, L., *J. Coat. Technol.*, **59**, (754), 35, 1987.
61. Patton, T. C., in *Paint Flow and Pigment Dispersion*, 2nd ed., John Wiley & Sons, New York, 1979, p. 377.
62. Frenkel, J., *J. Phys.*, **9** (5), 385, 1945.
63. Nix, V. G., and Dodge, J. S., *J. Paint. Technol.*, **45** (586), 59, 1973.
64. Kuczynski, G. C., and Neuville, B., *Conference on Sintering and Related Phenomena*, Paris, 1960.
65. Geguzin, J. E., *Physik des Sinters*, Leipzig, 1973.
66. Hann, H., and Jonach, B., *Plaste und Kautschuk*, **24** (2), 135, 1977.
67. Lange, P. G. de, *J. Coat. Technol.*, **56** (717), 23, 1984.
68. Orchard, S. E., *Appl. Sci. Res.*, **A11**, 451, 1962.
69. Spitz, G. T., *Am. Chem. Soc., Div. Org. Coat. Plast. Chem.*, **33**, 502, 1973.
70. Hannon, M. J., Rhum, D., and Wissbrun, K. F., *J. Coat. Technol.*, **48** (621), 42, 1976.
71. Nakamichi, T., *Prog. Org. Coat.*, **8**, 19, 1980.
72. Walz, G., and Kraft, K., *Proceedings of 4th International Conference on Organic Coatings Science and Technology*, Athens, 1978, p. 56.
73. Taylor, J. R., and Foster, H., *J. Oil Colour Chem. Assoc.*, **54**, 1030, 1971.
74. Quach, A., and Hansen, C. M., *J. Paint. Technol.*, **46** (592), 40, 1974.
75. Göring, W., *Farbe und Lack*, **83**, 270, 1977.
76. Greidanus, P. J., *12th Congress of the Federation of Scandinavian Paint and Varnish Technologists*, Helsinki, 8–11 May, 1988.
77. Scholtens, B. J. R., Linde, R. van der, and Tiemersma-Thoone, P. J. M. *DSM Resin—Powder Coating Resins Symposium*, Nordwijk, May 1988.
78. Henig, A., Jath, M., and Mohler, H., *Farbe und Lack*, **86** (4), 313, 1980.
79. Linde, R. van der, and Belder, E. G., *7th International Conference on Organic Coatings Science and Technology*, Athens, 1981, p. 359.
80. Franiau, R. P., *Paintindia*, September 1987, p. 33.
81. Pappas, S. P., and Hill, L. W., *J. Coat. Technol.*, **53** (675), 43, 1981.
82. Fink-Jensen, P., *Farbe und Lack*, **68**, 155, 1962.
83. Feyt, L., and Bauwin, H., *J. Coat. Technol.*, **52** (664), 87, 1980.
84. Nowak, E., Deutch, J. M., and Berker, A. N., *J. Chem. Phys.*, **78** (1), 529, 1983.
85. Gusman, S., *Off. Dig.*, May 1952, p. 296.
86. Gabriel, S., *J. Oil Col. Chem. Assoc.*, **58**, 52, 1975.
87. Hahn, F. J., and Steinhauer, S., *J. Paint Technol.*, **47** (606), 54, 1974.
88. Pesach, D., and Marmur, A., *Langmuir*, **3**, 519, 1987.
89. Kitayama, M., Azami, T., Miura, N., and Ogasawara, T., *Trans. ISU*, **24**, 743, 1984.
90. Hansen, C. M., and Pierce, P. E., *Ind. Eng. Chem., Prod. Res. Develop.*, **13** (4), 218, 1974.

91. Schwartz, A., *J. Polym. Sci.*, **12**, 1195, 1974.
92. Stachowiak, S. A., *7th International Conference on Organic Coatings Science and Technology*, Athens, 1981, p. 469.
93. Ghiljadow, S. K., *XII FATIPEC—Congress Book*, 1974, p. 617.
94. Crowley, J. D., Teague, G. S., Curtis, L. G., Foulk, R. G., and Ball, F. M., *J. Paint. Technol.*, **44** (571), 56, 1972.
95. Ghijsels, A., and Klaren, C. H. J., *XII FATIPEC Congress Book*, Garmisch-Partenkirchen, 12–18 May 1974, p. 269.
96. Galy, J., Sabra, A., and Pascault, J. P., *Polym. Eng. Sci.*, **26** (21), 1514, 1986.

5

Technology of Production of Powder Coatings

5.1	Premixing	226
	5.1.1 Tumbler Mixers.	228
	5.1.2 Double Cone Blenders	229
	5.1.3 Horizontal Mixers	230
	5.1.4 High-speed Blenders	233
	5.1.5 Conical Mixers	237
5.2	Hot Melt Compounding of Powder Coatings	238
	5.2.1 Batch Compounding by Z-blade Mixers	239
	5.2.2 Continuous Compounding	240
	5.2.2.1 Buss Ko-Kneader	240
	5.2.2.2 ZSK Twin Screw Extruder	244
5.3	Fine Grinding.	251
	5.3.1 Hammer Mills.	252
	5.3.2 Pin Disc Mills.	253
	5.3.2.1 Micro ACM Air Classifying Mills	253
	5.3.2.2 ZSP Circoplex Classifier Mills	257
	5.3.3 Opposed Jet Mills	258
5.4	Particle Size Classification.	261
	5.4.1 Classification by Sieving.	262
	5.4.1.1 Tumbler Screening Machines.	263
	5.4.1.2 Vibratory Screening Machines	264
	5.4.1.3 Pneumatic Tumbler Screening	266
	5.4.1.4 Centrifugal Sifters	268
	5.4.2 Centrifugal Air Classifiers	269
5.5	Powder Collection and Dedusting	279
5.6	Quality Control.	284
	5.6.1 Acid Number Determination of the Powder Coating Resins.	284
	5.6.2 Hydroxyl Number Determination	284
	5.6.3 Epoxy Equivalent Weight Determination.	285
	5.6.4 Determination of the Glass Transition Temperature	285
	5.6.5 Determination of the Melting Point and Melting Point Range	286
	5.6.6 Determination of the Softening Point	286
	5.6.7 Viscosity of Powder Coating Resins	287
	5.6.8 Colour of Powder Coating Resins	288
	5.6.9 Determination of the Gel Time.	288

TECHNOLOGY OF PRODUCTION OF POWDER COATINGS

5.6.10	Accelerated Stability Test	289
5.6.11	Density of the Powder Coating Materials	289
5.6.12	Compatibility of Powder Coatings	290
5.6.13	Particle Size Analysis	290
5.6.14	Measurement of Electrical Properties of Powder Coatings	291
	5.6.14.1 Measurement of Powder Resistivity	291
	5.6.14.2 Powder Coating Permittivity	292
	5.6.14.3 Change/Mass Ratio (q/m)	293
5.6.15	Inclined Plate Flow Test	294
5.6.16	Gradient Oven Test	295
5.6.17	Measuring the Thickness of Powder Coatings	295
5.6.18	Solvent Cure Test	296
5.6.19	Hardness of the Cured Film	297
5.6.20	Flexibility Tests	298
5.6.21	Impact Resistance of the Cured Coating	298
5.6.22	Adhesion of the Cured Film	299
5.6.23	Gloss	299
5.6.24	Salt Spray Resistance	300
5.6.25	Review of the Tests for Pipeline Coatings	300
References		300

With respect to the manufacturing process, the production of powder coatings is much closer to the plastic rather than the paint technologies. The manufacturing techniques are based on equipment widely used in the plastic and fine powder industries. Therefore, the design of the powder paint plant differs substantially from one producing wet coatings. The manufacture of powder coatings consists of different processes connected in a logical sequence, as presented in Figure 5.1.

Factors which influence the plant effectiveness are analysed by Scott by means of a mathematical model whose algorithms incorporate data readily obtainable from production records [1]. The model includes personal absence loss factors, plant unavailability loss factors, a premix preparation step, the main plant preparation time and time for quality control.

The process of production can be discontinuous which is very often the case with small plants. However, for plants with a high capacity the individual process steps are usually integrated into a continuous form of production.

Figure 5.2 represents schematically the design of a modern plant for continuous production of powder coatings [2]. Separately weighed-out components are thoroughly mixed in the premixer. The resulting premix is then uniformly metered into the extruder. Before entering the grinding mill, the material leaving the extruder has to be cooled and granulated. The last step of the production process includes particle classification with recirculation of the oversized product back to the grinding mill. This layout of the plant for powder coating production can be modified depending on the capacity of the production line and the types of products. Plants with a small capacity operate in a completely discontinuous way. Regardless of the type of process (continuous or discontinuous) every

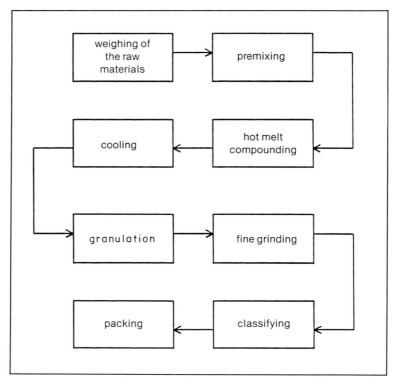

Figure 5.1 Schematic illustration of powder coating production procedures

production step involves specific equipment which will be described in detail in the following sections.

5.1 PREMIXING

Premixing of the ingredients of the powder coating prior to hot melt compounding is an operation whose importance must not be ignored. This step of the powder coating manufacture sometimes plays a decisive role in determining the performances of the coating. Insufficient premixing of the ingredients, especially of those which are employed in small concentrations such as additives and tinting pigments, cannot be compensated for later on in the second stage by the hot melt extrusion process. The hot melt compounding is a continuous process and a uniform feed of homogeneous material through the entire extrusion step is a prerequisite for good quality products. Insufficient premixing in the first step may

TECHNOLOGY OF PRODUCTION OF POWDER COATINGS

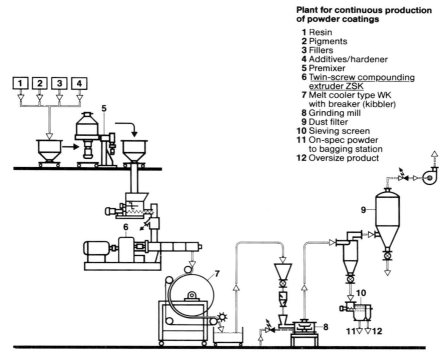

Figure 5.2 Plant for continuous production of powder coatings [2] (Reproduced by permission of Werner & Pfleiderer GmbH)

lead to inhomogeneous composition of the coating which is later manifested by inferior flow, bad mechanical properties, non-even colour of the coated surface, reduced gloss, surface defects, etc.

In principle the ingredients of the powder coating composition are solid materials. The best results in the premixing step are obtained when the particles of the constituents of the blend are of the same size. However, it is obvious that it can never be the case, since the difference in particle size between the pigments and the resins is several orders of magnitude. The problem of incorporation of small quantities of additives such as flow agents or catalyst is successfully overcome by using the master batch technique.

It is important to run the mixing cycle so that the binder is not broken down excessively. Ideally, the binder should have a particle size of 2–4 mm when mixing is completed. If the particles are too small the premix will tend to fluidize during feeding, and if they are too large excessive power will be needed to flux the premix in the compounder [3].

The choice of premixing equipment can be made on the basis of mixing intensity, colour change frequency and whether the process is continuous or a batch process. The variety of formulations in the production plant influences the type of premixer selected, because changeover and cleaning times become crucial in the production operation. For lines with a regular colour change high-speed mixers or container mixers are preferred, due to their easy cleaning and short mixing time of 2–3 minutes. For lines with very little colour changes, ribbon blenders are more advantageous due to lower investment and a larger batch size [4].

Apart from the process requirements, to which a great amount of attention should naturally be paid, it is necessary to have an adequate mechanical design in order to prevent inconsistent performance or even breakdown of the mixing equipment. This is especially important when the mixer has to be stopped in mid process either accidentally or intentionally in order to check the progress of the operation. It is essential that the drive in such cases must be able to withstand repetitive starting under full load. The most preferable driving construction is with direct on-line electric motors which provide two to three times additional torque than is normal on starting. When this cannot be tolerated, mixers with different direct driving arrangements such as slip rings, fluid couplings and centrifugal clutches or indirect arrangements in the form of V-belts are available. Different driving arrangements together with design details which provide safe loading of the mixer, vessel and agitator designs, construction of the mixer outlets and cleaning methods are discussed by Craddock [5].

Subsequent sections describe different types of mixing equipment typically used in the production of powder coatings.

5.1.1 TUMBLER MIXERS

Two types of tumbler mixers are used in the production of powder coatings. The simplest and cheapest example is the drum hoop mixer which consists of a frame with a strapping that is placed over the top of a previously charged drum, and the whole is turned sideways and placed over rollers close to the floor. The time of mixing is normally 10–15 minutes. The size of drum used in this type of mixer does not usually exceed 200 litres.

Drum hoop mixers are cheap machines suitable for small size discontinuous production characterized by a frequent colour change. They are easy to operate machines with many advantages. There is no delay between charging and mixing separate batches since several drums may be used and prepared for mixing. There are no delays due to colour change. The same drums may be used for storing the premix and then transporting it to the feed hopper of the extruder. After use the drums can be cleaned easily. However, they are not very effective in mixing small amounts of tinting pigments. Another disadvantage is the accumulation of fine pigments and extenders on the walls of the drums in layers with a thickness of several millimetres [6].

TECHNOLOGY OF PRODUCTION OF POWDER COATINGS   **229**

5.1.2 DOUBLE CONE BLENDERS

Much higher capacities of mixing space up to 4000 litres are available at the cone blenders which are very often obliquely angled with respect to the rotating shaft (Figure 5.3). During tumble blending of the ingredients, the spiral movement of

Figure 5.3 Double cone blender (Reproduced by permission of Kemutec Group Ltd; Macclesfield, UK. Manufacturers of 'Gardner' Mixers & Blenders)

the material due to the offset angle of the cones ensures a completely uniform distribution of the particles with minimum product breakdown in a totally closed environment. The blender can be supplied for operation under vacuum conditions or with a jacket for heating and/or cooling.

As in the case of drum hoop mixers these machines are very easy to operate and maintain. Because they do not contain baffles on the inside of the cones, cleaning the machine when the colour is changed is not difficult. Easy discharging of the mixed material is provided by a simple diaphragm valve. The same problems of low efficiency of mixing when small amounts of tinting pigments are present and a build-up of finely divided pigments on the walls of the cones are not successfully solved.

5.1.3 HORIZONTAL MIXERS

Horizontal mixers consist of two main parts: the horizontal body, which can be semi (so-called 'U' body) or full cylindrical in shape, and the horizontal rotating shaft carrying mixing blades. The design of the blades, the geometry and specific weight of the materials to be mixed determine the rotational speed of the shaft (Figure 5.4).

The traditional 'U' trough mixers have a long narrow shape, and as the agitator rotates, mixing is achieved by moving the product from both ends to the opposite ends at the same time in a countercurrent manner. The longer the mixer the longer the agitator ribbons or the greater the number of elements where the agitator is of the interrupted ribbon design. The new so-called 'HE' series of ribbon mixers produced by Gardner are characterized by a small length/width ratio, providing some special benefits. To complement the short aspect ratio of the mixer (length to width), a technically advanced double interrupted ribbon agitator has been developed which has only four helix segments and two scraper elements, irrespective of mixer capacity. The combination of the agitator design and short mixing trough ensures substantial reductions in mixing cycles. The minimal number of helix elements, together with reduced mixing times, result in less work being done on the product to achieve the desired mix, thus minimizing the product degradation in terms of particle size reduction. As will be seen, this is a rather important condition for good compounding efficiency of the single screw extruders. Having an easily accessible, short, wide trough and only four helix elements on the agitator, these mixers are easy to clean.

Standard HE series mixers are available with working capacities of 500–10 000 litres. Space requirements for the mixers are tabulated in Table 5.1.

The so-called turbulent rapid mixer of Draiswerke GmbH (Figure 5.5) offers the possibility that the important units of the machine are exchangeable, thus facilitating subsequent conversion to suit a change of production. The machine is designed for operation in various speed ranges to permit optimum utilization for a given application.

TECHNOLOGY OF PRODUCTION OF POWDER COATINGS

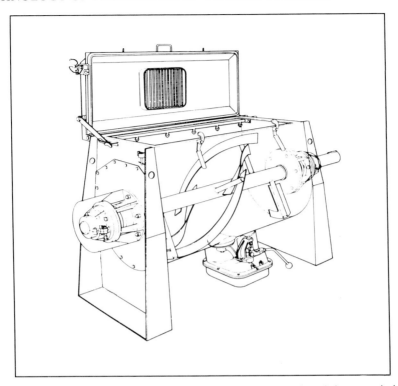

Figure 5.4 Model 500HE horizontal ribbon mixer (Reproduced by permission of Kemutec Group Ltd; Macclesfield, UK Manufacturers of 'Gardner' Mixers & Blenders)

Table 5.1 Space requirements for HE series ribbon mixers (Reproduced by permission of Kemutec Group Ltd; Macclesfield, UK. Manufacturers of 'Gardner' Mixers and Blenders)

Model	Working capacity (L)	Length (mm)	Width (mm)	Height (lid closed) (mm)
500HE	500	1560	905	1060
750HE	750	1710	1005	1170
1000HE	1000	1860	1105	1280
2000HE	2000	2280	1400	1540
2500HE	2500	2420	1520	1530
3000HE	3000	2580	1575	1750
4000HE	4000	2800	1720	1910
5000HE	5000	3024	1865	2090
7500HE	7500	3265	2025	2285
10000HE	10000	3530	2195	2500

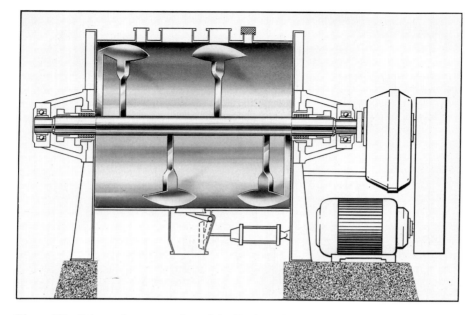

Figure 5.5 Schematic presentation of the Drais turbulent rapid mixer (Reproduced by permission of Draiswerke GmbH)

In the low or subcritical speed range the machine functions as a blender, with mixing elements pushing the material in the axial and radial directions, according to the countercurrent principle, to ensure intimate mixing of all the ingredients. The advantage of this system is that practically the entire internal space of the tank can be utilized and power requirements are small in relation to the volume of feed. On the other hand, the mechanical stressing of the material is minimized, so the braking of the particles of the binder during the mixing operation is not excessive, ensuring in this way problemless feeding of the extruder in the next step of the powder coating production.

If the mixing elements rotate at high speed close to the so-called supercritical speed range, then the mixing pattern differs from that produced at low speed rotation of the shaft, as presented in Figure 5.6. The work performed by the mixing elements in circulating the ingredients within the mix is of secondary importance. Mixing efficiency is primarily dependent on the effect of particle impact. The individual rotating 'T' elements plough through the mix alternatively, with varying angles of attack. Large chunks of material are thereby torn out of the mix and thrown into a series of crossing paths, which in turn cause the material to be uniformly scattered over the entire length of the mixing chamber before falling back into the bulk of the mix. Although in this way the mixing efficiency is much higher, the machine should operate in this centrifugal mixing

TECHNOLOGY OF PRODUCTION OF POWDER COATINGS

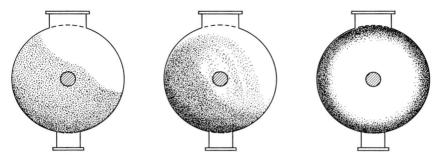

Figure 5.6 Schematics of mix pattern at different mixer speeds: (1) low-speed mixing with blades constantly pushing the material forward, (2) high-speed turbulent mixing, (3) turbulent peripheral layer produced by mixing at extremely high speed (Reproduced by permission of Draiswerke GmbH)

mode only when the particle disintegration does not cause problems with respect to the feeding of the extruder.

At very high speeds, the centrifugal action impels the mix towards the inner wall of the cylindrical mixing chamber, where it forms a turbulent annular layer of increasing density in the course of the outward travel. Interaction with the braking effect of friction at the wall of the mixing chamber results in large differences in particle velocity within the annular layer and thus intensive mixing turbulences. Consequently uniform disintegration and blending of the constituents is achieved in a matter of seconds.

Depending on the mode of operation, mixing elements with different designs are employed, as presented in Figure 5.7. Drais turbulent mixers are produced in various sizes ranging from 2.5 to 50 litres for research applications and from 100 to 40 000 litres for production purposes.

An advantage of the horizontal mixing machines is the high efficiency of mixing. Dispersion is adequate for most powder blending purposes, including the incorporation of small amounts of pigments. The easy removal of the shaft from the mixing chamber allows good cleaning access to the cylinder and the blades.

5.1.4 HIGH-SPEED BLENDERS

High-speed blenders are very effective mixing equipment for preparation of dry powder blends. A typical item of this type is presented in Figure 5.8. The mixer is designed for quick mixing normally between 10 seconds and 10 minutes of powders, pellets and viscous liquids with fillers, pigments or other additives. Its design allows the easily accessible mixing head to serve a number of batch containers, one at a time. The head of the mixing equipment can be easily and quickly cleaned when frequent colour change is required. The mixing rotor

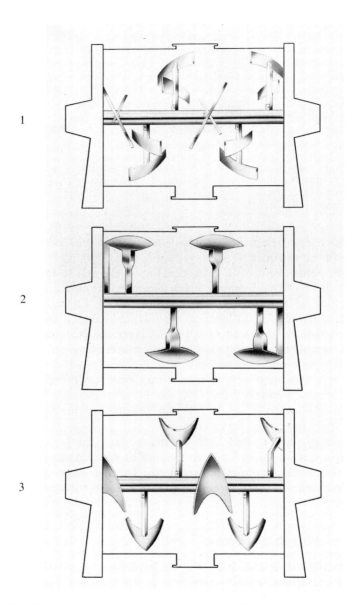

Figure 5.7 Basic types of mixing elements of the Drais turbulent rapid mixer: (1) paddle agitator universally applicable in medium to high speed ranges, (2) double helical mixing element suitable for low speed ranges, (3) spearhead paddle mainly employed for high-speed mixing (Reproduced by permission of Draiswerke GmbH)

TECHNOLOGY OF PRODUCTION OF POWDER COATINGS

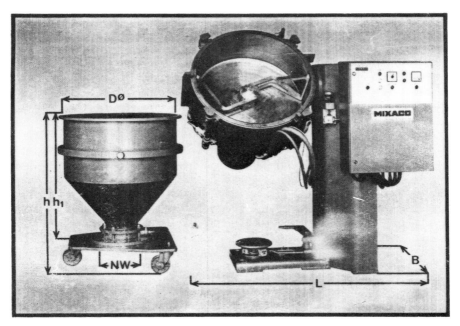

Figure 5.8 Mixaco container mixer (Reproduced by permission of Dr Herfeld GmbH & Co KG)

operates at 5 m/s circumferential speed. The mixer is also produced in a two-speed version with 5 and 10 m/s circumferential rotor speed.

The operational sequences of the premixing step are presented schematically in Figure 5.9. After adding the ingredients of the batch in the container it is pushed under the mixing head. The container is then pneumatically lifted and automatically locked to the mixer head by manual or pneumatically actuated clamps dependent upon the container size. By means of a gear motor the head/container assembly is inverted into the mixing position. In this position the head serves as a mixing chamber. The mixing rotor design, in its inverted operating position, centrifugally flings up the material against the mixing chamber and the container wall. As the material rises and reaches the conical area of the inverted container it falls back on to the mixing rotor. This flow pattern takes place in seconds and continues through the timed mixing cycle, thus creating a uniform mix. After the timed mixing cycle is completed the mixer is tilted down and the container is released. The mixer is now available for another container.

Technical data about the Mixaco container mixers are presented in Table 5.2.

Several advantages of the high-speed mixers compared to the slow-motion blenders make them very attractive mixing equipment for powder coating manufacture. A distinguished feature of these machines is the very short mixing

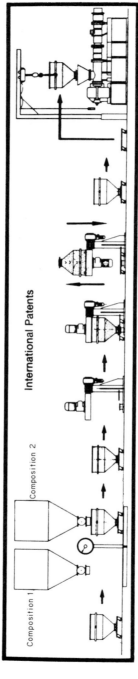

Figure 5.9 Schematic presentation of the preblending operation with the Mixaco container mixer (Reproduced by permission of Dr Herfeld GmbH & Co KG)

TECHNOLOGY OF PRODUCTION OF POWDER COATINGS

Table 5.2 Mixaco container mixers—technical data (Reproduced by permission of Dr Herfeld GmbH & Co. KG)

Container mixer CM	150	300	600	1000	2000
Capacity (L)	150/120	300/240	450/360	1000/800	2000/1600
Motor (kW)					
1 speed	3	5.5	11	18.5	45
2 speed	4.4/5.5	7.5/10	18.5/22	26/33	42/56
Tilting motor (kW)	0.37	0.37	0.55	2.2	4
Container lift (bar)	6	6	6	6	6
Weight (kg)	600	1250	2500	3500	5500
L (mm)	1380	1475	1760	2250	2600
B (mm)	1515	1760	2175	2555	3125
H (mm)	1740	2171	2425	2815	3500
NW φ (mm)	150	150	220	250	300
D φ (mm)	647	844	995	1376	1726
h (mm)	833	1071	1235	1575	1935
h_1 (mm)	655	861	1025	1365	1610

time accompanied by extremely high mixing efficiency. Even small amounts of tinting pigments can be easily incorporated into the premix, being homogeneously dispersed through the premix. Considerable savings can be achieved in this way, reducing the amount of pigments used to obtain pastel shades, since the rapid mixing action helps the development of full colour. Small amounts of liquids can be homogeneously dispersed, which eliminates the need to use the expensive master batching technique for incorporation of liquid additives. Finally, quick cleaning combined with the possibility of using several mixing containers with one mixing head and a short mixing time eliminates the problems of frequent colour change, at least in the preblending step.

5.1.5 CONICAL MIXERS

Conical mixers offer low energy consumption to a high product output ratio and make optimum use of fractional batch sizes without detracting from the mixing ability. They usually comprise a single or multicone static body with an orbital moving screw around the wall of the container, as presented in Figure 5.10. The main characteristic of the conical blenders is the three-dimensional mixing action as a combination of the rotational movement of the screw, causing horizontal rotation of the material, upward movement of the ingredients due to the helical shape of the screw and vertical downward movement due to the gravitational forces which begin when the material reaches the top wide part of the conical container.

Conical mixers require a minimum floor area. As they are rather tall machines in many cases they need a two-storey building to be installed. They provide

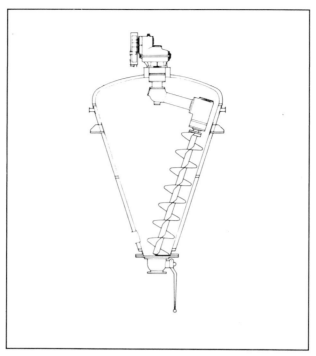

Figure 5.10 Scheme of a conical mixing machine (Reproduced by permission of Kemutec Group Ltd; Macclesfield, UK. Manufacturers of 'Gardner' Mixers & Blenders)

efficient mixing which is practically independent of the batch size. The cleaning of the conical mixers is not a very easy operation. Therefore they are more suitable for a large uniform production rather than a production characterized by frequent colour changes.

5.2 HOT MELT COMPOUNDING OF POWDER COATINGS

The compounding of powder coatings involves several processes which find a place in one single piece of equipment. During this step of the production the binder and crosslinker (not always) are molten and mixed on a molecular scale, the pigment agglomerates are broken to primary particles and homogeneously dispersed together with the additives in the molten binder/crosslinker mix and the surface of the pigment particle is wetted by the molten binder.

Three types of machines can be used in principle for performing the hot melt compounding. The main equipment for carrying out the compounding is the

TECHNOLOGY OF PRODUCTION OF POWDER COATINGS   239

extruder, although simple machines such as Z-blades and heated twin-roll mills may be used.

The energy necessary for melting the binder is supplied partly by heat transfer between the walls of the extruder and the material to be compounded and partly by the heat which is generated due to the high shear gradients in the machine.

The dispersive mixing which breaks down the pigment agglomerates is a process which requires high energy input. A relatively small amount of energy is required for the distributive mixing by which the primary pigment particles are homogeneously distributed throughout the molten binder.

The theory of deagglomeration of the pigment clusters during extrusion has been discussed by Winkler, Klinke and Dulog [7]. They point out that the efficiency of the dispersion process depends on the energy density present at the so-called effective extruder volume where the dispersion process takes place. Franz and Bolt have already indicated that for a given volume flow of the material through the extruder, the overall dissipated energy is directly proportional to the shear rate and the residence time [8].

The technical problems created by the new generation of fast curing powder coatings have almost completely excluded the Z-blade mixers from the machines used to produce contemporary powder coatings. The alternative spray drying technique starting from a liquid paint failed because of economical reasons. However, Z-blade mixers are still in use for batchwise production of thermoplastic powder coatings where the relatively long residence time at high temperature during production does not affect the rheological behaviour of the product.

5.2.1 BATCH COMPOUNDING BY Z-BLADE MIXERS

Z-blade mixers are steel heat-jacketed mixers that are charged with a solid or liquid component from the top and later discharge the homogeneous molten mass from the bottom. They are equipped with strong internal agitators that are shaped like the letter 'Z'.

Usually the solid resin is charged into the mixer and melted within several hours. Then the blades can be put into operation and the pigments, fillers and additives are charged without the curing agent. The components are processed at temperatures ranging from 110 to 130°C for one to four hours depending on the dispersion efficiency that has to be obtained. In the next step the temperature is lowered as much as possible and the hardener is incorporated in as short a period as possible. Before discharge the temperature is increased quickly to 130°C in order to reduce the viscosity for rapid run-out into cooling trays.

Although these machines are rather old and not widely used for production of powder coatings, they still possess some advantages. The preblending operation is completely avoided since the residence time in the machine is rather long. Liquid materials can be easily introduced without master batching techniques. It is easy to match the colour of the coating by addition of tinting pigments. In principle larger amounts of pigments and fillers can be incorporated, thus reducing the raw material costs. However, decisive disadvantages have contributed to the demise of this machine which has now almost disappeared from powder coating plants. Power consumption of the machine per weight unit produced of powder coating is very high. Compared to the extruders, Z-blade mixers have a much higher heat history which has certainly had a bad impact in the case of thermosetting powder coatings. It is very difficult to produce high gloss powder coatings by this technique. In this respect the extruder compounding technique is superior. As with any batch process, lot-to-lot uniformity is a constant quality control problem. An additional increase of the production costs is caused by low production rates, difficult cleaning, which in many cases employs solvents, and higher working power requirements.

5.2.2 CONTINUOUS COMPOUNDING

Two types of machines are used in the production of powder coatings: single screw extruders with a rotating and oscillating screw (Buss Ko-Kneader) and self-cleaning co-rotating double screw extruders (ZSK Werner and Pfleiderer and APV Baker Perkins extruders). Both extruders operate at pressure-free conditions by having a number of openings in the cylinder wall through which the materials can be added to the mix. The system is also characterized by a great variety in screw design. Application of transporting, neutral and contra-transporting knead elements in the ZSK extruder and transport, half transport or knead elements (varying from each other in the number of elements, gaps in the elements and pins through the walls) in the Buss Ko-Kneader influences the mixing and kneading operations [9].

5.2.2.1 Buss Ko-Kneader

Figure 5.11 represents schematically the inside of the kneading section of the Buss Ko-Kneader. In contrast to the other single screw and twin screw extruders, the screw of the Buss Ko-Kneader executes not only a radial movement but also an axial movement. In this way the kneading flights (1) positioned on the shaft (2) knead the mixture axially against the three rows of stationary kneading pins (3) in the barrel casing (4). The movement of the screw flights relative to the kneading pins produces a shearing gap in which the product is subjected to shearing with pronounced reorientation. In this way very high surface regeneration is achieved providing homogeneous mixing despite extremely short processing length. At the same time the process conditions are rather gentle due to balanced shearing and controllable temperature.

The process of producing powder coatings with the Buss Ko-Kneader is schematically presented in Figure 5.12. The container with the premix is placed on the metering unit and serves as a feedhopper for the metering unit. The

TECHNOLOGY OF PRODUCTION OF POWDER COATINGS

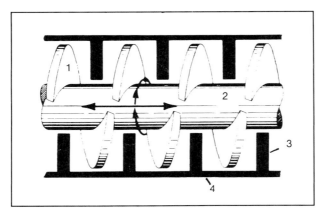

Figure 5.11 Schematic representation of the Buss Ko-Kneader [8] (Reproduced by permission of Buss AG)

metering unit ensures uniform, continuous, volumetric metering of the premix via the inlet screw into the extruder. To protect the downstream plant components, the premix is led through a metal detector after the metering unit. For all kneader sizes, the barrel zones and the screw are temperature controlled with water. Thus, start-up of the plant can be accelerated by preheating the extruder to temperatures of 10–20°C above the melting point of the resin to be processed. As soon as the molten material emerges from the die, the heat control is switched to direct cooling for removal of the heat. As a result, some energy introduced mechanically by dispersive mixing is removed. This in turn accelerates dispersive mixing and ensures a mild heat history typical for this type of machine. Temperature peaks, which can have an undesirable effect on the product quality, particularly in the case of rapidly curing powder coatings, are excluded by virtue of the principle of operation. Electronically controlled heating/cooling units keep the preselected temperature for each exactly constant.

In Figure 5.13 the temperature profile diagram of the Buss Ko-Kneader is presented. The barrel length can be divided into three zones: in-feed zone, plastification zone and kneading and homogenizing zone. The graph indicates the ideal mass temperature profile.

The self-purging principle which is one of the distinguishing features of this machine enables very easy cleaning of the barrel by flushing the kneader with resin or cleaning compound when colours are changed. Inspection of the machine can be carried out easily and quickly, since the two halves of the barrel are convenient to open and the mixing and kneading shaft is freely accessible without any dismantling.

The product emerging from the kneader is fed to a pair of cooling rolls. The water-cooled rolls spread out the cooled extrudate on the cooling belt. Additional

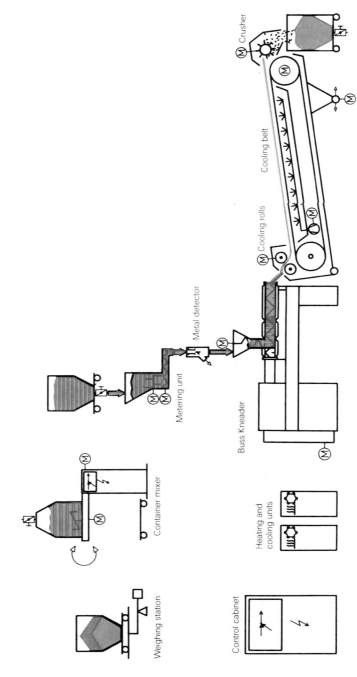

Figure 5.12 Buss concept for production of powder coatings [10] (Reproduced by permission of Buss AG)

TECHNOLOGY OF PRODUCTION OF POWDER COATINGS

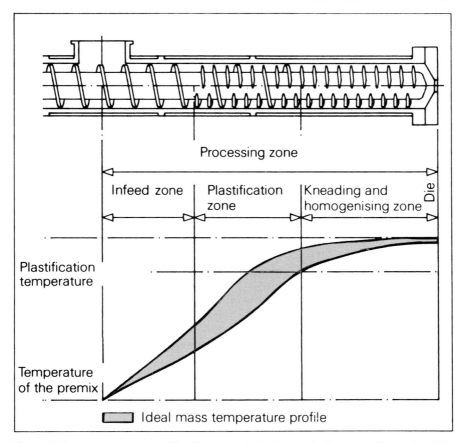

Figure 5.13 Temperature profile diagram of the Buss Ko-Kneader (Reproduced by permission of Buss AG)

cooling on the belt is provided by spraying with water underneath until the product becomes brittle before entering the crusher where it is cut in the form of flakes. The chips are collected in a container and conveyed to a sifter mill in the next production step.

Buss Ko-Kneaders are produced in different sizes and capacities starting from laboratory equipment with a capacity of 5 kg/h up to machines with a throughput of up to 2500 kg/h. Figures 5.14 and 5.15 represent the dependence of the kneader screw speed and the output of the machine (depending on the type of machine) and the residence time diagram of the material in the extruder casing.

Table 5.3 gives relevant data for the different types of Buss Ko-Kneader used for laboratory purposes or production of powder coatings.

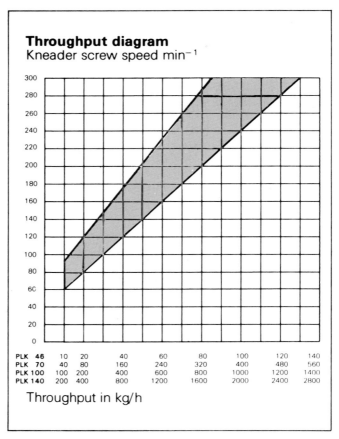

Figure 5.14 Throughput of the machine as a function of the screw speed (Reproduced by permission of Buss AG)

Harris [6] lists several advantages and disadvantages of the Buss Ko-Kneader. He counts the low power requirements, good temperature control, good compounding efficiency, short dwell time in the barrel, self-wiping action and easy way of cleaning among the strong points of this machine. The unsuitability for reprocessing finely divided powder and the reduced throughput when the premix entering the extruder is finely divided can be classed as disadvantages of the Buss Ko-Kneader.

5.2.2.2 ZSK Twin Screw Extruder

The Werner and Pfleiderer ZSK extruder (Figure 5.16) is based on a co-rotating and closely intermeshing twin screw system. The complete extruder consists of

TECHNOLOGY OF PRODUCTION OF POWDER COATINGS

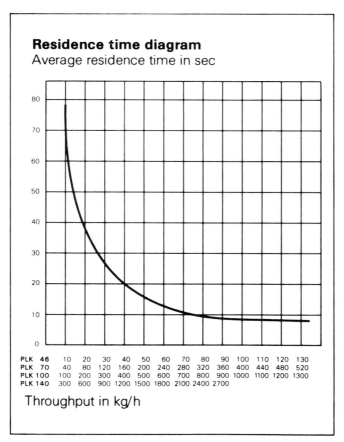

Figure 5.15 Residence time diagram of the Buss Ko-Kneader (Reproduced by permission of Buss AG)

the drive and the processing part. The drive is made up of a variable speed d.c. motor and a speed reduction gear with two co-rotating output shafts.

The processing section, i.e. the barrel and the two co-rotating screws, is designed on the building block principle. The barrel consists of separate sections bolted together to provide the required processing length.

The processing section is available in two distinct types: closed barrel or horizontally split barrel with hinge. The second easy access version can be operated without the use of tools and provides quick access to the screws. In both versions (closed or split) the screws can be easily withdrawn from the barrel. With two sets of screws, production can continue with the second set while the first is taken away from the extruder to be cleaned. In this way utilization of the whole plant significantly increases.

Table 5.3 Types of Buss Ko-Kneaders (Reproduced by permission of Buss AG)

Application	Laboratory		Production		
Type	PLK 46	PLK 46	PLK 70	PK 100	PLK 140
Throughput up to (kg/h)	40	120	500	1200	2500
Metering unit					
Hopper volume (L)	—	60	120	200	450
Connected load (kW)	—	0.38	0.49	0.67	0.87
Inlet screw					
Connected load (kW)	—	0.18	0.75	2.2	3.0
Feed hopper					
Hopper volume (L)	10	—	—	—	—
Connected load (kW)	0.55	—	—	—	—
Buss kneader					
Screw diameter (mm)	46	46	70	100	140
Screw length L/D	7	7	7	7	7
Maximum screw speed (r/min)	270	280	300	300	300
Connected load (kW)	7.5	11	32	65	130
Cooling belt					
Cooling zone length (mm)	—	2500	3000	5000	10000
Belt width (mm)	—	400	800	1000	1000
Cooling rolls diameter (mm)	—	160	250	400	400
Heating and cooling installations on the extruder					
Number of units	2	2	2	2	2
Heating capacity per unit (kW)	7.2	7.2	7.2	10.8	10.8

TECHNOLOGY OF PRODUCTION OF POWDER COATINGS

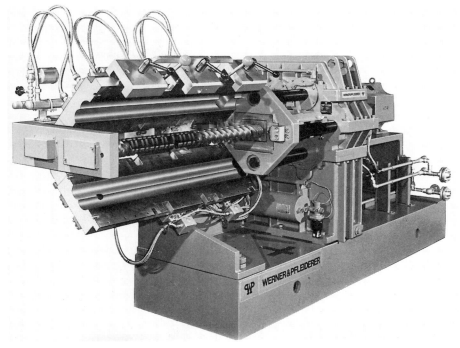

Figure 5.16 Twin screw extruder ZSK 70 M 175 for powder coatings with split barrel to provide easy access to the screws (Reproduced by permission of Werner & Pfleiderer GmbH)

The screws as presented in Figures 5.17 and 5.18 consist of a pair of one-piece screw shafts onto which a suitable sequence of deep-flighted screw and kneading elements, each for a specific effect, is arranged.

The cross-cut operational scheme (Figure 5.19) of the closed twin cylindrical barrel containing two screws with transverse lens-shaped kneading elements set at 90° shows how the screw geometry provides efficient conveying, pressure build-up and self-cleaning action. The material is continuously wiped from the agitator screw and the barrel wall and transferred from one zone created between the wall and the kneading element to another. These zones are the places where particularly effective dispersing and homogenizing take place.

An important feature of ZSK screws is their profile. Irrespective of the shape of the elements, they always intermesh and in operation the crest of one constantly wipes the root of the other while maintaining a very small uniform gap between them. This achieves a high degree of self-cleaning. The kneading elements and the conveying screws have identical cross-sections so that the self-cleaning effect is maintained along the full length of the screws.

The ZSK extruder is characterized by a high torque capability and a large free

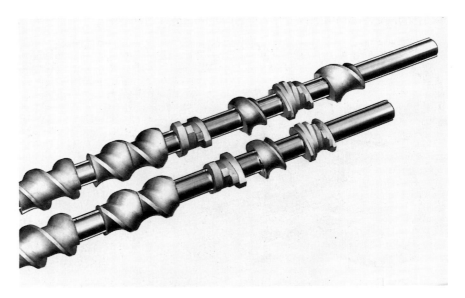

Figure 5.17 Building block principle of the ZSK screws (Reproduced by permission of Werner & Pfleiderer GmbH)

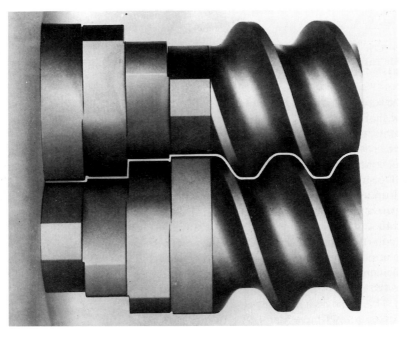

Figure 5.18 Sealing profile of the self-cleaning ZSK conveying and kneading elements (Reproduced by permission of Werner & Pfleiderer GmbH)

TECHNOLOGY OF PRODUCTION OF POWDER COATINGS

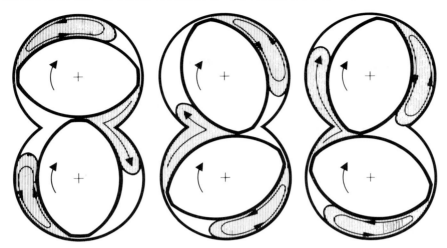

Figure 5.19 Function of kneading discs in ZSK machines (Reproduced by permission of Werner & Pfleiderer GmbH)

volume in the screw channels. Power and volume are matched to optimize processing parameters as well as the economics of the compounding operations.

The building block principle of the machine with interchangeable screw elements and barrel sections allows the user to vary and accurately control conditions anywhere along the process length, which can range from 9 to 48 L/D. The individual barrels (Figure 5.20) having an L/D ratio of 3 or 4 are produced in different designs, enabling multiple feeding of ingredients or venting of moisture or removal of other volatiles. In this way solid, liquid or even gaseous components can be introduced anywhere within the processing section. Both the barrel elements and screw components in the feed and melt zone have removable wear-protected liners (Figures 5.20 and 5.21). The rest of the machine is made of nitrided steel. This makes it possible for only defined sections of the total length of the processing part that need to be wear-protected to be replaced after long use.

The temperature control of the barrel can be achieved by electric heating combined with water cooling, by pressurized water or by heat transfer cooling. In addition, on several types of ZSK machines, there is a possibility for internal cooling of the screw shafts.

The production of powder coatings on ZSK twin screw extruders is characterized by an extremely short residence time which on average throughput rates ranges from 10 to 20 seconds. However, since the product is in the molten state in only approximately 65% of the barrel length, this means that the material is exposed to a correspondingly shorter thermal stress.

The molten compound emerging from the extruder falls between chilled rolls

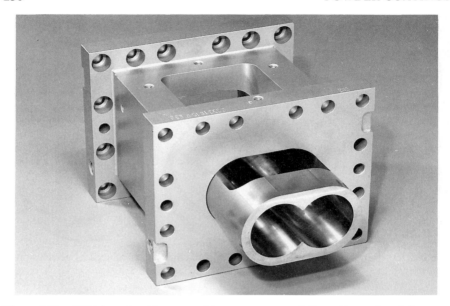

Figure 5.20 ZSK barrel with removable wear-protected liner (Reproduced by permission of Werner & Pfleiderer GmbH)

and is squeezed into a wide thin strip. The additional cooling takes place on an endless conveyor band before entering the braking roll. If the available floor space is limited, or for throughputs higher than 1 ton per hour, a special design of cooling roll device with a breaker is recommended, as presented in Figure 5.2.

Five out of nineteen different types of ZSK twin screw extruders whose technical specifications are listed in Table 5.4 are used for the production of powder coatings. The maximum throughput is in the range between 60 and 2200 kg/h.

Compared to the single screw extruders, twin screw extruders consume somewhat more energy per unit of produced coating. However, they have several

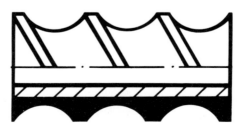

Figure 5.21 Wear-protected ZSK screw bushing (Reproduced by permission of Werner & Pfleiderer GmbH)

TECHNOLOGY OF PRODUCTION OF POWDER COATINGS

Table 5.4 ZSK twin screw extruders—technical data[a] (Reproduced by permission of Werner & Pfeiderer GmbH)

Type	ZSK 30	ZSK 40	ZSK 58	ZSK 70	ZSK 92
Maximum throughput (kg/h)	60	180	500	1200	2200
Drive power (kW)	5.4	20	32	108	150
Screw speed (r/min)	300	300	300	300	300
Screw diameter (mm)	30	40	58	70	92
Channel department (mm)	4.7	7.1	10.3	12.5	16.3
Length of processing section, maximum L/D	42	48	48	48	48
Length of screw shafts (above foundation) (mm)	1100	1100	810	1100	1100
Length of machine (mm)	2200	2600	3400	4000	6000
Width of machine (mm)	780	720	700	700	700
Height of machine (mm)	1400	1400	1400	1550	1450

[a] The machine dimensions are based on processing length $L = 24D$, excluding the discharge section but including the main drive.

advantages. The extremely short dwell time allows safe production of coatings with very high reactivity. The processing of finely powdered materials is less problematic and the dispersion is more efficient so that smaller amounts of strong colour are required due to the efficient colour development [6].

5.3 FINE GRINDING

The fine grinding process follows the hot melt compounding. After leaving the extruder die, cooling and squeezing between the chill rolls, additional cooling on the cooling band to a brittle, easily breakable continuous strip and flaking by a kibler unit, the material is usually collected in movable containers and transported to the grinding mills.

Powder coating is formulated in such a way that the glass transition temperature of the system is at least 40°C to ensure storage stability of the coating. If the grinding of the material is performed at room temperature by appropriate removal of the heat developed during the grinding process, the material is subjected to the mechanical forces action below its glass transition temperature. Polyester, epoxy or acrylic resins used for the manufacture of thermosetting powder coatings usually have molecular weights between 3000 and 6000. With such a molecular weight below their glass transition temperature, these binders are brittle materials with a low modulus which allows easy non-problematic grinding. High molecular weight thermoplastic binders, however, are extremely tough and resilient. Cryogenic grinding at low temperatures is the

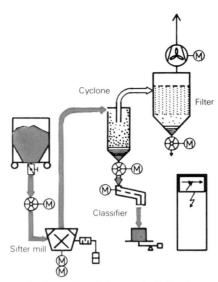

Figure 5.22 Scheme of grinding, classifying and dedusting unit for powder coating production (Reproduced by permission of Buss AG)

only method for reducing the size of the particles of the thermoplastic powder coatings.

Different types of grinding mills are used by the powder coating producers to bring the particle size of the coating within a desirable range. All of them utilize a process of attrition in which the size reduction is the result of impact of the solid granules or flakes with the grinding surface or with each other. Contemporary grinding systems are delivered in one set together with the particle size classifiers and dedusting devices, as presented schematically in Figure 5.22.

Since the 1960s efforts of the mill producers concentrated on developing a grinding machine that combines both the grinding and classifying procedures in one device which will simplify the plant design. Schwamborn [11] gives a survey of the inventions that occurred in the period 1968–1988 to improve the air classifying mill system with a compressed-air operated bag filter which is considered as a modern grinding system for powder coating production. More than sixteen patents referring to this system issued in a period of 20 years indicate the permanent improvements leading to a complete grinding/classifying line that provides clean and safe continuous grinding of powder coatings.

5.3.1 HAMMER MILLS

Hammer mills are among the first devices used for fine grinding of powder coatings. The construction of the mill is rather simple, as presented in Figure 5.23.

TECHNOLOGY OF PRODUCTION OF POWDER COATINGS

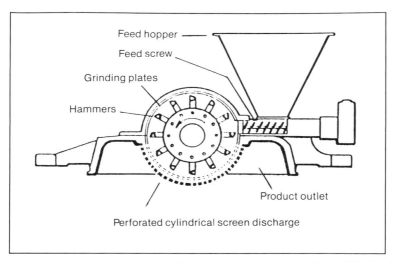

Figure 5.23 Scheme of a hammer mill (Reproduced by permission of Federation of Societies for Coatings Technology)

The granulate to be ground enters the milling chamber through an appropriate feeding unit. The grinding takes place as a result of impact between the granulated material and the rotating hammers. Contemporary hammer mills operate with a stepless speed variation of the rotor between 1000 and 5000 rev/min [12]. In this way it is possible to influence the output of the machine and the quality of the ground powder.

The outlet of the machine is separated from the milling chamber by an exchangeable screen. The type of screen also influences the output of the machine. For powder coating production, herringbone slotted or cross slotted screens are preferred because they give a finer particle size [6].

Hammer mills are produced with a water cooling option on request, so that if necessary the heat developed during the grinding process can be removed efficiently.

5.3.2 PIN DISC MILLS

5.3.2.1 Micro ACM Air Classifying Mills

The ACM classifying mill operates on the principle of impact milling and employs high rotor speeds with a striking edge velocity of up to 120 m/s. Size reduction is effected by the impact of the material particles on the rotating grinding surfaces and the fixed liner (3), as presented in Figure 5.24, as well as by the impact of the particles against each other. The rotating grinding surfaces

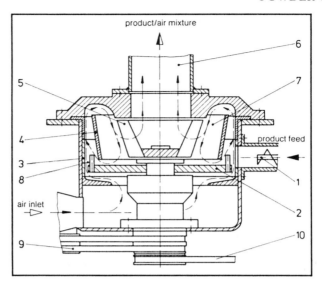

Figure 5.24 Scheme of ACM air classifying mill (Reproduced by permission of Hosokawa MikroPul GmbH)

1 feed screw
2 grinding rotor with pins or other grinding surfaces
3 liner
4 shroud ring
5 classifier wheel
6 discharge nozzle
7 separating zone
8 grinding zone
9 grinding rotor drive
10 classifier drive

perform the active crushing work and impart acceleration to the particles for further size reduction by rebound.

The powder coating granules or flakes which are to be fine ground are fed into the side of the grinding chamber by means of an adjustable speed screw (1). The air enters the grinding chamber at the circumference of the grinding disc (2) and accelerates the feed and ground material towards the blades of the shroud ring (4). The product/air mixture is aligned in the blades of the shroud ring and then enters the classifying section in a defined direction as a uniformly distributed flow. The fine material is carried with the air through the classifier (5) to the outlet

TECHNOLOGY OF PRODUCTION OF POWDER COATINGS 255

connection (6) whilst the rotating classifier throws the coarse material into the shroud ring. From here it falls onto the grinding disc and is admitted into the grinding zone (8) for further reduction.

In the classifying section two opposite forces act upon each particle: centrifugal force induced by rotation of the classifier and centripetal force or sweeping force produced by the air flow. The fine particles with small mass follow the sweeping force and pass with the air through the classifier. The coarse particles are rejected by the rotating classifier. The adjustment of the separation limit is effected by changing the speed of the classifier. Contrary to customary classifying mills with stators or synchronous classifiers, the Mikro ACM classifying mill incorporates a dynamic air classifier which is infinitely adjustable independently from the speed of the grinding rotor.

The complete rotor of the Mikro ACM mills, consisting of the classifier, grinding disc, bearings and V-belt pulleys, can be quickly and easily removed as a complete unit because it is attached to the mill chamber housing by only four bolts. The important components of the rotor are exchangeable, so that depending on the material to be ground and the desired type of finished product, the best configuration of the mill can be chosen. Figure 5.25 represents different types of grinding and classifying components and the complete assembled rotor.

Mikro ACM classifying mills are produced in five standard types whose technical data are given in Table 5.5. The layout of a complete Mikro ACM grinding installation is presented in Figure 5.26.

Table 5.5 Technical data of Mikro ACM classifying mills (Reproduced by permission of Hosokawa MikroPul GmbH)

Type	ACM 3	ACM 10	ACM 30	ACM 60	ACM 200
Capacity factor	0.3	1	2.7–3.1	5.4–6.2	16–20
Drive capacity					
Grinding rotor (kW)	2.2	7.5	22	45–55	132–200
Classifier (kW)	0.37	1.1	4	7.5	30
Speed (r/min)					
Grinding rotor (max)	11 800	7200	4700	2900	2000
Classifier (max)	5400	4000	2900	2600	2000
Classifier (min)	1000	700	500	450	400
Air flow (max)(m^3/h)	600	1500	4500	9000	27 000
Air flow (min)(m^3/h)	300	750	2250	4500	13 500
Weight (kg)	350	450	900	2000	4500
Overall dimensions (mm)					
Width	1200	1250	1900	2500	4000
Depth	850	850	1100	1700	1600
Height	1120	950	1250	1700	2400

Rotor

Most important grinding and classifying components

Classifier

24 blades-short. To obtain a narrow grain spectrum.

Classifier

24 blades-long. For maximum possible product fineness.

Impact lining

Smooth design. For abrasive products and those with a tendency to cake.

Impact lining

Notched design. For a high degree of fineness and non-caking products.

Grinding rotor
with grinding pins

For non-abrasive products, with high degree of fineness.

Grinding rotor
with grinding blocks

For abrasive products. The edges of the blocks are hard metal tipped.

Figure 5.25 (*caption opposite*)

TECHNOLOGY OF PRODUCTION OF POWDER COATINGS

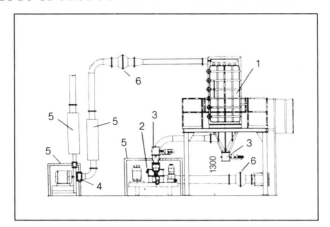

Figure 5.26 Layout of pressure resistant and sound insulated Mikro ACM grinding installation: 1. Mikro-Pulsaire fine dust filter; 2. Mikro ACM classifying mill; 3. Rotary air lock valve; 4. Mikro explosion protected caravent fans (90 basic types); 5. Noise protection ensuring sound level below 80 dB; 6. Explosion protection valves (Reproduced by permission of Hosokawa MikroPul GmbH)

5.3.2.2 ZSP Circoplex Classifier Mills

The ZSP Circoplex classifier mills developed by Alpine AG employ a similar grinding principle which has already been described for the Mikro ACM air classifying mills. However, classification of the ground product is done in a different way. The product to be ground is fed from above via a rotary airlock (1) (Figure 5.27) directly into the classifying chamber, where particles which are already of the required particle size are directly separated by an ultra-fine classifier (2). The design and the operating principle of the classifier sold under the trade name of Turboplex will be discussed in the next section.

The size reduction section of the mill consists of the impact beater system (3) and the stationary triangular-ribbed grinding track (4) which surrounds the grinding chamber. The exchangeable impact beaters are especially hardened to achieve a long service life. The design of the impact beater system allows a very gentle size reduction with minimum energy consumption. The triangular-ribbed grinding track is made from wear-resistant special cast iron and is split into individual segments. Air is supplied to the mill through the duct (6) and conveys the pulverized product to the horizontally mounted classifier. The fine material is discharged through the blow-out opening (5), while the coarse particles are rejected from the classifier and recirculated for further grinding.

The mill can be easily opened for cleaning by a crank unit for small types or by an electrohydraulic system with lifting cylinders for bigger machines. The impact beater system can be removed so that the grinding chamber is accessible for thorough and quick cleaning.

Figure 5.25 Most important grinding and classifying components of Mikro ACM mill and an assembled rotor (Reproduced by permission of Hosokawa MikroPul GmbH)

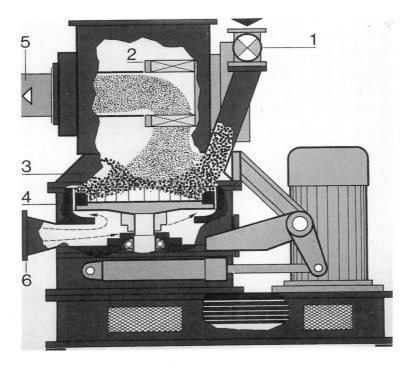

Figure 5.27 Circoplex classifier mills ZSP (Reproduced by permission of Alpine AG)

The particle fineness of the end product is exclusively set by adjusting the classifying wheel speed which is indicated on a digital speed display. However, the optimum speed of the impact beater system for low energy consumption is fixed by the producer on delivery of the machine and can be readjusted later on if required. A particle size distribution between 10 and 90 μm can be achieved in one pass, without inadmissible ultra-fine dust and oversize particles. Table 5.6 summarizes the data for the six different size ranges of Circoplex classifier mills.

5.3.3 OPPOSED JET MILLS

In conventional grinding machines the main grinding mechanism involves contact between the particles whose size has to be reduced, the grinding media (hammers, rotating pins, rods or discs) and the mill lining. In order to obtain superfine powder qualities, this type of mechanical grinding is associated with a rapid rise in energy consumption which makes the grinding uneconomical.

The contemporary tendency towards thin layer powder coatings with ultra-fine particle sizes makes the use of the opposed jet mills very attractive. With these

Table 5.6 Circoplex classifier mills—summary of the range available (Reproduced by permission of Alpine AG)

Mill type	100 ZSP	200 ZSP	315 ZSP	500 ZSP	630 ZSP	750 ZSP
Grinding fineness (d97 in μm)	8–100	8–120	8–120	8–120	10–180	10–200
Fine impact mill						
Chamber diameter (mm)	200	400	630	1000	1250	1530
Drive power (kW)	3	11	22	45	75	132
Maximum speed (rev/min)	11 200	5600	3550	2250	1800	1460
Drive motor speed (rev/min)	3000	3000	3000	3000	1500	1500
Fine classifier						
Turboplex type	100 ATP	200 ATP	315 ATP	500 ATP	630 ATP	750 ATP
Drive power (kW)	2.2	4.0	7.5	15.0	22.0	30.0
Speed (rev/min)	1150–11500	600–6000	400–4000	240–2400	200–2000	160–1600
Main dimensions						
Length (mm)	650	1300	2000	2900	3450	3850
Width (mm)	380	770	1450	1790	2220	2570
Height closed (mm)	700	1100	1410	2325	2850	3270
Height open (mm)	1000	1400	1850	2800	3800	4300

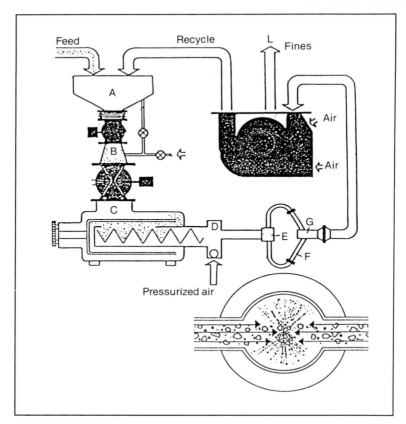

Figure 5.28 Operational principle of the opposed jet mill [14] (Reproduced by permission of OY Finnpulva AB)

mills particle sizes between 15 and 35 μm with 97% of the material being below 12 μm can be easily obtained [13]. The grinding mechanism is based on interparticle collisions between opposed suspension jets in the grinding chamber. This is usually combined with a pneumatic microclassifier in a closed circuit with the jet mill as presented in Figure 5.28 [14].

The material through the feed funnel (A) reaches the grinding pressure in an equalizing chamber (B), after which it goes into the feed chamber (C). The steady, uniform material flow coming from the feed chamber is dispersed in the pregrinding chamber (D). The feed suspension through the flow divider (E) is led into two nozzles (F) and then into the grinding chamber (G), where the working gas pressure and temperature decrease and the velocities of the gas and particles increase to reach the kinetic energy needed for grinding. In the grinding chamber the particles collide with each other and are ground into a product that,

TECHNOLOGY OF PRODUCTION OF POWDER COATINGS

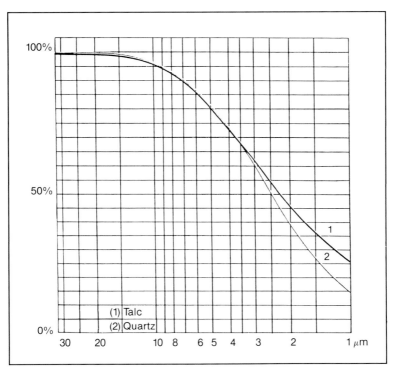

Figure 5.29 Granule size distribution of quartz and talc obtained by the opposed jet mill (Reproduced by permission of OY Finnpulva AB)

depending on the type of the powder coating and mill capacity, has 30–90% of fine product for classification. The residual pressure of the working gas carries the ground particles to the microclassifier (H) working in a close circuit with the jet mill.

Figure 5.29 represents the granule size distribution of milled talc and quartz obtained at jet mill FP3 of Finnpulva AB. The technical data of different types of opposed jet mills produced by OY Finnpulva AB are presented in Table 5.7.

5.4 PARTICLE SIZE CLASSIFICATION

The contemporary grinding equipment is usually delivered as a complete unit including a classifying device. The output of the mill is adjusted to the capacity of the classifier and maximum allowed particle size. However, the classifying mills are not the only equipment used by the powder coating producers and neither is the air stream classification the only way of separating the coarse particles from

Table 5.7 Technical data of Finnpulva opposed jet mills (Reproduced by permission of OY Finnpulva AB)

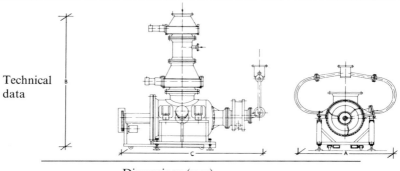

Type	Dimensions (mm)			Capacity (t/h)	Weight (kg)
	A	B	C		
FP 2	450	1700	1700	0.1	500
FP 3	600	2100	2100	0.3	800
FP 4	700	2300	2300	0.5	1500
FP 6	900	2600	2600	1.0	2000
FP 8	1200	3000	3000	2.0	3000
FP 10	1400	3400	3400	3.5	4500
FP 15	1900	4300	4300	7.0	8000

the fine powder. The hammer and some pin disc mills do not provide particle classification; this operation must be done in a separate step of the process. As seen in the previous section, the opposed jet mills have also to be accompanied by a classifier in order to obtain a desirable particle size range.

The screen is probably the best known type of classifier which is widely used by the industry for separation of powders according to size alone. For such an operation industrial screen classifiers are within a range of screen apertures between 10 mm and 75 μm. The capacity of the screen classifiers decreases rapidly when the particle size of the material to be classified is lower than 70 μm. For dry classification below 70 μm and particularly below 50 μm centrifugal air classifiers are most commonly used [15].

5.4.1 CLASSIFICATION BY SIEVING

Mechanical sieving or screening is one of the oldest known process techniques for classification still used in current practice in the production of powder coatings. The same technique can be applied not only to particle size classification but also to scalping, a term used for the removal of small quantities of oversize from a feed

TECHNOLOGY OF PRODUCTION OF POWDER COATINGS

which is predominantly fine. The screening method in principle is used in the dust removal filters in the final step in the powder coating production where the dedusting filter has to retain all particles leaving the dedusting cyclone in the exhausting air flow. Finally, the policing or removal of foreign contaminants accompanied by the breaking down of powder agglomerates which can block the spray gun nozzle is another possible application of the screening technique in the powder paint overspray recovery.

Mechanical sieving allows separation of particles as small as 50 μm although separations down to 20 μm may be achieved with an air-jet sieve [6]. The industrial sifting machines range in capacity from a few kilos per hour throughput up to hundreds of tons depending on the nature of the product to be classified. They can be designed for single or multiproduct splits. Both vibrational and centrifugal sifting machines for classifying the product into fine and coarse fractions are frequently used in the production of powder coatings.

5.4.1.1 Tumbler Screening Machines

The movement of the material to be classified across the screen is achieved by a three-dimensional rotation of the screening drum. This rotation is produced by a rigid crank drive which has an adjustable slope. A vertical movement is superimposed on the horizontal gyratory movement to result in a drum oscillation with the parameter speed n, eccentricity of the crank e, radial slope α and tangential inclination β. The tumbler screening machines are designed in such a way that all mentioned parameters are infinitely variable. The material to be screened can move over the screen deck at any required speed or remain on the screen deck for any desired period. The interaction of tangential and radial inclination results in a spiral movement of the material across the screen, as shown in Figure 5.30.

The tumbler screening machines can be equipped as single, double, triple or quadruple screens to provide continuous screening of up to five fractions. Assembly of the machine is rather simple, as presented in Figure 5.31.

Because of their circular construction these machines can be easily equipped with supplementary devices which provide continuous trouble-free screening of the material by permanent cleaning of the screen bottom from the accumulated fines. In its simplest construction it can be done by bouncing balls made out of highly wear-resistant rubber which are fitted between a ball tray and the underside of the screen. They clean the screen continuously by their impact and wiping action. Another accessory for the same purpose is the roller brush cleaning device composed of two or more brush bearing arms carried on a bowl mounted drive unit driven from the main shaft. Cleaning of the screen can be achieved by an air cleaning device which comprises nozzle arms that rotate beneath the screen deck and blow air through the screen to clean it. A combination of air cleaning

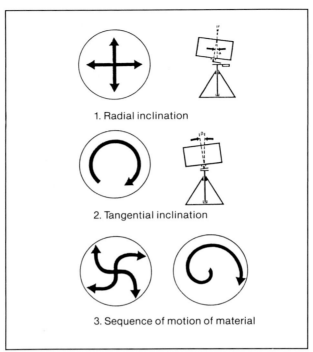

Figure 5.30 Spiral movement of the material across the screen in the tumbler screening machines (Reproduced by permission of Allgaier-Werke GmbH)

and rotational roller brushes is very effective for extremely difficult materials that stick to the underside of the screen.

5.4.1.2 Vibratory Screening Machines

These screening machines are specially suitable for removal of the oversize from the fines, the so-called scalping. Oversized particles are discharged at the side, while the discharge of the fines is done through the centre of the hopper shaped basic cylinder. The continuous discharge of the fine particles allows relatively high capacity.

The vibrational motion of the screen is achieved by the action of two unbalanced motors, each with a set of flyweights, offset against each other in order to obtain a circular motion forcing the particles to travel across the screen from the centre to the periphery. The intensity and direction of this oscillation are adjustable according to the characteristics of the material to be sieved.

The vibratory screening machines are constructed in modules, to give a single or double deck unit providing classification up to three fractions.

Figure 5.31 A five-unit construction system for a double screen tumbler sifter (Reproduced by permission of Allgaier-Werke GmbH)

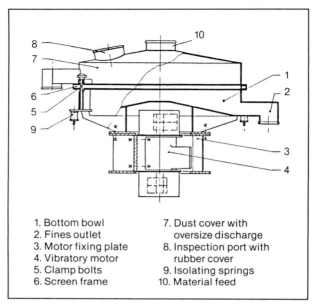

Figure 5.32 Vibratory tumbler screening machine (Reproduced by permission of Allgaier-Werke GmbH)

A combination of vibratory and tumbling action is realized in the vibratory tumbler screening machine which is schematically presented in Figure 5.32.

The sifter is available in a single or multideck unit depending on the number of fractions to be collected. Material is fed centrally and travels in a spiral screening pattern across the screen cloth to the periphery. Drive is by a vibratory motor arranged in the machine centre which generates circular tumbling vibrations by appropriate adjustment of the flyweights. The tumbling motion is easily adjustable in three dimensions in order to obtain an optimum machine setting, taking into consideration the flowing and screening characteristics of the material.

5.4.1.3 Pneumatic Tumbler Screening

Pneumatic tumbler screening installations are used to separate fine materials which are difficult to screen, especially in cases where the fine fraction ranges down to 30 μm. Even with high acceleration, the inertia forces of the individual particles are not sufficient to overcome the surface tensions that arise through electrostatic charges, moisture or van der Waals forces. In such cases the air-jet screening is the only solution for overcoming these forces which block the efficiency of the sieving machine.

267

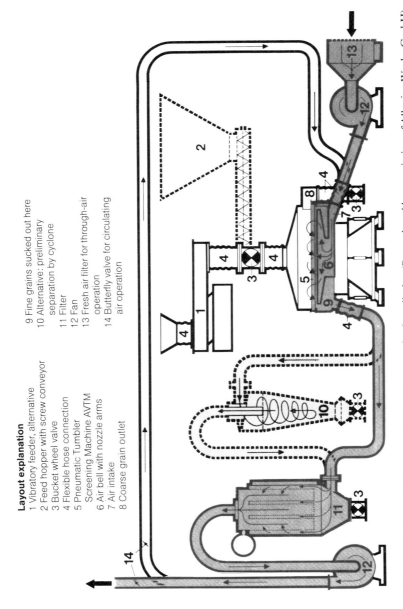

Layout explanation
1 Vibratory feeder, alternative
2 Feed hopper with screw conveyor
3 Bucket wheel valve
4 Flexible hose connection
5 Pneumatic Tumbler Screening Machine AVTM
6 Air bell with nozzle arms
7 Air intake
8 Coarse grain outlet
9 Fine grains sucked out here
10 Alternative: preliminary separation by cyclone
11 Filter
12 Fan
13 Fresh air filter for through-air operation
14 Butterfly valve for circulating air operation

Figure 5.33 Layout of a pneumatic tumbler screening installation (Reproduced by permission of Allgaier-Werke GmbH)

268 POWDER COATINGS

A pneumatic tumbler screening installation is presented in the diagram in Figure 5.33. The screen deck is cleaned from below by jets of air from a rotating nozzle system. This causes the material to be fluidized. The coarse material separates from the fine material and by means of the tumbling motion of the screen is continuously discharged. The fine material is drawn down through the screen and transported with the extracted air to an appropriate filter system. The pneumatic screening equipment can be fed either with fresh air or with recirculated air. Recirculation makes it possible to work with an inert gas which thus provides additional safety with regard to dust-explosion hazards.

5.4.1.4 Centrifugal Sifters

Centrifugal sifting machines, where the vibrating flat bed screen is replaced by a stationary cylindrical screen, have gained wide acceptance in powder coating manufacture. A sectional arrangement of a typical centrifugal sifting machine is presented in Figure 5.34. The material to be screened is fed via the sifter inlet by

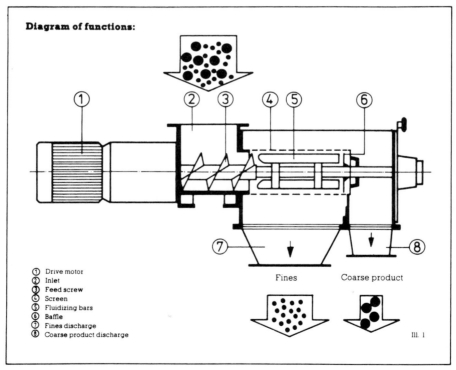

Figure 5.34 Centrifugal sifting machine (Reproduced by permission of AZO GmbH & Co. Postfach 11 20. Rosenberger Strasse, D-6960 Ostenburken)

TECHNOLOGY OF PRODUCTION OF POWDER COATINGS

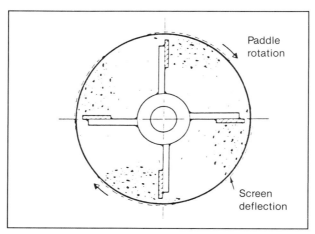

Figure 5.35 Uneven circumferential product loading in the centrifugal sifting machine [16] (Reproduced by permission of Paint & Resin Turret Group plc)

means of a rotary feeder, conveyor screw or vertical conveyor. The feed screw conveys the material into the cylindrical sifting chamber where it is picked up by the rotating fluidizing bars and thrown centrifugally against the screen. The helix configuration of the fluidizing bars ensures that the powder particles are traversed axially along the whole length of the machine. The fines penetrate the screen while the coarse particles, after passing round the end baffle, are discharged continuously.

The rotating fluidizing bars provide efficient means of distributing material over the screen periphery. However, powdered material to be sifted is not uniformly distributed in either the circumferential or axial direction. The uneven circumferential product loading is presented schematically in Figure 5.35 [16]. Each fluidizing bar carries a wave of semifluidized material in front of it, leaving an empty space between the front of the wave and the next bar. This produces a very useful secondary vibrating effect which helps in the self-cleaning of the screen. The sack fibres are screened without a beard forming at the screen, which adversely affects the screening process. In order to avoid the appearance of electrostatic charges during the sifting of powder coatings, the use of carbon fabrics is recommended.

The benefits and limitations of the centrifugal sifting machines as an alternative to flat bed sifting discussed by Tunnicliffe (16) are summarized in Table 5.8.

5.4.2 CENTRIFUGAL AIR CLASSIFIERS

Operating principles of the centrifugal air classifiers are based on the action of a vertical flow of air in both rotary and cyclonic devices. In the case of cyclones, this

Table 5.8 Benefits and limitations of centrifugal sifting machines compared to flat bed sifters [16]

Benefits	Limitations
Higher throughput in relation to screen area for a given separation efficiency	Design limited to single split type, i.e. output of two product streams only; cascade of machines necessary for multisplit operation
5–20% higher separation efficiency compared to other mechanical screening machines	Not suitable for very friable and very abrasive products
Vibration-free operation, thus less demanding with respect to supporting structures	Limited to medium size capacity; however, not relevant in the case of powder coating manufacture
Less noisy and dusty operation compared to vibrating sieves	
Quite strong deagglomerating action, less fines contaminating the coarse fraction	

is obtained by the tangential air inlet as presented in Figure 5.36. Rotary centrifugal air classifiers make use of a high-speed rotor located inside the classifier.

The vortex formed in the cyclone causes an inward spiral movement of the air stream from the tangential inlet to the central outlet. Similar streamlines can be drawn for the air which enters at the rotor periphery in the case of rotary centrifugal air classifiers.

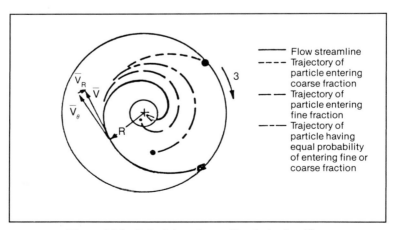

Figure 5.36 Principles of centrifugal air classifiers

TECHNOLOGY OF PRODUCTION OF POWDER COATINGS

The tangential velocity component (V_t) at any radial position (R) in the separating device [15] is given by

$$V_t = KR^n \qquad (5.1)$$

where K and n are constants whose values depend on the geometry of the equipment.

There is a difference in the tangential velocity component profiles between particles exposed to the action of free vortex formed by the tangential air inlet in the case of cyclones and forced vortex which exists in the rotary centrifugal air classifiers. The exponent n has a value of -1 in the case of cyclonic classifiers and 1 in the ideal case for the rotary types. The actual tangential velocity profile for the rotary centrifugal air classifiers is a combination of these two extreme cases, as presented in Figure 5.37 [15].

Particles which enter the classifier together with the air stream are subjected to the action of two forces: centrifugal force, which is proportional to the cube of the particle diameter and the square of the particle tangential component velocity, and drag force, which is proportional to the particle diameter and the ratio between the velocities of the air and the particle in the radial direction. Consequently, the trajectories of the large particles are dominated by the centrifugal forces and will be flung out to the wall of the container. In the case of cyclones they will sink in a circular path through the bottom exit. In the case of rotary classifiers they will be collected as a coarse fraction at the periphery of the

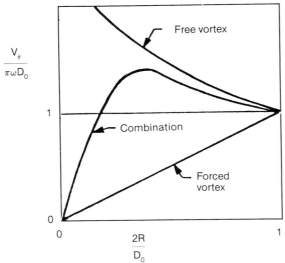

Figure 5.37 Air tangential velocity profiles [15] (Reproduced by permission of Federation of Societies for Coatings Technology)

rotor. The small particles exposed to a large drag per unit mass will follow the air stream and will leave the cyclon or the rotary classifier through the central outlet.

Depending on the rate of volumetric air flow (Q), the diameter (D_o) and the width (H) of the high speed rotor used to obtain the spiral-vortex air flow path, the rotational speed of the rotor (w), viscosity of the air stream (μ) and the particle density (δ), there is some critical particle diameter ($D_{50\%}$) known as the cut size where the drag and centrifugal forces are equal [17]:

$$D_{50\%} = \left(\frac{3k}{\pi D_0 w}\right)\left(\frac{\mu Q}{\pi H \delta}\right)^{1/2} \tag{5.2}$$

In the above equation k is a constant dependent on the flow pattern and has to be determined for a specific classifier design. The particles with this critical diameter have an equal chance to enter either the coarse or the fine fraction.

Even a superficial qualitative analysis of the equation defining the critical particle diameter shows that it will be more difficult to remove the fine particles from unpigmented powder coatings having a low particle density. Removal of the fine particles will also be adversely affected by increased air flow rates. On the other hand, increasing the rotor speed decreases the cut size.

In the ideal case when the operating conditions of the system completely comply with the mathematical description it will be a sharply defined boundary between the fine and the coarse fraction as presented in Figure 5.38 (the dashed line). In practice, however, such a sharp classification can never be achieved. Several reasons contribute to a deviation of the practical results from the ideal. Powder particles are not of the spherical shape that is assumed in the derivation

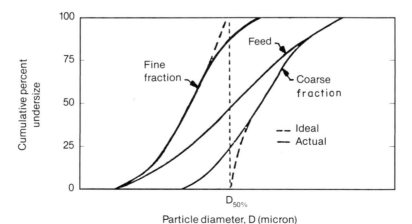

Figure 5.38 Representation of the classification results by fraction size distribution curves [15] (Reproduced by permission of Federation of Societies for Coatings Technology)

TECHNOLOGY OF PRODUCTION OF POWDER COATINGS

of equation (5.2). The interactions between the particles in the air stream are also neglected. The collision between the particles affects the particle velocity during its residence in the classifier which can, for example, decrease the velocity of the tangential component of the coarse particles, resulting in contamination of the fine fraction with particles having a larger diameter than the critical. On the other hand, fine particles can enter the classifier in the form of agglomerates behaving as having a larger diameter and are carried to the coarse fraction. This phenomenon is especially emphasized when extremely fine powders have to be classified. Decreasing the particle size rapidly increases the surface/volume ratio, resulting in increasing interaction between the particles due to the surface forces such as van der Waals forces. For these reasons it is very difficult to obtain a particle size lower than 5 μm by centrifugal air classifiers [15].

It is difficult to choose which criteria should be considered as relevant in the evaluation of the classifier efficiency since the cut size, the sharpness of the cut and the output of the machine are important elements in judging the equipment performances to be mutually dependent. The geometrical method suggested by Eder [18] seems to be widely accepted [15, 19]. It uses the following equation:

$$\Phi = \frac{D_{75\%}}{D_{25\%}} \tag{5.3}$$

in which $D_{75\%}$ is the particle size diameter corresponding to 75% grade efficiency and $D_{25\%}$ is the particle size diameter corresponding to 25% efficiency (Figure 5.39). The grade efficiency which is plotted against the particle diameter

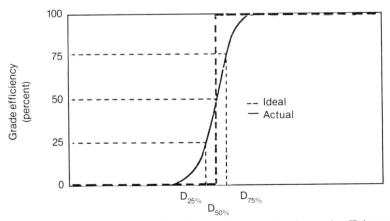

Figure 5.39 Representation of classification performance by the grade efficiency curve [15] (Reproduced by permission of Federation of Societies for Coatings Technology)

in Figure 5.39 is defined as

$$n_{(D)} = 100 \frac{\text{weight of size } D \text{ entering coarse fraction}}{\text{sum of weights of size } D \text{ in fine and coarse fraction}}$$

The choice of the most suitable classifier depends on the desirable cut size and the output of the machine. High capacity units are usually used in closed circuit grinding systems and have a cut size range between 20 and 70 µm. This is the usual particle size range for powder coatings. For classifying below 20 µm units with a moderate capacity are used. Laboratory centrifugal units with a capacity of a few grams per hour can achieve classifications even below 1 µm.

The principal construction of a widely used centrifugal air classifier with a horizontal arrangement of the classifying wheel shaft is depicted in Figure 5.40. The machine consists of a horizontally mounted classifying wheel (1) with an

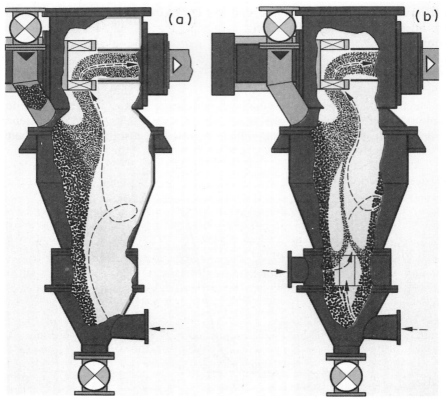

Figure 5.40 Turboplex ultra-fine classifier (Reproduced by permission of Alpine AG)

outlet for the fine particles and classifying air flow (2). The classifying air injected into the machine base (3) flows inwards through the classifying wheel and discharges the fine material, whereas the coarse particles, being rejected by the classifying wheel, leave the classifier through the coarse material outlet (4). The coarse material is permanently cleaned from the fine particles by intensive rinsing with air on its way to the outlet before it leaves the classifier. In this way additional separation of the fine material is achieved. The product can be fed from above by a rotary valve (5) or directly through the inlet of the classifying air flow (3). In the latter case the machine acts as an air stream classifier and can be operated directly in-line with pneumatically discharging mills. The fineness of the end product is adjusted simply by varying the classifying wheel speed.

Turboplex classifiers differ from similar types of equipment with respect to the horizontal arrangement of the classifying wheel shaft. This gives a longer service life, higher fines yield and improved cleanliness of the coarse material from fine particles. This construction helps to eliminate the formation of deposits in the fine material discharge when handling materials which are sticky or tend to build up. Another distinguished characteristic of the classifier is a reliable rinsing air supply between static and moving parts which ensures a very precise sharp top size limitation of the fine material produced. Since the material feed is independent from the air supply, even when the feed rate is variable fine material with a constant quality is produced.

A modified version of the single wheel classifier (Figure 5.40b) with an additional air flow inlet provides an even higher fines yield and optimum cleanliness of the coarse material. This construction also helps in classifying products than are difficult to disperse in air.

Additional improvement in the centrifugal air classifiers is the multiwheel version presented in Figure 5.41. It consists of several horizontally mounted classifying wheels each with a separate drive (1). The wheel speed is controlled via one adjustable electrical frequency inverter. The product is introduced either by an airlock feeder (2) or directly into the classifier air flow, when it acts as an air stream classifier. The fine material is discharged through the outlet (3) while the coarse fraction leaves the classifier from the bottom (4). A distinguished feature of the multiwheel classifier is a very high output rate of superfine products with a particle size range down to 3–6 μm. With increasing classifier capacity the cut point and precision of the cut change in the case of superfine products can usually only be obtained at the expense of the machine capacity. By installing several, small, equal-sized classifying wheels in the classifier head, ultra-fine products with a high output rate are obtained in a single step classifying operation. The feed rate of the multiwheel classifiers ranges approximately from 150 to 7500 kg/h, depending upon the end product fineness, the nature of the product and the size of the classifier.

Table 5.9 summarizes technical details of the Turboplex ATP classifiers.

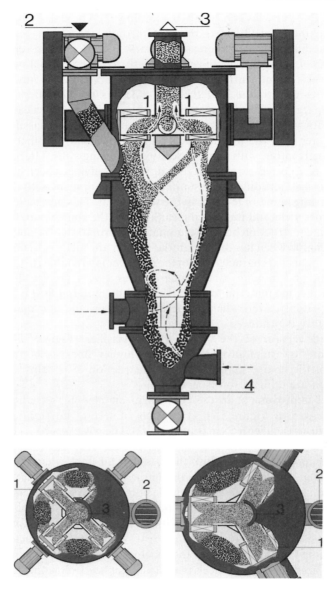

Figure 5.41 Turboplex multiwheel classifier (Reproduced by permission of Alpine AG)

Table 5.9 Turboplex ATP centrifugal air classifiers—technical data (Reproduced by permission of Alpine AG)

Turboplex ATP	100	100/4	200	200/4	315	315/3	500
Fineness (d97)[a]	4–100	3–60	5–120	4–70	6–120	6–120	8–120
Feed rate (kg/h)[b]	50–200	150–400	200–1000	600–3000	500–2500	1500–7500	1250–8000
Classifying wheel							
Number	1	4	1	4	1	3	1
Diameter (mm)	100	100	200	200	315	315	500
Speed (rev/min)	1150–11500	1150–11500	600–6000	600–6000	400–4000	400–4000	240–2400
Drive power (kW)	4	16	5.5	22	11	33	15
Classifier design							
Single wheel	a		a		a		a
Multiwheel		a		a		a	
Air stream		a	a	a	a	a	a

[a] Fineness range d97 = particle size through which 97% of the product passes. Data are based on a material with a density of 2.7 g/cm^3.
[b] Dependent on fineness d97 and fine material capacity. Data are based on classification with a very high separation sharpness.

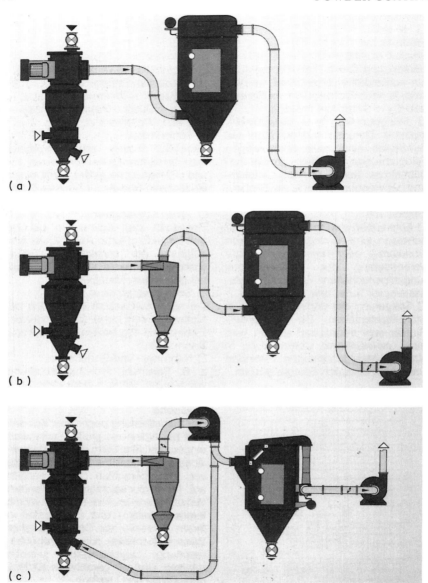

Figure 5.42 (*caption opposite*)

TECHNOLOGY OF PRODUCTION OF POWDER COATINGS

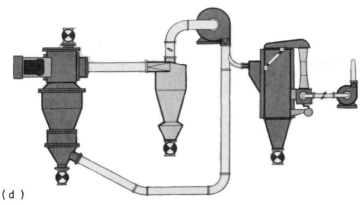

(d)

Figure 5.42 Typical standard plant designs for collecting the end product (Reproduced by permission of Alpine AG)

5.5 POWDER COLLECTION AND DEDUSTING

The collection of fine particles leaving the outlet of the air classifier is usually done by so-called side removal filters or more often by a combination of a cyclone for preliminary fine material separation and a side removal filter. Several typical plant designs for the last stage of the powder production process are presented in Figure 5.42.

Through-air operation without preliminary fine material separation via cyclone is presented in Figure 5.42(a). This is the most economical design from the investment and maintenance standpoint, but certainly not the most economical for exploitation.

Secondly, Figure 5.42(b) represents through-air operation with preliminary separation of the fine material via a cyclone. Cyclones have become the essential component in modern grinding systems for powder coatings. They reduce considerably the cleaning expense in comparison with filters and permit great availability of the system, even in the case of frequent colour changes. The collecting efficiency of the cyclone, however, decreases with increasing dimensions. This makes the use of cyclones for capacities higher than 1000 kg/h inefficient. Figure 5.43 represents the cyclone collecting efficiency in terms of product loss per hour as a function of the cyclone size and capacity [20].

Figure 5.42(c) depicts an operation in a partly closed circuit, where 50% of the classifying air is recirculated back into the classifier after separation from the fine material in the cyclone. The costs for the final dust collection are very favourable with this mode of operation.

Finally, Figure 5.42(d) presents a closed circuit operation where it is necessary to take only a small air bleed from the closed air circuit. To dedust this bleed air, a

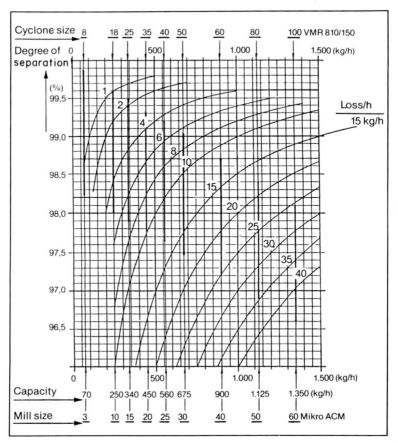

Figure 5.43 Cyclone collecting efficiencies in Mikro ACM systems for powder coatings (Reproduced by permission of Hosokawa MikroPul GmbH)

relatively small baghouse filter is sufficient. In bigger plants even this cost is avoided by ducting the bleed air into a central dedusting system.

The combination of a cyclone with a side removal filter permits the lowest product loss regardless of the capacity of the plant. In large batch production the side removal filter is more efficient than the cyclone. For medium and small batches the cyclone can almost completely replace the side removal filter. The choice between cyclones as powder coating collectors and side removal filters depends on the plant capacity and the diversity of powder coatings produced with respect to the types and colours. Table 5.10 gives a comparison between these two systems.

Very often residual dust filters are used as safety devices for the side removal

TECHNOLOGY OF PRODUCTION OF POWDER COATINGS

Table 5.10 Comparison between cyclone and side removal filters

Cyclone advantages versus side removal filters	Cyclone disadvantages versus side removal filters
Quick access	Lower product recovery rate
Easier, faster and perfect cleaning	High loss of additives (fumed silica, for example)
Less powder deposits	Decreasing efficiency when the ultra-fine portion increases
No fibre contamination from the filter media	Dropping efficiency in the case of air leakage
Stable pressure drop and continuous powder flow	No performance control while in operation; product losses are realized when the batch is finished
Constant granulometry	More sensitive to deviations in the operating conditions; supervision by trained personnel necessary for correct cyclone use
Lower investments, no spare parts, no maintenance	
No frequent change of filter bags and no risk of malfunction	

filters. In such cases the dust emission of the grinding system can be reduced to less than 5 mg/m^3. Dust removal filters also function as safety equipment in the case of malfunction of the side removal filter.

Residual dust removal filters require almost no maintenance work, increase the reliability of the grinding and classifying operation and improve the pollution control.

Among several commercial residual dust removal filters the so-called Mikro Pulsaire bag filters are very often in use for production of powder coatings. They consist of removable filter bags placed in a housing with the different sizes and form dependent upon the required filter capcity (Figure 5.44).

The air contaminated by the powder particles enters the air plenum and flows from the outside to the inside of the filter bags. The fastening of the bags, which are slipped over bag retainers, can be done either by means of bag clamps or by clampless twist venturi, as shown in Figure 5.45. With both versions the blow tubes can be moved out in groups to allow quick access to the filter bags.

The dust collects on the outside of the filter bag, while the clean air passes through the filter media into the inside of the bag and leaves the filter through the outlet nozzle.

Cleaning of the filter bags is accomplished by very short compressed air jets coming from the nozzles of the blow tubes, which are mounted above the filter bags. During the injection of compressed air into the venturis, extra air is added using the jet pump principle. Relaxation occurring when the cleaning air leaves the venturi tube produces a pressure wave that continues over the whole filter bag

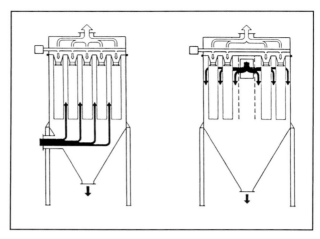

Figure 5.44 Principle of operation of Mikro Pulsaire dust collectors (Reproduced by permission of Hosokawa MikroPul GmbH)

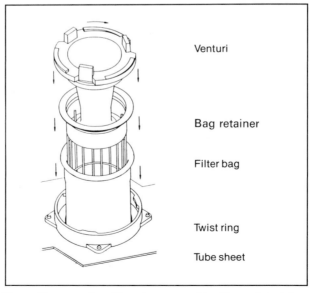

Figure 5.45 Venturi/bag retainer/filter bag assembly (Reproduced by permission of Hosokawa MikroPul GmbH)

TECHNOLOGY OF PRODUCTION OF POWDER COATINGS

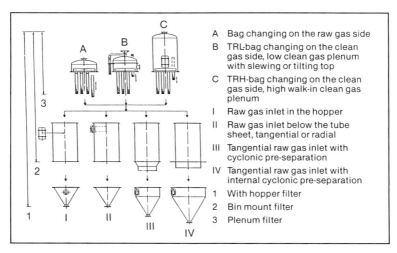

Figure 5.46 Principal design of cylindrical filters (Reproduced by permission of Hosokawa MikroPul GmbH)

area. This flexes the filter bags. The accumulated dust is thrown off into the hopper and from there is discharged through the discharging device, which is usually an airlock rotary valve.

The cleaning sequence is automatically controlled by an electronic control instrument and by the diaphragm and solenoid valves fitted to each blow tube.

Different designs of the cylindrical types of Mikro Pulsaire dust collectors, as presented in Figure 5.46, allow bag changing on the raw gas side (A), clean gas side by means of slewing or tilting top (B) or on the clean gas side having a high walk-in clean gas plenum (C). The raw gas inlet can be in the hopper (I), which is a suitable construction for heavy dusts prone to agglomeration. For light non-agglomerating and very fine dusts, as in the case of powder coating production, the raw gas inlet is below the tube sheet in the tangential or radial version (II). Constructions with the tangential raw gas inlet and cyclonic preseparation (III) and the tangential raw gas inlet and internal cyclonic preseparation (IV) are also available.

The clampless bayonet–venturi technique permits quick and secure installation of filter bags and bag retainers. A metal ring is sewn into the collar of the filter bags, serving as a seal and preventing the bag from dropping through the tube sheet opening. If the bags need to be dropped into the hopper during the replacement, the metal ring is replaced by a flexible one.

Mikro Pulsaire dust filters are highly effective devices for gas cleaning. They are produced in various models and sizes so that it is easy to choose for a model that best fits the plant capacity. The cleaning operation is fully automatic. The filter operates at a constant pressure drop between 8 and 15 mbar due to continuous efficient cleaning of the filter media. The only moving parts of the filter are the diaphragms of the externally mounted compressed air pulse control valves, which require a minimum of maintenance.

5.6 QUALITY CONTROL

This section refers to the methods used to assess quality of powder coatings by tests performed either by the powder producers or by the end users. Since many of the test procedures are standardized there is no need to spend much time describing the details. In such cases the principle of the test method will be briefly described with reference to the standardized test procedures. Somewhat more detail will be given for the test methods typical for powder coatings only, which have not yet been the subject of standardization. The testing procedures for quality assessment of powder coatings have been a matter of official standardization [21].

5.6.1 ACID NUMBER DETERMINATION OF THE POWDER COATING RESINS [22]

The acid number or the acid value is defined as the number of grams of potassium hydroxide required to neutralize the carboxyl groups of one gram of resin. Since most of the powder coating resins have a relatively low solubility, the toluene/ethanol solvent blend normally used for acid value determination of most of the resins for conventional solvent based coatings is commonly replaced by tetrahydrofuran/water blend in the ratio of 10:1. This solvent blend is suitable for dissolving most of the powder coating resins.

The following formula is used to calculate the acid number:

$$\text{Acid number} = \frac{56.1 AN}{W} \qquad (5.4)$$

where

A = potassium hydroxide solution used for titration (ml)
N = normality of potassium hydroxide solution
W = weight of the resin (g)

5.6.2 HYDROXYL NUMBER DETERMINATION [23]

The normal procedure for determination of the hydroxyl number of powder coating resins involves dissolution of the sample in a blend of cyclohexanone and pyridine (ratio of 1:1), reaction with acetic anhydride, addition of water to react with the excess of acetic anhydride, addition of tetrahydrofuran to redissolve the sample and titration of the reaction product with 0.5 N potassium hydroxide.

TECHNOLOGY OF PRODUCTION OF POWDER COATINGS 285

This procedure which slightly differs from the standardized methods has been specially developed for polyester powder resins and related products in the laboratories of DSM Resins BV [24]. The hydroxyl number can be calculated from the formula:

$$\text{Hydroxyl number} = \frac{56.1(B - A)N}{W} + \text{acid number} \qquad (5.5)$$

where

A = potassium hydroxide solution spent for titration of the sample (ml)
B = potassium hydroxide needed to titrate the blank (ml)
N = normality of the potassium hydroxide solution
W = weight of the resin sample (g) (recommended to be calculated as $W = 280/$expected hydroxyl value)

5.6.3 EPOXY EQUIVALENT WEIGHT DETERMINATION [25]

The perchloric acid method widely used for this purpose is based on the stoichiometrical reaction of 1,2-epoxy group with hydrogen bromide, which is generated during the reaction of perchloric acid ($HClO_4$) with tetraethylammonium bromide (NEt_4Br). The procedure involves titration of a solution of epoxy resin dissolved in methylene chloride containing tetraethylammonium bromide and crystal violet indicator with a standardized solution of perchloric acid in glacial acetic acid. When the epoxy groups are completely consumed, the free hydrogen bromide causes a colour change in the crystal violet indicator.

The epoxy equivalent weight (EEW) of the resin is calculated by the following formula:

$$\text{EEW} = \frac{100W}{AN} \qquad (5.6)$$

where

W = epoxy resin sample (g)
A = perchloric acid solution used for titration (ml)
N = normality of the perchloric acid solution

5.6.4 DETERMINATION OF THE GLASS TRANSITION TEMPERATURE

Differential scanning calorimetry is the most commonly used method for determination of T_g of powder coating resins and powder coatings. This technique was extensively described in Section 3.2.1 and will not be a matter of consideration in this section. Since T_g is dependent on the heating rate, the report on T_g should always be accompanied by the data relating to the heating rate used in the test. For powder coating resins it is common practice to use a heating rate of 5°C/min [26].

5.6.5 DETERMINATION OF THE MELTING POINT AND MELTING RANGE [27]

Polymers used for production of thermoplastic powder coatings exhibit a certain degree of crystallinity and a more or less well-defined melting temperature. This temperature determines both the processing conditions during production of powder coatings and the stoving condition during their application. On the other hand, polymers which are used for production of thermosetting powder coatings are mainly of an amorphous nature. They do not exhibit defined melting temperatures like the crystalline polymers, but rather a broad melting range in which the material begins to flow. This melting range should be considered as a minimum temperature interval for processing powder coatings during extrusion.

Although the d.s.c. technique can be used to determine the melting point of the crystalline polymers, a much simpler and cheaper technique employs the so-called capillary method. This method can be used to determine the melting point, melting range, sintering point and fusion point of the resins

The resin which has to be tested is powdered with a mortar and pestle and the powdered material is filled in a capillary tube to a depth of 1.5–2.0 cm. The capillary is then placed in a bath with a controllable spread of heating. The temperature is raised quickly to about 20°C below the expected melting range. Thereafter the heating rate is slowed to 1–2°C/min up to the melting point. The report may include the following:

1. The melting range is defined as the two temperatures in between which the sintering and fusion of the material takes place.
2. The melting point is defined as the temperature at which all the material just becomes molten.
3. The sintering point is defined as the temperature at which the first signs of contraction of the powdered column in the capillary occurs and at which the material column frees itself from the capillary wall.
4. The fusion point is defined as the temperature at which the material shows a fusion or translucency at any point in the capillary tube.

For practical reasons laboratories which possess a gradient oven can make use of this apparatus to evaluate the melting range.

5.6.6 DETERMINATION OF THE SOFTENING POINT [28, 29]

A method that has been well established for years is the so-called ring and ball method. The procedure is rather simple and consists of melting and pouring the molten resin into a preheated metal ring with standardized dimensions. After cooling the excess resin is cut with a preheated knife and removed. The ring with the solid resin is then placed in a horizontal position in a bath with glycerol or other heat transfer liquid. A ball with a standardized dimension and weight is placed on the surface of the solid resin and the glycerol bath is then heated with a rate of 5°C/min. The softening point of the resin is recorded as the temperature at which the material touches the lower horizontal plate.

5.6.7 VISCOSITY OF POWDER COATING RESINS [30, 31]

Two instruments are typically used for determination of the resin viscosity: the Emila rheometer with a special measuring cup at elevated temperatures and the ICI cone and plate viscosimeter. Although different in construction, both instruments operate on the same principles.

In the case of the Emila rheometer a hollow measuring cylinder is rotated by a motor at a constant speed in the measuring cup filled with the molten resin. The measuring cup is heated by a stream of oil coming from a thermostat or directly by electronically controlled electric heating elements. The rotation of the cylinder causes laminar flow in the liquid. The resistance against this rotation is proportional to the dynamic viscosity of the molten resin and can be read from the torque engine.

The operating principle of the ICI cone and plate viscosimeter is similar. It differs from the Emila rheometer in the construction of the measuring head which consists of a thermostatted plate and rotating cone instead of a cup and cylinder (Figure 5.47).

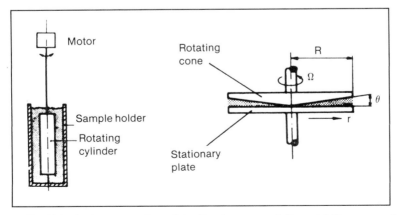

Figure 5.47 Principal construction of Emila rheometer (left) and ICI cone and plate viscosimeter (right)

The ICI cone and plate viscosimeter is a handler instrument, easily cleaned and requiring less material for testing. However, the temperature control is not as accurate as with the Emila rheometer. Moreover, in the case of high viscosities, the resin creeps up and detaches from the plate, thus reducing the contact surface between the rotating cone and the stationary plate realized through the resin layer, which leads to inconsistent results in high viscosity regions.

The Emila rheometer is widely used for melt viscosity determination of powder coating resins in Europe. The ICI cone and plate and the procedure for measuring melt viscosity is recommended by the Powder Coating Institute in the United States [30].

5.6.8 COLOUR OF POWDER COATING RESINS [32]

The most widely used method for determination of the colour of the powder coating resins is the 'Gardner 1953' using the Gardner–Hellige varnish comparator. The resin is placed in a clean Gardner viscosity tube in a 50% solution in dimethyl formamide. Determination of the colour is done visually by comparing the colour of the sample with the reference colour from the colour disc that best matches the colour of the sample.

5.6.9 DETERMINATION OF THE GEL TIME

The gel time of powder coating resins mixed in a given proportion with the suitable crosslinker gives an indication of the reactivity of the system. It is the standard test for resins and for powder coatings. The method is simple, and although it seems to be very subjective it gives useful results.

By the very common manual method for gel time determination described in the DIN Standard 55990, ca. 200 mg of the material is placed on a preheated thermostatted plate with a convex shaped cup. The stop watch is started immediately after pouring the material into the cup. The material is stirred gently with a thin needle over the total surface of the cup. After a certain time the material tends to creep up the stirring needle detaching at a certain moment from the cup bottom. This time is registered by the stop watch as the gel time. A procedure differing only in small detail is recommended by the Powder Coating Institute [33]. The range for any gel time determined by this procedure is generally $\pm 10\%$ compared to the control material.

Automated versions of the gel time testers eliminate the errors due to differences in individual operators. Weder [34] reports a standard deviation of 2.1 seconds at an average gel time of 185 seconds with an automatic gel time tester. Compared to this the manual operation provides a standard deviation of 11.6 seconds at an average gel time of 193 seconds. The instrument operates on thermomechanical principles. The same amount of probe (200 ± 10 mg) is heated to the desired temperature in a heating block. A precision measuring dolly made

TECHNOLOGY OF PRODUCTION OF POWDER COATINGS   289

of glass or stainless steel dips into the specimen and is pulled out at programmable intervals. The movement of the dolly which is directly dependent on the viscoelastic properties of the molten powder coating is plotted against the setting time to produce a penetration curve. All functions of the dolly are set on a central control unit which evaluates at the same time the measuring data and transmits them to either a graphical recorder or small computer, depending on the configuration. In this way the reciprocating movement of the dolly is recorded as a measure of the degree of hardening against time. Compared to the manual methods, the automatic methods for determination of the gelation time offer much better reproducibility.

5.6.10 ACCELERATED STABILITY TEST [35,36]

When the glass transition temperature of the coating composition is lower than the storage temperature, powder particles undergo agglomeration which in extreme cases can completely destroy the free flowing properties of the powder. Such a product is unsuitable for application by either fluidized bed or electrostatic spraying. In the case of rapid curing thermosetting powder coatings a premature reaction can take place during storage that has an adverse effect upon the film forming properties during curing.

A procedure for the accelerated stability test recommended by the Powder Coating Institute allows rather quick evaluation of both the physical and chemical stability of powder coatings [37]. The method suggests storage of a small quantity of powder coating at 40 or 45°C and periodical visual testing of free flowing properties of the material followed by determination of the gel time. The test is discontinued when the powder is no longer free flowing and/or the formed lumps cannot be easily broken.

The number of days before physical properties of the powder coating are affected is reported as a measurement for physical stability of the product.

Chemical stability of powder coatings is usually expressed in terms of a decrease in gel time compared to the original readings. It is very difficult to establish certain norms which will define chemical stability of the coating. Therefore it is recommended that these results be used in comparison against standard thermosetting powder coatings with a well-established chemical stability. To obtain reliable results care should be taken that the standard powder coating employs the same curing chemistry.

5.6.11 DENSITY OF THE POWDER COATING MATERIALS [38-40]

The method is recommended by the Powder Coating Institute and can be used to determine the density of powder coatings, pigments, resins, crosslinkers and other solid ingredients of powder coatings. The method uses a 50 ml volumetric flask in which part of the liquid (in most cases hexane) is displaced by the solid material.

The density of the powder coating material can be calculated by the following equation:

$$D_p = \frac{W_{fp} - W_f}{50 - (W_{fpl} - W_{fp})/D_l} \tag{5.7}$$

where

W_f = weight of the empty flask
W_{fp} = weight of the flask containing the powder sample
W_{fpl} = weight of the flask containing the powder sample and the liquid up to the 50 ml mark
D_l = density of the liquid used

The precision and accuracy of the method have not been determined statistically, but a limited survey of testing performed by various laboratories indicates that results appear to differ by less than 0.05 g/ml.

5.6.12 COMPATIBILITY OF POWDER COATINGS [41]

This test can indicate possible problems caused by contamination of the coating during its production or application by other powder coatings or related raw materials. Two powder coatings which are expected to be applied on the same line are mixed in proportions indicated below:

100/0; 99.5/0.5; 95/5; 75/25; 50/50; 25/75; 5/95; 0.5/99.5; 0/100

After spraying on the test panels and curing at the recommended temperature the panels are checked for cratering or gassing, changes in gloss, colour contamination and changes in physical properties after curing.

5.6.13 PARTICLE SIZE ANALYSIS [42]

Particle size is an important factor influencing powder coating application performances. Not only the average particle size but also the particle size distribution becomes more important, especially in areas where powder coatings are expected to meet sophisticated requirements.

Many different methods for particle size determination are used in practice. The list includes microscopic analysis, optical arrays, flow ultramicroscopy, light scattering, sieving, filtration, sedimentation methods, electrozone sensing, turbidimetry, optical diffraction, optical particle counters, velocimetry, neutron scattering, centrifugation, etc. Excellent articles review the methods developed up to 1975 [43] and 1989 [44]. The reader interested in the methods typically used to determine the particle size of powder coatings can consult the book of Harris [45]. Because of space reasons, only the principles of the technique based on laser diffraction particle sizing which has been available for over a decade will be

TECHNOLOGY OF PRODUCTION OF POWDER COATINGS

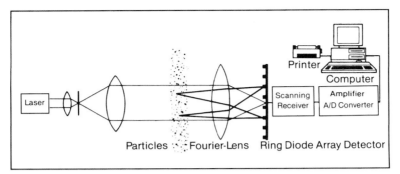

Figure 5.48 The layout of a typical laser diffraction particle size analyser (Malvern Series 260 Oc) (Reproduced by permission of FMJ International Publications Ltd)

described in this section. This technique is at the moment certainly the best method offering speed, performance, ease of use and reliable results, and is unmatched by other techniques.

The layout of the diffraction particle sizer produced by Malvern Instruments Ltd is presented in Figure 5.48 [46].

A small helium–neon laser provides a parallel monochromatic laser beam which is expanded in diameter to ensure that a large number of particles will be in the measuring zone at any one time. The sample is placed in a special feeder attachment. It is dispersed in a fast air stream which carries the particles in a controlled flow through the measuring zone. As the sample passes through the beam, individual particles scatter light at angles related to their respective sizes; small particles scatter light at large angles and vice versa. The scattered light is collected by a Fourier transform lens and sent onto a silicon photodiode array detector. The detector measures the light scattered over a range of angles and produces a signal related to the composite light energy pattern of all particles in the sample. This signal is digitized and analysed by the computer. The particle size distribution is then displayed or printed in tabular or graphical form.

5.6.14 MEASUREMENT OF ELECTRICAL PROPERTIES OF POWDER COATINGS

5.6.14.1 Measurement of Powder Resistivity

Resistivity of powder coatings is an indicator of the chargeability of the particles and their behaviour during application by electrostatic guns. Hughes [47] describes one of the simplest methods for measuring resistivity of powder coatings which makes use of the so-called resistivity cell as illustrated in Figure 5.49.

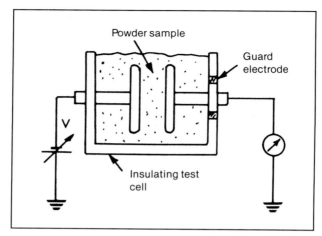

Figure 5.49 Schematic of a simple powder resistivity test cell (Reproduced by permission of the Oil and Colour Chemists' Association)

A sample of powder coating is subjected to a steady direct current potential while the steady state conduction current is measured. Knowing the cell geometry, the resistivity of the powder coating (r) can be easily calculated by the following equation:

$$r = \frac{VA}{Id} \quad (\Omega\,\text{m}) \tag{5.8}$$

where

A = electrode surface
d = distance between the electrodes
V = cell potential
I = steady state electrical current

Contrary to what can be expected, evaluation of different resistivity cells indicates that resistivity is minimally affected by powder compression. Therefore, prior to commencing measurements, the powder is gently tapped so that particle packing gives reproducible results [47].

5.6.14.2 Powder Coating Permittivity

As will be seen in the following chapter, the relative permittivity of the powder (ε_r) might be an important parameter influencing the charge acceptance of the powder particles. Measurement of the permittivity can be done with the same resistivity cell using a standard capacitance measuring instrument. The relative permittivity can be determined as a ratio between the capacitance of the

TECHNOLOGY OF PRODUCTION OF POWDER COATINGS

resistivity cell filled with powder coating (C_r) and the air capacitance of the empty cell (C_0):

$$C_0 = \frac{\varepsilon_0 A}{d} \qquad (5.9)$$

$$C_r = \frac{A\varepsilon_0 \varepsilon_r}{d} \qquad (5.10)$$

where

ε_0 = permittivity of the free space = 8.854×10^{-12} F/m
ε_r = relative permittivity of the powder particle

From (5.9) and (5.10) follows

$$\varepsilon_r = \frac{C_r}{C_0} \qquad (5.11)$$

5.6.14.3 Charge/Mass Ratio (q/m)

Measurement of this very important charging characteristic of powder particles is relatively simple, although certain precautions have to be taken in practice. The so-called Faraday cup arranged as presented in Figure 5.50 is the most commonly used device for determining the charge/mass ratio [47].

A small metallic cup is enclosed by a larger grounded metallic cup. The inner cup is connected to an electrometer instrument so that any charge associated with the collected sample sprayed by the electrostatic gun onto the test cell can be

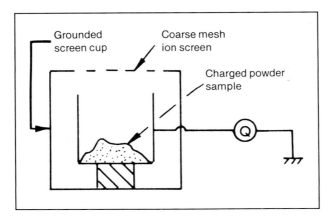

Figure 5.50 Schematic of a Faraday cup test cell (Reproduced by permission of the Oil and Colour Chemists' Association)

recorded. Dividing the recorded charge by the difference in weight of the cup before and after the test leads to the charge/mass ratio for the powder in question.

This simple technique is quite suitable for determining q/m values of powder coatings when applied by triboelectric guns. However, the technique is vulnerable to errors when the efficiency of high-voltage corona guns has to be assessed. In such a case the flying powder particles are always accompanied by a stream of free ions emanating from the gun nozzle. The charge of the free ions entering the inner cup will be registered by the instrument, although they do not contribute to the charge/mass ratio. As a result much higher measurements than the true values are obtained. An improvement on the measuring technique involves an attached grid screen across the inlet of the Faraday cup which traps the free ions before they enter the inner cell. In practice this technique does not perform satisfactorily. The grid itself is very soon coated with powder particles which initiate back-ionization, thus discharging further oncoming particles and adversely affecting the accuracy and validity of the obtained results. An improvement on the technique using a water irrigated grid has been proposed by Moyle and Hughes [48]. In this method powder particles accumulating on the grid trap are permanently washed out by a stream of water, allowing an accurate measurement to be taken of the charge/mass ratio of a corona-charged powder cloud.

Details about the construction and practical use of a commercially available electrostatic powder coating diagnostic kit are given by Hearn [49].

5.6.15 INCLINED PLATE FLOW TEST

The flow of powder coatings is too complex a phenomenon to be determined by a single test. However, the inclined plate flow test can be considered as a useful indicator of the degree of flow that may occur during the curing cycle, contributing considerably to the surface appearance of the cured film. The test procedure is rather simple. A pellet of powder coating prepared by means of a mould press is placed horizontal to the glass panel that has been preheated to the test temperature. The panel stands in an oven on a metal plate rack assembly which is capable of being maintained in both the horizontal and inclined positions by means of a lever without opening the oven door. When the pellet is slightly molten in the flat position for a time of 30 seconds, the panel is tilted to 65° and kept in this position for 15 minutes. The amount of flow is measured as the length of the trace that is made by the molten powder. Since the oven design, angle of inclination and pill variations affect the results, making interlab reproducibility difficult to correlate, a procedure established by the Powder Coating Institute recommends a test temperature of 177°C (350°F), pill thickness of 6 mm, pill weight one half of the specific gravity of the test powder, angle of inclination of 65° and time for keeping the pellet in a horizontal position of 30 seconds [50].

TECHNOLOGY OF PRODUCTION OF POWDER COATINGS 295

5.6.16 GRADIENT OVEN TEST

The gradient oven laboratory apparatus manufactured by Byk-Labotron is a very useful instrument for evaluation of the stoving characteristics of powder coatings. The essential part of the oven is the heating bank which consists of 45 heating elements embedded in aluminium blocks. The elements are arranged side by side, and are separated by insulation material so that each element may be programmed with a different temperature. The microprocessor through which the control of the heating bank is effected allows a temperature rate control of between 2 and 30°C/min and simulation of real time–temperature profiles of industrial ovens. The oven can produce a linear gradient with a maximum span of 100°C within the operating range of the instrument between 40 and 250°C and a stepwide gradient with two, three or four steps. In all cases the maximum difference between the steps can be 50°C.

The coating to be tested is applied on a panel which is placed on the heating bank. An electrically driven pressing device ensures an efficient contact between the panel and the heated surface, so that the time taken for the sample to reach the programmed surface temperature varies between 20 and 120 seconds dependent on the thickness of the test panel.

The gradient oven can be used to determine the melting point or melting range of the resins used in the manufacture of powder coatings or the coatings themselves. However, the most important use of the gradient oven is in the evaluation of the curing characteristics of the thermosetting coating. The possibility to set up different curing temperatures on a single test panel results in considerable savings in experimental time since several tests may be conducted simultaneously. For example, in one single experiment with a fixed curing time one can easily determine the minimum curing temperature needed to reach good mechanical properties. At the same time an impression of the coating behaviour at undercure and overcure can be obtained with respect to the mechanical properties, together with the temperature dependence of the yellowing tendency. The use of the stepped gradient facility allows larger test areas of a single temperature to be produced in cases where the optimum curing temperature is more closely defined. Finally, the instrument allows a simulation of the temperature profile of industrial ovens. In such a case the temperature of the heating bank is uniform along the whole length at any given time following the profile of the real industrial oven, as presented in Figure 5.51 [51].

5.6.17 MEASURING THE THICKNESS OF POWDER COATINGS [52–54]

The simplest method for measuring the thickness of powder coatings involves causing mechanical damage to the film in order to reach the substrate and measurement of the depth of the niche by micrometer.

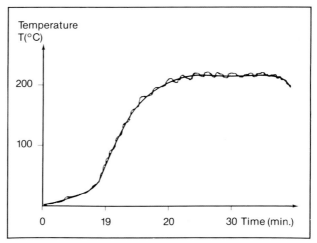

Figure 5.51 Recorded temperature profile together with its simulation made by the gradient oven (Reproduced by permission of FMJ International Publications Ltd)

Modern methods for measuring the film thickness of powder coatings are based upon high-frequency eddy current and electromagnetic induction.

The eddy current method is used to determine the thickness of coatings applied on non-ferrous metals such as copper and copper alloys, zinc and zinc alloys, aluminium and aluminium alloys and other magnetic alloys including some stainless steels. For coatings on magnetic steel, such as iron and iron alloys including some stainless steels, the electromagnetic induction method is virtually always used since it has been proved to be the most reliable [55]. Instruments that are commercially available are produced in both hand-held and bench located versions. Contemporary hand-held instruments automatically calculate the mean result from either a single series or a multiple of series or batches. They provide the range or thickness measured and often a whole series of statistical data. Many of them can store the measurements data for later off-loading into a printer or computer. Bench located versions which are more suitable for laboratory use have all the facilities of the hand-held instruments. However, the larger size permits addition of options that cannot be incorporated in the hand-held version, such as continuous monitoring of the film thickness when large numbers of simply shaped components are to be controlled.

5.6.18 SOLVENT CURE TEST

The methods for monitoring the curing of powder coatings described in Chapter 3 are strong tools in the hands of the paint chemist that are necessary in

TECHNOLOGY OF PRODUCTION OF POWDER COATINGS   297

order to develop a proper powder coating formulation. However, they are too complicated, time consuming and therefore unsuitable for a quick evaluation of the state of cure that is very often necessary for the end user to carry out. A solvent cure test method has been developed by which one can differentiate between cured, partially cured and uncured thermosetting powder coatings [56]. The method resembles the well-known acetone resistance test used in solvent borne coatings. A 3 × 3 inch soft cloth pad is attached to the ball end of a 2 lb ball peen hammer. The hammer is used to provide a constant force on the test panel and eliminate operator dependence. The pad is saturated with the testing solvent which is methyl ethyl ketone (MEK) for pure epoxy powders or a blend of 10% MEK and 90% xylene for hybrids, TGIC or polyurethane coatings. The pad is slid on the test panel looking for signs of coating failure. Although it is stated that less than 25 double rubs indicate insufficient curing, 25–100 rubs partial curing and more than 100 rubs full curing, results obtained by this test should always be compared with known cured panels of the same type of coating.

5.6.19 HARDNESS OF THE CURED FILM [57–62]

The common test methods used to determine the hardness of organic surface coatings are also used to measure the hardness of cured powder coatings. These are: indentation hardness, pencil hardness and pendulum hardness. The test methods are standardized and the procedures are described in detail in the corresponding standards.

The Tukon indentation tester is the device used to determine the hardness of the films with an indenter that is a pyramidal diamond of prescribed dimensions, known under the name Knoop indenter [57]. After placing a fixed weight on the indenter for a fixed time period the indentation of the diamond is determined by a microscope. Dividing the applied weight by the projected area of the indentation the so-called Knoop hardness number is obtained. Test results are rather inconsistent for rubbers and materials with a T_g lower than the ambient testing temperature because of partial recovery of the deformations. Fortunately, this is not the case with powder coatings where rather reproducible results can be expected. A similar indentation hardness method uses the so-called Pfund indenter which is an hemispherical quartz or sapphire indenter of prescribed dimensions. The result of the test is reported in terms of the Pfund hardness number. Another standardized method for determination of the hardness of powder coatings is the Bucholz indentation test [58].

Determination of the hardness by means of a square shaped graphite pencil is a very popular test method because of its simplicity. A set of calibrated wood pencils having the following scale of hardness: 6B, 5B, 4B, 3B, 2B, B, HB, F, H, 2H, 3H, 4H, 5H, 6H, which increase from left to right, is used as a tool for determination of the hardness of the film. The hardness is reported as the grade of the hardest pencil that does not cause any marring of the surface [59].

Pendulum hardness test methods make use of the Persoz pendulum, the König pendulum and the Sward rocker [60–62]. The principles of the tests are similar although the constructions of the testers differ considerably. In the pendulum testing devices, the pendulum is in contact with the coated surface through two steel balls. In the case of the Sward rocker the contact between the tester and the film surface is assured through two flat metal rings. In both cases the testing element (the pendulum or the ring) is replaced from its equilibrium position and oscillates freely, swinging back and forth through a small angle. The oscillation amplitude decreases with time due to absorption of mechanical energy by the coating. Harder coatings give longer damping times while soft coatings are characterized by short damping times. In the case of the Sward rocker the damping is the result of both absorption of mechanical energy by the coating and energy loss due to the rolling friction between the rings and the surface. Test results in the case of pendulum hardness are reported as the time in seconds required for reduction of the oscillation amplitude from a specified starting angle to a specified ending angle. Similarly, the test results with the Sward rocker are reported as twice the number of swings required for a fixed amount of dumping.

5.6.20 FLEXIBILITY TESTS

Mandrel bend tests being a matter of standardization [63–66] consist of bending the coated metal panels over metal supports of cylindrical or conical shape. In the case of cylindrical mandrels the test results are reported as the smallest diameter of bending at which a crack of the coated surface is not detected by the unaided eye. When a conical mandrel is used as a support the test results are reported as the distance from the small end of the cone to the end of the crack in the panel. Both methods provide formulas according to which the elongation at the back of the coating can be easily calculated and reported when necessary.

5.6.21 IMPACT RESISTANCE OF THE CURED COATING

Impact resistance of powder coatings is determined in a simple way by dropping a weight down a guide tube onto a hemispherical indenter which rests on a coated panel. Due to the hole opposite the indenter, the panel together with the coating undergoes very rapid deformation during which cracking of the coating may occur. The Gardner impact tester having 2 and 4 lb weights and a 40 inch guide tube with a maximum impact of 160 inch lb is a widely used instrument for performing the tests. The results are reported as the maximum inch-pound value obtained at which there is no cracking of the film. The report must contain an indication of a direct or reverse impact depending on whether the coated surface is in an up position in contact with the indenter or down. As a rule, the reverse impact test is more severe than the direct impact test. Impact resistance depends very much on the thickness of the film. Therefore for comparison purposes the impact test should be carried out at the same coating thickness. Adhesion to the substrate also plays an important role. Insufficient adhesion can lead to a misinterpretation of the results.

5.6.22 ADHESION OF THE CURED FILM

The so-called crosscut adhesion test is the most common test method for assessment of the adhesion of a powder coating to the substrate. A series of parallel cuts are made with a special knife in the film deep enough to touch the metal surface. A second series of parallel cuts is then made at right angles to the first. By means of pressure sensitive adhesive tape the squares which are lost from the surface are pulled off. The result is reported as the number of squares remaining on the surface after this treatment or with a number rating from 5B to 0B, depending on the percentage of the affected area of the lattice. The method as such, although not very accurate, has been standardized [67, 68] and is widely used in practice.

More reliable results which can be directly correlated to the adhesion properties of the film provide the pull-off test method [69]. This method is based on determination of the greatest perpendicular force that a coated surface area can bear before a plug of material is detached. However, since this method makes use of tensile stress as compared to shear stress applied by scratch or knife adhesion, the results obtained by both methods may not be comparable.

5.6.23 GLOSS

The gloss of powder coatings is determined as specular gloss, which is defined as the relative luminous reflectance factor of a specimen in the mirror direction. The relative luminous reflectance factor, on the other hand, represents the ratio of the luminuous flux reflected from a specimen to the luminous flux reflected from a standard surface under the same geometric conditions. According to ASTM D 523 [70] the standard surface is a polished glass. Measurements are performed by means of commercially available optical instruments. Commercially available gloss meters consist of a light source furnishing an incident beam, means for locating the surface of the specimen and a receptor of the reflected light, which is a photosensitive device responding to visible radiation. Measurements are made usually with 60° and 20°, and sometimes with 85° geometry. The 60° geometry is normally used for determination of the gloss. However, if the gloss rates are higher than 70 at an incident angle of 60°, then 20° geometry is more suitable. In contrast, when the gloss rates are lower than 10 at 60°, then 85° geometry gives more consistent and suitable results for comparison.

5.6.24 SALT SPRAY RESISTANCE

The salt spray resistance test is very often performed in order to evaluate the performance of the coated ferrous and non-ferrous metals under heavy corrosion conditions. The conditions in salt spray testing, the apparatus required, the preparation of the test specimens and the reporting of the results have been a matter of standardization [71–73]. The test method consists of spraying 5% water solution of sodium chloride with a pH between 6.5 and 7.2 over the coated panel which is placed in the salt spray chamber at a temperature of 35°C. When it is desired to determine the development of corrosion from an abraded area in the coating, a scratch is made with a sharp instrument so as to expose the underlying metal before testing. The period of test is usually a matter of mutual agreement between the supplier and end user of the coating or it complies with already existing standards depending on the end use of the material. Regardless of the testing time, the test is continuous for the duration of the entire test period except for short interruptions necessary to inspect, rearrange or remove the test specimens.

The salt spray test is useful in rating the relative resistance of the protective coatings in atmosphere containing salt. However, the reproducibility of the test is highly dependent on the type of substrate. It should also be noted that the obtained results should not be related to the corrosion resistance in other media.

5.6.25 REVIEW OF THE TESTS FOR PIPELINE COATINGS

Powder coatings are widely used for protection of pipelines. The requirements with respect to the coating properties in this application area differ considerably from those which are common for very standard powder coatings. A number of test procedures has been developed specifically for testing coating systems intended for use as pipeline coatings. Those standard methods applicable to the powder coatings are listed in the references [74–84] without summarizing the testing principles in this section.

REFERENCES

1. Scott, J. A., *Polymers Paint Colour Journal*, **179** (4239), 407, 1989.
2. Werner & Pfleiderer GmbH, Brochure WER 05 066/3-1. 5-VIII. 86.
3. Franz, P., in *Buss Powder Coating Technology*, Buss Brochure MK-T3P, October 1988.
4. Walty, O., *Production Process for Thermoset Powder Coatings*, DMS Resins Symposium, Singapore, 1986.
5. Cradock, I., *Manufacturing Chemist*, January 1989, p. 24.
6. Harris S., *The Technology of Powder Coatings*, Portcullis Press Ltd, Redhill, 1976.
7. Winkler, J., Klinke, E., and Dulog, L., *J. Coat. Technol.*, **59**, (754), 35, 1987.
8. Franz, P., and Bolt, A., *Powder Coatings*, **4** (1), 3, 1981.

TECHNOLOGY OF PRODUCTION OF POWDER COATINGS 301

9. Meijer, H. E. H., *The Modelling of Continuous Mixers, DSM Resins Powder Coating Symposium*, Nordwijk, May 1988.
10. Buss AG, Brochure MK/832/03/1088/001/E, p. 4.
11. Schwamborn, K. H., *Buss Powder Coating Training Course*, November 1988.
12. Frewitt M. G. H., Hammermills Brochure, October 1987.
13. 'Process, engineering and systems for mechanical processing technology', Alpine AG, Brochure 017/9e, p. 12.
14. United Nations, Economic and Social Council, Report ENV/WP. 2/5/Add. 153, 5 July 1988.
15. Schaller, R. E., and Lapple, C. E., *J. Paint Technol.*, **44** (571), 87, 1972.
16. Tunnicliffe, G., *Paint and Resin*, December 1987, p. 8.
17. Newton, H. W., and Newton, W. H., *A Study of Classification Calculation, Rock Prod.*, **1932**, 35.
18. Eder, Th., *Aufbereitungs Technik*, **4**, 140, 1961.
19. Rumpf, H., and Leschonski, K., *Chemie Ing. Technik*, **39**, 1231, 1967.
20. Schwamborn, K. H., *Tendencies of Development in Grinding Systems for Powder Coatings, Buss Powder Coatings Training Course*, Lucerne, November 1988.
21. ASTM D 3451: 1987, *Standard Practices for Testing Polymeric Powders and Powder Coatings*, 1987.
22. ASTM D 1639: 1983, *Standard Test Method for Acid Value of Organic Coating Materials*, 1983.
23. DIN 4629: 1979, *Paint Media; Determination of Hydroxyl Value by Titrimetric Method*, November 1979.
24. DSM Resins, Powder Coating Resins Brochure, July 1988, p. 62.
25. ASTM D 1652: 1988, *Standard Test Method for Epoxy Content of Epoxy Resins*, 1988.
26. DSM Resins, Powder Coating Resins Brochure, July 1988, p. 6.3.
27. DIN 53 181: 1981, *Binders for Paints and Varnishes; Determination of the Melting Range of Resins by the Capillary Method*, March 1981.
28. DIN 4625: 1981, *Binders for Paints and Varnishes; Determination of the Softening Point; Ring and Ball Method*, June 1981.
29. ASTM E28: 1967, *Standard Test Method for Softening Point of Resins by Ring-and-Ball Apparatus*, 1967.
30. *Recommended Procedure for Measurement of Melt Viscosity of Powder Coating Materials*, no. 5, Powder Coating Institute, Alexandria, VA 22314, USA.
31. DIN 53 214: 1982, *Testing of Paints and Varnishes; Determination of Rheograms and Viscosities by Rotational Viscosimeters*, February 1982.
32. ASTM D 1544: 1980, *Standard Test Method for Determination of Color of Synthetic Resins*, 1980.
33. *Recommended Procedure for Gel Time Reactivity*, No. 6, Powder Coating Institute, Alexandria, VA 22314, USA.
34. Weder, J. A., *Finishing*, **13**(4), 41, 1989.
35. DIN 55 990, pt 7: 1980, *Testing of Paints, Varnishes and Similar Coating Materials; Powder Coatings; Assessment of Resistance to Caking*, June 1980.
36. DIN 55 990, pt 8: 1980, *Testing of Paints, Varnishes and Similar Coating Materials; Powder Coatings; Assessment of the Chemical Storage Stability*, June 1980.
37. *Recommended Procedure for Gel Time Reactivity*, No. 1, Powder Coating Institute, Alexandria, VA 22314, USA
38. *Recommended Procedure for Density of Powder Coating Materials*, No. 4, Powder Coating Institute, Alexandria, VA 22314, USA.
39. DIN 55 990, pt 3: 1979, *Testing of Paint, Varnishes and Similar Coating Materials; Powder Coatings; Determination of Density*, December 1979.

40. ASTM D 1475: 1985, *Standard Test Method for Density of Paint, Varnish, Lacquer, and Related Products*, 1985.
41. *Recommended Procedure for Compatibility of Coating Powders*, No. 2, The Powder Coating Institute, Alexandria, VA 22314, USA.
42. DIN 5590, pt 2: 1979, *Testing of Paint, Varnishes and Similar Coatings Materials; Powder Coatings; Particle Size Distribution*, December 1979.
43. Collins, E. A., Davidson, J. A., and Daniels, C. A., *J. Coat. Technol.*, **47**(604), 35, 1975.
44. Barth, G. H., and Sun, S. T., *Anal. Chem.*, **61**, 143, 1989.
45. Harris, S. T., *The Technology of Powder Coatings*, Portcullis Press Ltd, London, 1976, pp. 143–53.
46. Ward, J. (Ed.), *Thermoset Powder Coatings*, FMJ International Publications Ltd, Redhill, Surrey, 1989, p. 39.
47. Hughes, J. F., *J. Oil Col. Chem. Assoc.*, **4**, 145, 1989.
48. Moyle, B. D., and Hughes, J. F., *IEEE Trans. Ind. Applic.*, **IA-20**(6), 1631, 1984.
49. Hearn, G. L., in *Thermoset Powder Coatings*, Ed. J. Ward, FMJ International Publications Ltd, Redhill, Surrey, 1989, p. 116.
50. *Recommended Procedure for Inclined Plate Flow*, No. 7, The Powder Coating Institute, Alexandria, VA 22341, USA.
51. Cauchois, G., in *Thermoset Powder Coatings*, Ed. J. Ward, FMJ International Publications Ltd, Redhill, Surrey, 1989, p. 128.
52. DIN 50 986: 1979, *Measurement of Coating Thickness; Wedge Cut Method for Measuring the Thickness of Paints and Related Coatings*, March 1979.
53. ASTM D 1186: 1981, *Standard Methods for Nondestructive Measurement of Dry Film Thickness of Non-magnetic Coatings Applied to a Ferrous Base*, 1981.
54. ASTM D 4138: 1982, *Standard Method of Measurement of Dry Film Thickness of Protective Coating Systems by Destructive Means*, 1982.
55. Latter, T. D. T., in *Thermoset Powder Coatings*, Ed. J. Ward, FMJ International Publications Ltd, Redhill, Surrey, 1989, p. 124.
56. *Recommended Procedure for Solvent Cure Test*, No. 8, The Powder Coating Institute, Alexandria, VA 22341, USA, 1987.
57. ASTM D 1474: 1985, *Standard Test Methods for Indentation Hardness of Organic Coatings*, 1985.
58. DIN 53 153: 1977, *Testing of Paints, Varnishes and Similar Coating Materials; Bucholz Indentation Test on Paint Coatings and Similar Coatings*, November 1977.
59. ASTM D 3363: 1974, *Standard Test Method for Film Hardness by Pencil Test*, 1974.
60. DIN 53 157: 1987, *Testing of Paints, Varnishes and Similar Coating Materials; Assessment of the Mechanical Damping Behaviour of Coatings Using Konig Pendulum Apparatus*, January 1987.
61. ASTM D 4366: 1984, *Standard Test Methods for Hardness of Organic Coatings by Pendulum Damping Tests*, 1984.
62. ASTM D 2134: 1966, *Standard Test Methods for Softening of Organic Coatings by Plastic Compositions*, 1966.
63. ASTM D 1737: 1985, *Standard Test Method for Elongation of Attached Organic Coatings with Cylindrical Mandrel Apparatus*, 1985.
64. ASTM D 522: 1985, *Standard Test Method for Elongation of Attached Organic Coatings with Conical Mandrel Apparatus*, 1985.
65. DIN 53 152: 1985, *Testing of Paints, Varnishes and Similar Coating Materials; Bend Tests on Coatings; Method Using Cylindrical Mandrel*, October 1985.
66. DIN 6860: 1988, *Testing of Paints, Varnishes and Similar Coating Materials; Bend Tests on Coatings; Method Using Conical Mandrel*, March 1988.
67. ASTM D 3359: 1983, *Standard Methods for Measuring Adhesion by Tape Test*, 1983.

TECHNOLOGY OF PRODUCTION OF POWDER COATINGS

68. DIN 53 151: 1981, *Testing of Paints, Varnishes and Similar Coating Materials; Crosscut Test on Paint Coatings and Similar Coatings*, May 1981.
69. *Standard Method for Pull-off Strength of Coatings using Portable Adhesion Testers.*
70. ASTM D 523: 1985, *Standard Test Method for Specular Gloss*, 1985.
71. ASTM B 117: 1985, *Standard Method of Salt Spray (Fog) Testing*, 1985.
72. ASTM D 1654: 1984, *Standard Method for Evaluation of Painted or Coated Specimens Subjected to Corrosive Environments*, 1984.
73. DIN 53 167: 1985, *Paints, Varnishes and Similar Coating Materials; Salt Spray Tests on Coatings*, December 1985.
74. ASTM G 6: 1983, *Standard Test Methods for Abrasion Resistance of Pipeline Coatings*, 1983.
75. ASTM G 8: 1985, *Standard Test Method for Cathodic Disbonding of Pipeline Coatings*, 1985.
76. ASTM G 9: 1982, *Standard Test Method for Water Penetration Into Pipeline Coatings*, 1982.
77. ASTM G 10: 1983, *Standard Test Method for Specific Bendability of Pipeline Coatings*, 1983.
78. ASTM G 11: 1983, *Standard Test Method for Effect of Outdoor Weathering on Pipeline Coatings*, 1983.
79. ASTM G 13: 1985 and G 14: 1983, *Standard Test Method for Impact Resistance of Pipeline Coatings* (Limestone Drop Test, 1985, and Falling Weight Test, 1983.
80. ASTM G 17: 1983, *Standard Test Method for Penetration Resistance of Pipeline Coatings (Blunt Rod)*, 1983.
81. ASTM G 18: 1983; *Standard Test Method for Joints, Fittings, and Patches in Coated Pipelines*, 1983.
82. ASTM G 19: 1983, *Standard Test Method for Disbonding Characteristics of Pipeline Coatings by Direct Soil Burial*, 1983.
83. ASTM G 20: 1983, *Standard Test Method for Chemical Resistance of Pipeline Coatings*, 1983.
84. ASTM G 42: 1985, *Standard Test Method for Cathodic Disbonding of Pipeline Coatings Subjected to Elevated Temperatures*, 1985.

6

Powder Coatings Application Techniques

6.1	Surface Preparation				305
	6.1.1	Mechanical Methods of Cleaning			305
	6.1.2	Chemical Cleaning and Pretreatment			306
		6.1.2.1	Cleaning and Pretreatment of Cold- and Hot-rolled Steel		306
			6.1.2.1.1	Alkaline Cleaners and Acid Cleaners	307
			6.1.2.1.2	Conversion Coating of Steel	308
				6.1.2.1.2.1 Iron Phosphate	308
				6.1.2.1.2.2 Zinc Phosphate	311
		6.1.2.2	Cleaning and Pretreatment of Aluminium		316
			6.1.2.2.1	Alkaline Cleaners	316
			6.1.2.2.2	Acid Cleaners	318
			6.1.2.2.3	Conversion Coatings of Aluminium	318
				6.1.2.2.3.1 Iron Phosphate Coating	318
				6.1.2.2.3.2 Zinc Phosphate Coating	319
				6.1.2.2.3.3 Chromium Phosphate	319
				6.1.2.2.3.4 Chromate Conversion Coatings	321
		6.1.2.3	Cleaning and Pretreatment of Galvanized Steel		321
			6.1.2.3.1	Alkaline and Acid Cleaners	322
			6.1.2.3.2	Conversion Coatings for Galvanized Steel	322
				6.1.2.3.2.1 Zinc Phosphate Conversion Coating	322
				6.1.2.3.2.2 Chromium Phosphate Conversion Coatings	323
6.2	Application of Powder Coatings				324
	6.2.1	Electrostatic Spraying Technique			325
		6.2.1.1	Corona Charging Guns		326
			6.2.1.1.1	Problems Associated with Corona Spray Guns	331
		6.2.1.2	Tribo Charging Guns		335
		6.2.1.3	Alternative Guns		337
		6.2.1.4	Factors Affecting the Spraying Process		339
	6.2.2	Fluidized Bed Process			344
	6.2.3	Electrostatic Fluidized Bed			346
	6.2.4	Flame-spray Technique			347
	6.2.5	Comparison between Different Application Techniques			349
6.3	Design of the Spraying Booths				349
	6.3.1	Colour Change in the Powder Coating Process			352
6.4	Troubleshooting				354
	References				360

6.1 SURFACE PREPARATION

The substrate to which the powder coating is applied is in most cases metallic. Metallic articles are easily coated in a one-spray operation with a film thickness of 50–75 μm. This is quite an advantage in comparison with the multicoat wet paint system when thicknesses of this range have to be achieved. With powder coating there is no flash-off period, no running away from the sharp edges and no sagging due to excessive film thickness. However, a powder coating system does not avoid the necessity of surface preparation. The importance of surface treatment prior to powder coating has been stressed in numerous papers [1–12]. Treatment of the metal substrate prior to coating includes cleaning of the surface followed by application of an inorganic layer of so-called conversion coating. Chemical conversion coatings are applied to the metal substrate because of three reasons: (a) they provide temporary in-process corrosion protection to the metal prior to application of powder coating, (b) they promote good adhesion of the powder coating to the substrate during film formation and (c) they afford additional corrosion protection to the metal during the lifetime of the product. The cleaning of the surface cannot be avoided, even in cases when a special chemical pretreatment of the surface is not considered.

The methods used in the cleaning of the surface can be classified into two groups: mechanical and chemical methods. Since chemical methods of cleaning depend very much on the type of surface, they will be discussed together with the surface pretreatment specifically for different substrates.

6.1.1 MECHANICAL METHODS OF CLEANING

Mechanical methods of cleaning the surface before coating are widely used in practice. The principle is the same, using an abrasive action to remove the dirt in the form of tightly adhering contaminants on the steel surface. At the same time certain etching of the surface takes place, improving in this way the adhesion of the coating. The following forms of mechanical treatment are used: scratch-brushing by means of a rapidly rotating wire disc-brush; sanding by means of abrasive discs, wheels, cloths and papers; and dry or wet blasting with abrasive materials.

The most important method for mechanical preparation is airblasting. This is often called sandblasting because of the use of quartz sand as an abrasive medium, which in the form of sand jet is directed at the surface of the article through a special nozzle by compressed air.

A method that also works by throwing the abrasive material on the surface to be treated, without the use of an air stream, is the airless centrifugal wheel blast.

The airless centrifugal wheel blasting is a more economical method from an energy point of view. In the airblast systems high volumes of clean, dry compressed air are used to accelerate the abrasive particles. Therefore the energy which is consumed by the airless systems is only 10% of that used in the airblast operation [13]. On the other hand, by airblasting it is very easy to position the blowing nozzles to direct the blasts into nooks that are shadowed from the blast pattern of the centrifugal wheel.

The airblasting or airless blasting techniques are the most appropriate for removal of mill scale from the mild steel surfaces, removal of oxides and corrosion products, removal of other dry soils such as mould sand residues, residues of paints, etc. It also removes the scratches and other surface irregularities, providing uniform roughness, which ensures good coating adhesion. Coating must take place immediately after the blasting because the cleaned surface is in a highly reactive state and corrosion occurs very soon.

6.1.2 CHEMICAL CLEANING AND PRETREATMENT

Surfaces on which oil or grease are present cannot be blasted directly, because these substances cannot be removed completely by blasting. Moreover, they contaminate the abrasive itself. Therefore the oil or grease deposits must be removed before sandblasting.

A chemical method that can be used for all types of surfaces is one employing solvents as degreasing agents. For degreasing the surface a simple solvent wiping process can be used. Vapour degreasing by halogenated hydrocarbons like trichloroethylene is also used. It should be kept in mind that vapour cleaning does not remove the solid dirt from the surface, and therefore if the subsequent blasting does not take place, then wiping or liquid/vapour degreasing has to be used. Trichloroethylene is still widely in use despite the health and safety concerns and the necessary heating installations in order to provide adequate cleaning action at elevated temperatures.

Safety, health and environmental considerations work in favour of the aqueous cleaners. Therefore the most common cleaning method, which is also the most economical, uses emulsified cleaners. These are cleaning agents such as organic solvents which have been incorporated in the water phase by means of emulsifiers. Since the surface on which the powder coating is applied can be of a different nature, cleaning agents specially suitable for cold- or hot-rolled steel, aluminium or galvanized steel have been developed.

The next operation which follows the cleaning step is the application of the so-called conversion coatings, which also differ for different substrates. In the following sections the substrate preparation will be discussed separately for different types of material.

6.1.2.1 Cleaning and Pretreatment of Cold- and Hot-rolled Steel

Cold-rolled or the so-called mild steel is free from heat scale and, when properly packed, even from rust.

POWDER COATINGS APPLICATION TECHNIQUES

Hot-rolled steel has a surface covered with a blue/black mill scale which is very brittle in nature and will flake off upon bending. Scale is a product of gaseous corrosion at high temperature and consists of the anhydrous oxides FeO, Fe_3O_4 and Fe_2O_3. Whereas ferric oxide, Fe_2O_3, is chemically stable, ferrous oxide, FeO, is less stable and readily changes to hydrated ferric oxide under the influence of water and oxygen. The cracked mill scale is cathodic to the underlying steel and promotes corrosion.

Hot-rolled pickled and oiled steel is also available on the market and is a steel that has been pickled to remove the hot scale and oiled to prevent rusting during storage.

All of these types of steel have to be cleaned before powder application. Several different types of chemical cleaners can be used for this purpose.

6.1.2.1.1 Alkaline Cleaners and Acid Cleaners

A wide variety of different alkaline cleaners is available on the market formulated for each specific application. In general they are combinations of sodium phosphates, silicates and carbonates. Often they contain chelating agents, solvents and surfactants.

The resistance of steel to alkaline attack gives a great degree of freedom for adjusting the properties of the alkaline cleaners to the type of material undergoing the cleaning operation. A contemporary tendency in the development of alkaline cleaners is to lower the operating temperature by means of a proper choice of surfactants and chelate systems providing benefits of lower operating costs. The cleaned substrate must undergo subsequent rinsing operation in order to remove the metasilicates present in the cleaning agent. Significant progress has been made by use of the so-called inhibited cleaners containing certain titanium salts [5]. These salts act as crystal size regulators for the conversion coatings, which are applied in the following pretreatment step.

New surfactant systems also enable efficient cleaning action at less alkaline conditions than was the case with the aqueous cleaning agent of the first generation. This is especially important for cases where zinc phosphate conversion coatings are used in the next pretreatment step. Zinc phosphate conversion coatings applied on a steel surface being treated with strong alkaline cleaning solutions in general have a larger size of crystal deposited on the surface. This decreases the mechanical properties of the conversion coating and the gloss of the powder coating applied as a top layer because of the rough nature of the surface. Research efforts in this area are also in the direction of development of cleaners/conversion coatings blends, or more specifically cleaners/iron phosphates [5].

The cleaning process may vary from hand wiping cold cleaning to an immersion or spraying process. All processes have to end up with a rinsing procedure in order to remove the cleaning agent from the surface. Dependent on the amount and type of dirt, the cleaning procedure may be repeated several times. The best results are obtained at automatic hot-spraying cleaning lines where the cleaning stage at a temperature of 60–70°C within 45–60 seconds is followed by rinsing within 30–45 seconds. In the case where the immersion cleaning method is used, residence times between two and five minutes are necessary for efficient cleaning [13, p. 49].

Rust can be very effectively removed by using mineral acids like hydrochloric or sulphuric acid. Also the heat scale can be removed in the same manner. Strong mineral acids for rust removal are used in principle only in cold operations, and except in highly specialized operations almost never in spray equipment. The reactions between iron and its oxides and sulphuric acid proceed in accordance with the following equations.

$$Fe + H_2SO_4 \rightarrow FeSO_4 + H_2 \tag{6.1}$$

$$FeO + H_2SO_4 \rightarrow FeSO_4 + H_2O \tag{6.2}$$

$$Fe_2O_3 + H_2SO_4 \rightarrow Fe_2(SO_4)_3 + 3H_2O \tag{6.3}$$

$$Fe_3O_4 + H_2SO_4 \rightarrow FeSO_4 + Fe_2(SO_4)_3 + 4H_2O \tag{6.4}$$

Moderate or mild acid cleaners based on phosphoric acid or organic acids are used in spray applications.

On the market there are various acid cleaners convenient for hand wiping. Usually they contain detergents and thickening agents, making them suitable also for brush removing applications.

6.1.2.1.2 Conversion Coating of Steel

The chemical pretreatment of the metal surface is the following operation, which prepares the surface after the dirt, metal oxides, oil and grease have been removed. Conversion coatings on the metal surface consist of insoluble metal oxides or metal salts. Three main types of conversion coatings are used for a steel surface: iron phosphate, zinc phosphate and chromate conversion coatings. Due to pollution control problems connected with the disposal of chromium and the additional equipment requirements the latter coating has never become commercially important. The characteristics of the main types of conversion coatings for steel are listed in Table 6.1 [14].

6.1.2.1.2.1 Iron Phosphate
The compositions that are used to produce iron phosphate conversion coatings on steel are principally based on monosodium phosphate, although even the simplest treatment with phosphoric acid as such gives acceptable results. These mildly acidic compositions are available on the market in powdered or liquid form. The rest of the composition are various accelerating agents in the form of nitrites, nitrates or chlorates. When, at the same time, a cleaning action is considered next to the coating, surface active agents may

POWDER COATINGS APPLICATION TECHNIQUES

Table 6.1 Characteristics of different types of conversion coatings for steel [14] (Reproduced by permission of Spruyt, van Mantgem & de Does NV)

Type of phosphate coating	Colour	Deposition (g/m^2)	Thickness (μm)	Porosity (%)	Pencil hardness
Iron phosphate Fe$_3$(PO$_4$)$_2$·8H$_2$O	Blue	0.1–0.5	0.1–0.5	0.5–1	H
Zinc–iron phosphate Zn$_2$Fe(PO$_4$)$_2$·4H$_2$O	Medium grey	10–30	5–15	0.05–0.4	HB
Zinc phosphate Zn$_3$(PO$_4$)$_2$·4H$_2$O	Grey	2–10	1–5	0.05–0.5	HB→H
Zinc calcium phosphate Zn$_2$Ca(PO$_4$)$_2$·2H$_2$O	Light grey	1.5–6	1–3	0.05–0.4	HB→H
Manganese phosphate MnFe$_9$5H$_2$(PO$_4$)$_4$·4H$_2$O	Dark grey	8–40	3–25	0.5–3	B→HB

be included. Depending on the acidity of the composition, rust removal can also take place. In this case the compositions have moderate or strong acidity.

The iron phosphate conversion coating can be obtained by simple treatment of the steel surface with solutions containing 0.2–6 vol% of phosphoric acid. Usually the composition of the conversion coating is designed in such a way that the rinsing operation after the treatment is not necessary. Good results are obtained by air drying the treated material, but even better results are obtained by drying the surface in a stream of warm air. During the drying process the steel surface reacts with the phosphoric acid to form a layer of iron phosphate according to the following reaction scheme:

$$Fe + 2H_3PO_4 \longrightarrow Fe(H_2PO_4)_2 + H_2 \quad (6.5)$$

$$Fe(H_2PO_4)_2 \longrightarrow FeHPO_4 + H_3PO_4 \quad (6.6)$$

The layer of FeHPO$_4$ is rather thin, having a thickness between 0.2 and 0.5 μm.

If the concentration of phosphoric acid in the conversion coating solution is too high then a sticky layer of unreacted phosphoric acid remains on the surface. This hygroscopic layer decreases the corrosion resistance of the coated article and the adhesion of the powder coating on the substrate. Conversely, too low a concentration of phosphoric acid produces as a consequence the formation of a thin layer of porous powder such as iron phosphate which has a rather poor corrosion resistance.

Depending on the bath temperature, concentrations between and 4 and 6% by volume of phosphoric acid are used for operation at room temperature and 0.2–2% at 70–90°C.

For good phosphatization it is necessary to control the concentration of iron ions in the bath. Usually in continuous operation the concentration of iron in the bath increases and should be kept below 5 g/L.

Monosodium phosphate conversion coatings solution has a lower acidity than phosphoric acid solutions. The pH is usually in the range between 4.5 and 5.8.

Several different reactions simultaneously take place during treatment of the surface with monosodium phosphate based conversion coatings. In the first reaction a ferrophosphate is formed according to

$$Fe + 2NaH_2PO_4 + H_2O + \tfrac{1}{2}O_2 \longrightarrow Fe(H_2PO_4)_2 + 2NaOH \qquad (6.7)$$

The formed ferrophosphate reacts with sodium hydroxide and oxygen according to the following reactions, producing insoluble ferriphosphate or ferrihydroxide:

$$2Fe(H_2PO_4)_2 + 2NaOH + \tfrac{1}{2}O_2 \longrightarrow 2FePO_4 + 2NaH_2PO_4 + 3H_2O \qquad (6.8)$$

$$2Fe(H_2PO_4)_2 + 6NaOH + \tfrac{1}{2}O_2 \longrightarrow 2Fe(OH)_3 + 2NaH_2PO_4 + 2Na_2HPO_4 + H_2O \qquad (6.9)$$

During the drying of the surface insoluble ferrioxide is formed:

$$2Fe(OH)_3 \longrightarrow Fe_2O_3 + 3H_2O \qquad (6.10)$$

The oxygen necessary for the oxidative action during the phosphating process can be supplied by the air, especially in the case where phosphatization takes place by the spraying method. However, contemporary phosphating preparations contain accelerators which are oxidizing compounds, e.g. nitrites, nitrates, chlorates, bromates or borates. The accelerating action of the oxidizers consists in oxidizing the hydrogen formed during dissolution of the iron. Additionally, the accelerators oxidize divalent iron ions in the solution, promoting formation of insoluble ferric phosphate which precipitates in the form of a sludge.

Due to formation of dinatrium phosphate, the pH of the bath in continuous operations increases permanently. Where a pH of 6.0 is reached, the phosphating efficiency of the bath is practically zero. Therefore, permanent control and correction of the pH of the phosphating solution in the bath with sodium phosphate or phosphoric acid is necessary to ensure high quality of the conversion coating.

The weight of the formed conversion coating in the case of monosodium based solutions is between 0.3 and 0.7 g/m², with thickness not greater than 1 μm. Depending on the ratio between the formed ferriphosphate and ferrioxide the colour of the coating ranges from grey to deep blue.

The advantage of monosodium phosphate conversion coatings over the simple phosphoric acid based solution is that phosphatization and cleaning of the surface take place at the same time. Usually the conversion coating solution contains surface active agents, mostly of a non-ionogenic type.

The conversion coatings solutions can be applied in a very simple way using a rug or mop. The solution can be reapplied several times until the surface is clean and free of rust. Since this is a mainly hand applied technique the solution is in principle not preheated.

To speed up the phosphating action of the solutions they are very often applied by a steam gun. The solution is mixed with wet steam and applied by the gun when preheated to ca. 70 °C. Rinsing is usually done by the same gun using fresh steam only.

Iron phosphate conversion coatings can be applied by immersion application methods. In this case care should be taken that the substrate is extremely clean before coating, since during the immersion coating technique the cleaning action of the composition is not very good; this is also the case for spray or hand applications. The cleaning action during application of the coating can be improved by agitation. In this case substrates with a light soil can be treated in a one-stage process. The temperature of the bath can vary from 30 to 70°C dependent on the concentration of the active material in the immersion solution, the desired coating thickness, and capacity and design of the coating line.

The most widely used method for applying iron phosphate conversion coating is spray washing. Depending on the nature of the soil and the quality level requirements, spray systems operate in at least two stages (clean/coat in one stage and rinse with water in the second) up to even nine, including cleaning, rinsing, coating, postrinsing and drying steps subsequently distributed in a logical manner. Typical time schedules for an automatic line for complete pretreatment is 45–60 seconds for the cleaning steps, 45–60 seconds for the treatment and 15–30 seconds for the rinsing step. The temperatures in different zones of the line vary between 30 and 70°C [13, p. 52].

Iron phosphate conversion coatings are used for steel products destined for normal indoor use, since they do not provide outstanding corrosion protection. However, they are highly appreciated for their superior paint adhesion characteristics. The control of the phosphating operation is simplified by using liquid coating formulations which permit the use of automatic metering systems. For limited cleaning performances one-pack phosphating/cleaning solutions are available on the market. Twin pack formulations, on the other hand, allow the cleaner and the coater to be dosed separately to the same tank. This system provides greater flexibility concerning the treatment of materials with a different degree of contamination by oil, grease and other impurities [4].

6.1.2.1.2.2 Zinc Phosphate Compositions for obtaining zinc phosphate conversion coatings on the steel surface are in principle acid solutions containing zinc dihydrogen phosphate, $Zn(H_2PO_4)_2$, accelerating agents like zinc nitrate and additives for grain control and coating weight control.

In water solution zinc dihydrogen phosphate exists in several forms according to the following equilibrium [15]:

$$Zn(H_2PO_4)_2 \rightleftharpoons ZnHPO_4 + H_3PO_4 \qquad (6.11)$$

$$3ZnHPO_4 \rightleftharpoons Zn_3(PO_4)_2 + H_3PO_4 \qquad (6.12)$$

The equilibrium constants of these two reactions are dependent on temperature. At higher temperatures the equilibrium is shifted to the right, which as a consequence causes precipitation of the less soluble zinc hydrogen phosphate, $ZnHPO_4$, in the form of slurry. This can be avoided by adding small quantities of phosphoric acid which shifts the reaction equilibrium towards formation of zinc dihydrogen phosphate [14, p. 62].

In practice the concentration of tertiary zinc phosphate in the bath is almost close to saturation. The metal surface, when in contact with such a bath, is attacked by the phosphoric acid giving as a product ferrodihydrogen phosphate according to the already mentioned reaction (6.5). In this way, by consumption of phosphoric acid, the pH at the metal/solution interface is increased and the solubility products of zinc phosphate and iron hydrogen phosphate are exceeded. They precipitate and adhere well to the metal surface at the very reaction sites, building up an insoluble layer of considerable thickness according to the following overall reaction:

$$Fe + 3Zn(H_2PO_4)_2 \longrightarrow Zn_3(PO_4)_2 + FeHPO_4 + 3H_3PO_4 + H_2 \qquad (6.13)$$

In the solution, on the other hand, the pH remains quite low, thus preventing the formation of insoluble sludge of zinc phosphate salts.

The phosphating action can be accelerated by increasing the bath temperature. Reasonably high phosphating rates can be achieved at 90–100°C. This of course is connected with relatively high energy consumption, especially in the case where massive articles with high heat capacity have to be treated.

A much cheaper and more efficient method is one that employs accelerating agents. In this way acceptable phosphating rates can be reached at relatively low temperatures.

There are many different types of accelerators described in the patent literature, such as oxidizing agents (nitric acid, nitrates, nitrites, chlorates, borates), reduction substances (hyposulphites phosphites, benzaldehyde, hydroxylamine), organic nitrogen containing compounds, copper salts, etc. The most efficient and the most widely used accelerators are those of oxidizing types. The accelerating action of the oxidizing agents will be outlined in general terms since the detailed mechanism is beyond the scope of this book. These materials oxidize the hydrogen present in a layer of molecular size thickness over the metal surface. At the same time the ferro ions are oxidized to ferri ions which precipitate because of their low solubility. Both reactions promote further solubilization of the iron since the equilibrium of reaction (6.5) is shifted to the right.

The soluble iron salts in the solution have a very bad influence on the bath performance. In the first instance the phosphating rate decreases, but what is even worse, the obtained conversion coating has a lower corrosion resistance because of its porous nature. Therefore, control of the iron content in the bath is essential if good results are to obtained. In the case of too high a concentration of ferro ions, the addition of a certain amount of oxidizing agents such as sodium nitrate, sodium nitrite, hydrogen peroxide or chlorates can help by oxidizing the ferro ions to insoluble ferri ions which are than removed from the bath in the form of sludge of ferri phosphate. Where the sludge is heavy, normal operation of the phosphating process becomes quite difficult unless the whole bath is thoroughly cleaned.

During phosphatization in the presence of iron, nitrate is reduced to nitrite or to a nitroso compound which can only then react with the hydrogen liberated by dissolution of iron by phosphoric acid [15, p. 138]. Therefore direct addition of nitrites in the form of sodium nitrite is more effective in accelerating the phosphating action of the bath. Sodium nitrite has a double action in accelerating the process speed. It helps the formation of zinc phosphate by transforming the zinc dihydrogen phosphate and at the same time oxidizes ferro into ferri ions according to the following equation [14, p. 65]:

$$3\ Zn(H_2PO_4)_2 + 2\ NaNO_2 + 2\ Fe \longrightarrow$$
$$Zn_3(PO_4)_2 + 2\ FePO_4 + N_2 + 2\ NaH_2PO_4 + 4\ H_2O \qquad (6.14)$$

Sodium nitrite is a very strong oxidizing agent. Therefore, the phosphating rate in a nitrite containing composition based on zinc phosphate is very high. At bath temperatures of 40–80°C the residence time in the immersion operating baths is only 3–10 minutes depending on the bath temperature and the desired layer thickness. Using spray phosphatization the phosphating time is only 1–2 minutes [14, p. 66].

Sodium nitrite is not stable in acidic media. This instability which is manifested by the formation of nitrogen oxides is even more emphasized at higher temperatures. Therefore, a low operating bath temperature is desirable to achieve good bath stability. Temperatures higher than 80°C should be avoided when working with nitrite accelerated zinc phosphate conversion coatings.

If the concentration of the nitrite in the bath decreases because of its consumption during oxidizing the ferro salts in the solution or because of its decomposition, the concentration of ferro ions in the bath increases, lowering the efficiency of the bath operation. In continuously operating phosphating baths, the concentration of the nitrite in the bath solution must be kept on a certain level by subsequent additions of fresh sodium nitrite. On the other hand, too high a nitrite concentration will cause a heavy slurry of ferri phosphate to form. In extreme cases, when the addition of sodium nitrite is much too high, the bath becomes alkaline which causes precipitation of zinc hydrogen phosphate, $ZnHPO_4$. In such cases the slurry formation is even heavier. The other negative influence of an excess of sodium nitrite is the formation of a very thick irregular zinc phosphate layer on the metal surface. The optimal concentration of sodium nitrite in the bath providing good operation is between 0.1 and 1.0 g $NaNO_2$/L.

At good controlled operation conditions, the use of the nitrite accelerated bath is practically unlimited. It goes without saying that the slurry formed during the phosphating operation must be removed periodically from the bath.

Sodium chlorate is another oxidizing agent that has wide use in accelerating the phosphating process. The advantage of sodium chlorate over sodium nitrite is that it is possible to add this material to the concentrated zinc phosphate/phosphoric acid solution, since it is stable in acidic conditions. This allows ready-made one-component zinc phosphating solutions to be prepared, which is especially convenient when the concentration of the bath has to be corrected by addition of fresh phosphating agent. In this way the proportion of zinc phosphate/accelerator is kept constant, thus providing the deposition of a layer of the same quality by time. For continuous operating lines this is of great importance. A disadvantage of sodium chlorate accelerator is the formation of sodium chloride during the bath operation, which remains in the bath permanently increasing its concentration.

Metal salts in combination with oxidizing agents are often used as accelerators for phosphating steel. In principle copper, nickel, cobalt, silver, gold and platinum can be used as accelerators, but because of obvious commercial reasons the practical use is limited to the first two metal salts. During the phosphating process these metals precipitate on the steel surface and serve as a cathode, accelerating in this way the dissolution of the ferro phosphate. At the same time they catalyse nitrate decomposition and promote oxidation.

The protective function of the phosphate coatings depends very much on the crystalline character of the film. In general, protection increases with decreasing grain size. The size of the crystals of the zinc phosphate conversion coatings is dependent to a great extent on the type of cleaning agent used in the previous step. Especially when strong alkaline cleaning solutions are used, which is very often the case in immersion cleaning processes, the crystals of deposited zinc phosphate are of rather an undesirable size.

Refinement of the crystalline structure can be obtained by increasing the bath temperature. However, by using suitable additives the temperature can be reduced with no adverse effect on the grain size. This can be achieved in two ways: by adding regulators of the crystal growth in the bath before phosphatization takes place or adding crystal growth regulators directly in the phosphating solution.

Addition of titanium salts in concentrations of $1-10\,mg/L$ in the alkaline cleaning solution decreases the size of the crystals of the deposited zinc phosphate. Good results can be obtained with a special rinsing operation. For example, rinsing by immersing the article in a bath containing oxalic acid helps considerably to obtain a fine crystalline structure in the subsequent pretreatment with conversion coating. The action of the oxalic acid is a simple neutralization of the alkalies which remain after the cleaning and degreasing step.

A very fine crystalline structure can be obtained by modifying the phosphating solution by the addition of so-called regulators of crystal growth. These are calcium or barium phosphates and organic complexing agents like citric acid or tartaric acid. The best results are obtained using calcium ions. It is supposed that the composition of the deposited fine crystals on the steel surface is zinc calcium phosphate expressed by the formula $CaZn_2(PO_4)_2 \cdot 2H_2O$. Because of the fine crystalline structure, the deposited conversion coating is more compact and less porous than the common zinc phosphate layer. Therefore it provides superior corrosion protection. The thickness of the layer is of the same range as zinc phosphate.

In principle the application methods used for iron phosphate can be used in this case also. The brush-on treatment by a cold phosphating solution produces an acceptable coating thickness within 2–5 minutes.

The steam gun application method has not found commercial importance because of the poor quality of the coating and short life time of the operating equipment.

The immersion application method can be successfully used for zinc phosphate conversion coatings. The typical immersion technique involves five stages: clean, rinse, treatment, rinse and postrinse. The time required to develop the zinc phosphate immersion coating is 2–5 minutes. The working temperature varies in the same range as iron phosphate between 30 and 70°C.

The spray washer application accounts for the largest proportion of the zinc phosphate treatment process. Depending on the surface and quality requirements the treatment can involve five to nine steps which include cleaning, rinsing, treatment with the coating solution and drying, subsequently arranged in the most suitable way. Time requirements for each step are the same as those for the iron phosphate coatings.

The zinc phosphate pretreatment produces films of a greater thickness than the iron phosphate one ($1-5 \, g/m^2$ in comparison with $0.1-0.5 \, g/m^2$). Therefore it is favoured for articles which will be subjected to the most severe corrosion conditions. On the other hand, because of its crystalline structure, the conversion coating absorbs the powder coating which comes on top of it during the film formation and improves the adhesion of the powder coating to the substrate. The best corrosion protection of the surface is obtained by thick conversion coatings. On the other hand, zinc phosphate that is too thick can be disadvantageous for the flexibility of the complete system because of the brittle structure of the crystals.

A conversion coating can reduce the gloss of the organic coating applied on top of it, since the crystalline surface has an uneven microstructure. This is not very important for powder coatings, because in most cases they are applied in relatively thick layers, diminishing in this way the undesirable influence of the pretreated surface on the final gloss of the coating.

6.1.2.2 Cleaning and Pretreatment of Aluminium

Aluminium is one of the metals which is characterized by a certain degree of self-protection towards corrosion by building up a thin layer of aluminium oxide on its surface. However, this protection is not good enough to provide a long service life to aluminium articles exposed to normal weathering conditions. Moreover, aluminium oxide itself is not a very good base on which to apply powder or any other coating. The adhesion of organic coatings is in general very bad on aluminium unless the surface is specially treated in the ways described below.

The treatment of the aluminium surface before applying powder coatings includes phosphatization, chromium phosphatization and chromating of the surface. The other method is electrochemical oxidation of the surface which in most cases is used as the sole method, providing anticorrosion and decorative characteristics of the treated surface. It is almost never used as a pretreatment step in covering aluminium articles with powder coatings.

As in the case with steel, the chemical pretreatment of the aluminium is preceded by a necessary cleaning step in order to remove dirt, oil, grease and the corrosion products from the surface. Alkaline, acid and neutral types of cleaners are used for this purpose.

6.1.2.2.1 Alkaline Cleaners

The aluminium surface is sensitive to alkaline attack. Therefore alkaline cleaners for aluminium are milder than those used for cleaning steel. Alkaline salts such as sodium carbonate, sodium phosphate, potassium carbonate and sodium silicate are suitable for preparation of cleaning solutions. Only in extreme cases when heavy dirt has to be removed can alkaline cleaners containing small to moderate amounts of caustic soda be used very carefully.

Depending on the pH of the cleaners and their compositions, they can be classified into two groups: weak alkaline or almost neutral cleaners without any or with a very weak etching action on the surface and strong alkaline cleaners which dissolve the oxide layer on the surface which results in a noticeable loss of weight.

The first type of cleaners include solutions of sodium silicate, blends of sodium silicate, carbonate and pyrophosphate, trisodium phosphate, sodium pyrophosphate, and blends of sodium chromate, carbonate and pyrophosphate. Very often wetting agents are also included in order to improve the cleaning action.

Sodium silicate and sodium chromate have no etching action on the surface, and the weight loss is zero or almost negligible. The reason is that sodium silicate and sodium chromate do not react with aluminium as a metal. On the other hand, they undergo chemical reaction with the aluminium oxide layer, but the result is a transformation of the aluminium oxide into aluminium silicate or aluminium chromate, thus having almost no effect on the total weight of the treated material [14, p. 87].

$$2\,Al_2O_3 + Na_2SiO_3 \longrightarrow Al_2SiO_5 + 2\,NaAlO_2 \qquad (6.15)$$

$$4\,Al_2O_3 + 3\,Na_2CrO_4 \longrightarrow Al_2(CrO_4)_3 + 6\,NaAlO_2 \qquad (6.16)$$

$$NaAlO_2 + 2\,H_2O \longrightarrow NaOH + Al(OH)_3 \qquad (6.17)$$

$$Al + NaOH + H_2O \longrightarrow NaAlO_2 + 1.5\,H_2 \qquad (6.18)$$

The weight loss of the aluminium surface treated with sodium silicate or chromate solutions is in the range from 0 to $3\,mg/m^2$.

A weak etching effect is produced by treatment with sodium pyrophosphate and trisodium phosphate, where the weight loss between 40 and $1700\,mg/m^2$ is the result of the following chemical reactions:

$$Al + Na_4P_2O_7 + 3\,H_2O \longrightarrow NaAlO_2 + 3\,NaOH + 1.5\,H_2 \qquad (6.19)$$

$$Al + Na_3PO_4 + 3\,H_2O \longrightarrow AlPO_4 + 3\,NaOH + 1.5\,H_2 \qquad (6.20)$$

$$Al + NaOH + H_2O \longrightarrow NaAlO_2 + 1.5\,H_2 \qquad (6.21)$$

Sodium carbonate or its combination with sodium triphosphate which belongs to the etching types of alkaline cleaners removes the aluminium oxide layer, but at the same time these solutions attack the metal surface too, causing a weight loss in the range between 8 and $12\,g/m^2$:

$$Al_2O_3 + Na_2CO_3 \longrightarrow 2\,NaAlO_2 + CO_2 \qquad (6.22)$$

$$2\,Al + Na_2CO_3 + 3\,H_2O \longrightarrow 2\,NaAlO_2 + 3\,H_2 + CO_2 \qquad (6.23)$$

Two methods are generally used for cleaning aluminium articles with alkaline cleaners: the powder spray method, in which the articles pass through a tunnel while the cleaning solution is pumped from an outside storage tank under high pressure onto the parts, and the simple immersion method. The temperature of the cleaning bath varies from 40 to 70°C depending on the concentration of the cleaning agents, type and amount of dirt and the desired residence time of the articles in the cleaning station.

A special version of alkaline cleaning is electrocleaning. The articles are immersed in a bath in which a direct electrical current is passed through the solution. The parts to be cleaned serve as an anode, while other hanging electrodes are cathodes. The cleaning is improved by the scrubbing action of the oxygen bubbles which develop on the anode as a result of the electrolysis of water.

The simplest hand-wiping method can also be used, deriving benefit from the physical act of removing the soil from the surface by means of a cloth or sponge. This method is restricted to small discontinuous operating units or to cases were heavy massive articles have to be cleaned. For practical reasons it is not convenient in such cases to carry on a continuous cleaning operation with the article hanging on a conveyor chain or belt.

From an efficiency point of view electrocleaning is the best method followed by the spray, immersion and hand-wiping methods. On the other hand, the electrocleaning method is the most expensive due to the higher concentrations of active components and the cost of the electricity. The next most expensive is the immersion method, followed by the hand-wiping method, while the cheapest is the spray-wash cleaning method.

6.1.2.2.2 Acid Cleaners

These cleaners are composed of phosphoric acid or mild acidic salts of phosphoric acid alone or in combination with surfactants, or surfactants and solvents. The acid cleaners will remove the oxide film on the aluminium surface. On the other hand, they are not as effective as the alkaline cleaners in removing common soils from the surface.

In some cases acidic salts of sulphuric acid in low concentrations are also used. So-called activators, which are salts of hydrofluoric acid, are often used in combination with phosphoric or oxalic acid or pH regulators. Depending on the composition of the cleaner, the following reactions take place during cleaning of the aluminium surface with acid cleaners [14, p. 88]:

$$Al + H_3PO_4 \longrightarrow AlPO_4 + 1.5\,H_2 \tag{6.24}$$

$$2Al + 6(NH_4)HSO_4 \longrightarrow Al_2(SO_4)_3 + 3(NH_4)_2SO_4 + 3H_2 \tag{6.25}$$

$$2\,Al + 3\,Na_2HPO_4 + H_3PO_4 + 3\,Na \longrightarrow$$
$$AlPO_4 + AlF_3 + 3\,Na_3PO_4 + 3\,H_2 \tag{6.26}$$

In the case when aluminium alloys containing copper or silicium have to be cleaned, the alkaline cleaners produce black coloured sludge which is very difficult to rinse out. With the help of nitric acid, sulphuric acid, fluorides or chromates these materials are transformed into a soluble form and can be easily removed from the aluminium surface.

The same methods used for alkaline cleaners, with the exception of the electrocleaning method, can also be used for acid cleaners. The less expensive method again is the spray method, followed by the hand-wiping and immersion methods. The spray method is also the most effective concerning the cleaning action.

6.1.2.2.3 Conversion Coatings of Aluminium

6.1.2.2.3.1 Iron Phosphate Coating An iron phosphate solution can produce a conversion coating on aluminium which is not iron phosphate but aluminium phosphate. This is not an ideal system for aluminium but is very suitable for cases where articles made out of steel and aluminium have to be coated on the same line.

Aluminium phosphate coating produced out of iron phosphate is a very good base for a powder coating. The process is very easy to control and operate, but it can never be used exclusively for aluminium because the effective life of the bath is then very short. It is economical only when aluminium is processed together with steel.

6.1.2.2.3.2 Zinc Phosphate Coating A processing solution containing zinc dihydrogen phosphate, $Zn(H_2PO_4)_2$, and fluoride produces a very good conversion coating on aluminium. The chemical reactions involved are similar to those for the chemical treatment of steel.

Since during the phosphatization of aluminium articles the aluminium from the surface, like the iron in the case of steel, is dissolved and transferred in a soluble form into the bath solution, its concentration with time increases, which hinders the process. Addition of silicofluorides or fluorborates greatly helps the precipitation of solubilized aluminium in the form of an insoluble sludge. In continuous operation the zinc phosphate processing solutions will produce a large amount of insoluble sludge over a period of time, which may deposit on the heat transfer elements in the bath decreasing their efficiency or plug up the nozzles in the spray operation equipment. To maintain good operating conditions the process line must be cleaned very often. This is a rather weak part of this process, which otherwise produces a zinc phosphate layer that is a very good base for powder coatings. It has to be pointed out that this process can be used for lines which operate exclusively with aluminium articles.

As in all other cases, the process is accelerated by temperature; on an industrial scale the operating line temperature is in a range between 40 and 70°C. The coating produced in this way has a grey colour with a thickness between 1 and 5 μm and a deposition weight of $2-6\,g/m^2$. It provides very good corrosion resistance and greatly improves the adhesion of the powder coating applied at the final stage of the coating process.

6.1.2.2.3.3 Chromium Phosphate Chromium phosphate conversion coatings are obtained from solutions containing chromates, phosphates and fluorides. Chromate pretreatment of aluminium with solutions of the above compounds results in depositions of yellow, colourless or green layers depending upon the choice of chemicals.

During treatment of the surface, aluminium reacts with the hydrogen ions, reducing them to molecular hydrogen which exists in the bath in a gaseous form. However, a portion of the hydrogen atoms is consumed in reaction with the six-valent chromium, reducing it to a valency of three. The latter reacts with phosphoric acid to form an insoluble chromium phosphate which deposits on the aluminium surface.

The presence of fluorides accelerates these reactions by preventing formation of passive phosphate or a chromate layer. At low fluoride concentrations the deposition rate is slowed down because of passivation of the surface, while at high fluoride concentrations undesirable high etching of the surface takes place.

The optimum fluorides/chromates ratio depends on the concentration of phosphoric acid in the bath. For a very wide range of concentrations of phosphoric acid from 50 to 300 g/L the optimum hydrofluoric acid/chromic acid mole ratio falls in a rather narrow interval between 0.22 and 0.28. Concentrations of phosphoric acid higher than 300 g/L lead to too strong an etching of the aluminium surface and so are avoided in practice. The same holds for the hydrofluoric/chromic acid ratio. For a given concentration of phosphoric acid there is a hydrofluoric/chromic acid mole ratio above which there is no formation of conversion coating due to strong etching of the surface. For 300 g/L of phosphoric acid the critical hydrofluoric/chromic acid mole ratio is ca. 0.2, increasing in a linear manner to 0.4 for a bath with 50 g/L of phosphoric acid content.

Too low a fluoric/chromic acid ratio, on the other hand, slows down the deposition rate. Below some critical value there is no formation of conversion coating within the processing times used in practice. This critical ratio is between 0.19 and 0.13 for concentration of the phosphoric acid in the bath between 50 and 300 g/L.

The following reactions can be considered as a description of the process of precipitation of amorphous chromium phosphate conversion coating [14, p. 90]:

$$2Al + 2H_3PO_4 + H_2CrO_4 + 3HF \longrightarrow AlPO_4 + CrPO_4 + AlF_3 + 4H_2O + 1.5H_2 \quad (6.27)$$

$$AlF_3 + 3HF \longrightarrow H_3AlF_6 \quad (6.28)$$

The process is accelerated considerably by temperature. Depending on the desired deposition rate the bath temperature varies between 35 and 50°C with a residence time between 5 seconds and 3 minutes. After drying at room temperature, a coating layer with an approximate composition of 50–55% $CrPO_4$, 17–23% $AlPO_4$ and 22–23% water incorporated in the coating structure with traces of chromium fluoride is obtained [14, p. 91].

In continuous operation, the aluminium content in the bath solution increases permanently. There are two possible ways to control and keep the level of aluminium ions in the bath. The use of ion exchangers is the best solution, while a relatively simpler way is to add sodium or potassium fluoride to the bath which causes precipitation of the aluminium in the form of insoluble K_2NaAlF_6. A disadvantage of the latter method is the formation of a slurry of sodium potassium aluminium fluoride, which has to be removed by filtration.

The chromium phosphate conversion coating is the best conversion coating base that can be applied on aluminium. It is more effective compared to the phosphate pretreatment where aluminium, zinc and steel articles are treated on the same line. It exhibits better salt spray resistance and gives improved adhesion.

The process can be easily controlled and is less expensive to maintain. However, it must be remembered that its use is restricted exclusively to aluminium surfaces.

6.1.2.2.3.4 Chromate Conversion Coatings

Chromate conversion coatings for aluminium are solutions containing chromic acid or chromates and fluorides. Hydrochloric acid is often used to regulate the pH of the solution. Composed in this way, they deposit on the treated aluminium surface colourless to brown coloured conversion coatings, which have a very complex structure that can be described by the general formula $x Cr_2(CrO_4)_3 \cdot y Al_2O_3 \cdot z H_2O$. The composition of the layer by weight is: 24–28% Cr^{3+}, 0.4% Cr^{6+}, 1.5–7% Al^{3+} and 0.3–4% F^- [14, p. 88].

Although complex, reactions involved in the deposition of chromate conversion coatings can be expressed in the following simplified way:

$$3\,Al + 6\,H_2CrO_4 + 6\,HF \longrightarrow AlF_3 + Al_2O_3 + Cr_2(CrO_4)_3 + CrF_3 + 9\,H_2O \tag{6.29}$$

$$AlF_3 + 2\,KF + NaF \longrightarrow K_2NaAlF_6 \tag{6.30}$$

Depending on the application method, a layer with an average weight between 200 and 300 mg/m^2 can be obtained within 5–120 s at room temperature. The thickness of the layer depends on the temperature and the method of application. The most effective method is the spray application at temperatures not higher than 60°C. The colourless to light brown coating obtained is of an amorphous nature, providing outstanding corrosion protection. The colourless amorphous chromate layers obtained from diluted chromate solutions at a low pH between 3 and 4 have a lower corrosion resistance than the coloured coating obtained from more concentrated chromate solutions.

6.1.2.3 Cleaning and Pretreatment of Galvanized Steel

In a similar way to aluminium, the surface of galvanized steel undergoes self-passivation when exposed to atmospheric conditions by formation of a thin layer of zinc oxide or zinc carbonate. However, this layer of so-called 'white rust' does not give good protection in the long term, especially not in an environment of aggressive industrial atmosphere. The corrosion proceeds slowly, forming a visible white/grey powder-like layer.

Depending on the method of production, the thickness of the zinc layer in the galvanized steel varies from 3 to 200 μm. Even in the case where the coating is applied directly on a freshly zinced steel surface not containing zinc oxide or zinc carbonate, a reaction between the zinc and the carboxyl groups of the binder takes place, forming zinc soaps that reduce adhesion of the coating to the surface. As usual, the cleaning step precedes conversion coating deposition.

6.1.2.3.1 Alkaline and Acid Cleaners

Alkaline cleaners for galvanized steel differ from those used for steel in the strength of the alkali. These are in principle mild alkaline salts that prohibit attack of the zinc surface. When heavy dirt has to be removed from the zinc surface, small or moderate amounts of caustic soda can be added to improve the cleaning action. This addition will also cause a certain amount of etching of the surface which results in improved adhesion of the powder coating later on.

The same cleaning techniques used for aluminium, including electrocleaning, can also be used for zinc, with the same remarks concerning the efficiency and the costs of the separate methods being appropriate.

The acid cleaners for galvanized steel are not as effective as the alkaline cleaning agents. Moreover, the zinc surface is very sensitive to acidic conditions. These are the reasons why the acid cleaners are not widely used for cleaning galvanized steel. However, there are some special types of acid cleaners which are used to remove the white corrosion product from galvanized surfaces.

As with alkaline cleaners, an electrocleaning operation can also be performed with the acid cleaners. The parts to be cleaned serve as the anode. The scrubbing action of the oxygen bubbles developing on the anode as a result of electrolysis of the water increases the cleaning efficiency.

6.1.2.3.2 Conversion Coatings for Galvanized Steel

6.1.2.3.2.1 Zinc Phosphate Conversion Coating The zinc phosphate coating is the most important conversion coating applied on galvanized steel. The common conversion coating process known as iron phosphate does not produce a coating on galvanized steel. On the other hand, the chromium phosphate process is considered to provide a conversion coating of inferior quality compared to that obtained by zinc phosphate, especially with respect to the adhesion of the powder coatings which will be applied in the next step [13, p. 43].

Chemical reactions which take place during the deposition of insoluble zinc phosphate on the galvanized steel surface are very similar to those in the treatment of a steel surface with zinc phosphate. The main reaction is between the elementary zinc with zinc dihydrogen phosphate in a solution that contains free phosphoric acid as the pH regulator:

$$2\,Zn + Zn(H_2PO_4)_2 \longrightarrow Zn_3(PO_4)_2 + 2\,H_2 \qquad (6.31)$$

As in the case of phosphatization of steel, oxidizing agents such as nitrates and nitrites are widely used in order to accelerate the process and/or to reduce the bath temperature. Salts of nickel and copper are also effective accelerators and regulators of the growth of crystals. To obtain specially fine crystals, the phosphating step is preceded by treatment of the surface with solutions containing titanium salts. These titanium salts are usually added to alkaline types of rinsing solutions after cleaning of the surface has been completed.

POWDER COATINGS APPLICATION TECHNIQUES

Zinc–aluminium alloys are very difficult surfaces to phosphate. In principle dissolved aluminium is a hindrance for the phosphating process. Even amounts as low as 0.5 g/L may completely prevent the formation of a phosphate coating [15, p. 200]. In such cases addition of sodium fluoride or sodium fluorosilicate precipitates aluminium from the bath and improves the efficiency of the process. An alternative process is preliminary treatment of the surface with solutions of sodium or potassium hydroxide. As aluminium is more soluble, the surface layer consists almost entirely of zinc.

Due to the higher reactivity of zinc, application of the conversion coatings can be done even at room temperature. This can improve the process efficiency in the treatment of a zinc surface containing aluminium. At room temperature the rate of aluminium dissolution is slowed down to a level that provides satisfactory results in many cases. The reaction rate rapidly increases with an increase in the bath temperature for pure zinc layers. However, for obvious reasons this is not valid for zinc–aluminium alloys.

Application methods such as spray-wash, immersion or hand-wiping can be used for phosphatization of galvanized steel. The line operation steps include cleaning, rinsing, treatment with the conversion coating solution, rinsing, eventual posttreatment, rinsing with deionized water and drying. Line operating temperatures are the same as in the previous cases ranging from 30 to 70°C. The advantage of the zinc phosphate conversion coating is the possibility of running articles of galvanized and cold rolled or heat rolled steel on the same line.

6.1.2.3.2.2 Chromium Phosphate Conversion Coatings Chromium phosphate conversion coatings are usually blends of sodium bichromate and mineral or organic acid which form a colourless, yellow, brown or green layer on the zinc surface. The amount of deposited material ranges from 500 to 700 mg/m^2.

The following reactions take place in the chromating process:

$$Zn + H_2SO_4 \longrightarrow ZnSO_4 + H_2 \tag{6.32}$$

The hydrogen formed as a result of this reaction partly reduces the six-valent chromium to a three-valent chromium:

$$2\,Na_2Cr_2O_7 + 3\,H_2 \longrightarrow 2\,Cr(OH)_3 + 2\,Na_2CrO_4 \tag{6.33}$$

The final product which is formed on the metal surface is believed to be the result of the following reaction [16, 17]:

$$2\,Cr(OH)_3 + H_2CrO_4 \longrightarrow Cr(OH)_3 \cdot Cr(OH)CrO_4 + 2\,H_2O \tag{6.34}$$

After drying, the inorganic conversion coating with a film thickness of about 0.5 μm is obtained. Its composition can be described as $Cr_2O_3 \cdot CrO_3 \cdot 2H_2O$ [18, 19]. When in contact with water the chromate film hydrolyses and free hexavalent soluble chromium ion is formed. The normal colour of the coating is yellow. However, when the hexavalent chromium compounds are leached out, the yellow colouration disappears and the film takes on a greenish tint. In practice, chromate coatings on galvanized steel show colourations that range from yellows and golds to olive greens.

The residence time of the article in the chromating solution depends on the concentration of the active materials in the bath, the bath temperature and desired thickness of the deposited layer. In principle it is relatively short and in a bath temperature range between 15 and 40°C the residence time varies between 5 and 60 seconds. Because of the speed of the process, the usual practice is to perform passivation of the articles at room temperature.

Regardless of the type of cleaning agent or conversion coating, four application methods are in common use in treatment of the surface. Table 6.2 summarizes the advantages and disadvantages of the particular methods [13, p. 54].

6.2 APPLICATION OF POWDER COATINGS

Because of their different nature compared to liquid coatings, powder coatings are applied on the object to be coated by techniques that have little in common with the well-known methods of application of conventional solvent borne coatings.

At first sight it seems that application of powder coatings should be more difficult than application of wet paints. However, this is only a prejudice emanating from the fact that people (even those who are not professionally involved in painting) are quite familiar with the use of liquid coatings. Difficulties with powder coatings generally arise only in geometrically complicated objects, where among other things electrostatically screening effects may occur. These difficulties should not, however, give an impression that application of powder coatings is always problematic. Compared to the liquid coatings, powder coatings even have several quite essential advantages.

One of the greatest limitations associated with wet paint spraying is that the particles that miss the object are irretrievably lost. Only those hitting the workpiece are used for their purpose. The ratio between these two quantities, the so-called overspray, can sometimes be so high that, depending on the object geometry, it is this factor alone that determines whether the coating process is economic or not. Very often this is the reason for switching to another wet application technique such as curtain or dip coating. Although electrostatic paint spraying decreases the overspray rate, it can never be neglected as in the case of powder coatings, where the coating is recycled in the spraying booth.

The 100% utilization of powder coatings together with elimination of the solvents are among the elements which have enormously influenced the economy as well as the ecology of the painting process. These advantages far outweigh the technical difficulties encountered in the early stages of development of powder paint application techniques.

POWDER COATINGS APPLICATION TECHNIQUES

Table 6.2 Comparison between different application methods in surface treatment [13, p. 54] (Reproduced by permission of the Society of Manufacturing Engineers. Copyright 1985, edited from *Users Guide to Powder Coatings*)

Method	Advantages	Disadvantages
Brush-on, wipe-on	No capital investments Parts do not have to be moved Large heavy parts can be handled Can be started/stopped quickly Suitable for intermittent work schedules	Labour intensive Low production rates
Steam gun	Low capital investments Parts do not have to be moved Large heavy parts can be handled Can be started/stopped quickly Suitable for intermittent work schedules	Labour intensive Low production rates
Immersion	Can be automated Moderate production rates	Moderate capital investments Must be able to lift parts Requires start-up time
Power spray washer	Normally automated High production rate	High capital investment Requires start-up time

Four different powder coating processes have been developed during the last thirty years: electrostatic spraying, fluidized bed, electrostatic fluidized bed and flame spray, the last one being developed recently and not yet widely used [20]. These application techniques which will be described in the following sections offer the following advantages over the wet coating systems: self-regulating thickness of the applied powder coat, absence of sagging problems, good edge covering, low working power costs, low overall operating costs, low costs for training of the personnel, 100% of utilization of the coating material, meeting the most stringent environmental compliances [21].

6.2.1 ELECTROSTATIC SPRAYING TECHNIQUE

Electrostatic spraying is the most common process used for application of powder coatings in metal finishing. A diagram of the conventional powder coating spray system is shown in Figure 6.1. The basic principle of the process

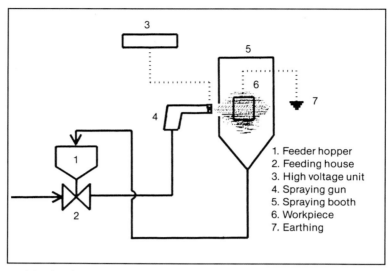

Figure 6.1 A schematic diagram of an electrostatic spray powder coating system

concerns propulsion of the dry powder by means of compressed air through a spray gun, in which it becomes electrically charged. The movement of the particles between the charged gun and the substrate to which the powder is applied is governed by a combination of electrical and mechanical forces. The electrical forces are the result of interaction between the charged powder particles and the electric field between the substrate and the gun, while the mechanical forces are derived from the air that blows the powder through the gun. Powder particles as electrically insulating material retain their charge and adhere to the workpiece. While the workpiece is in front of the gun the particles are held on the surface by the electric field from the gun. When the field is removed, the charged particles are still held to the surface, attracted by image charges in the substrate. The charged particle and its image form a positive and negative pair and can be considered to be opposite plates of a capacitor.

The oversprayed powder can be collected and reused. More than 98% overall material usage is claimed in the industrial powder coating installations. Usually the recovery system consists of a cyclone, followed by a filter to remove the last traces of dust particles from the air.

6.2.1.1 Corona Charging Guns

These are the most widely used application devices used to charge powder particles. Charging of the powder particles is imparted by high voltage applied to at least one electrode located inside the sprayer. As well as charging the particles,

POWDER COATINGS APPLICATION TECHNIQUES

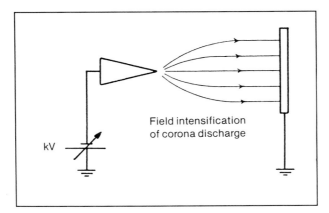

Figure 6.2 Electrostatic field lines between the corona gun and the workpiece [22] (Reproduced by permission of John Wiley & Sons, Inc.)

the corona electrostatic spraying gun functions to shape and direct the flow of the powder supplied by the feeder, to control the pattern size, shape and density of the powder cloud emitted by the gun and to control the deposition of the powder by electrostatic voltage levels.

In its simplest design, the corona gun may be a tube with a charging electrode in the form of a sharp pointed needle-like electrode or a fine gauge wire [22]. The electrode is connected to a high-voltage generator. Depending on the electrode geometry and the potential, a local electric field of about 3 MV/m can be created with field lines as presented on Figure 6.2. This is the field strength at which electrical breakdown or discharge of the air itself will occur in the vicinity of the pointed needle. This is done in a controlled low energy continuous process which manifests itself as a bluish glow. As a result of the local discharge, the air will be ionized, producing negatively charged ions, as is commonly the case with corona electrostatic spray commercial installations. The gun is constructed in such a way that powder particles carried by compressed air are directed to pass through this charged space, picking up the negative ions on their way to the workpiece (Figure 6.3).

As can be expected, not all powder particles leaving the gun nozzle are charged and not all negative air ions are picked up by the powder particles. Typically only 0.5% of the ions produced by a corona are associated with powder particle charging. The remaining 99.5% exist as free ions in the powder cloud [23]. In other words, the material leaving the electrostatic powder spray gun is composed of negatively charged powder particles, uncharged particles and negative ions.

The mechanism of charging powder particles by negative ions is not well understood. The situation is especially complicated bearing in mind that the powder paints are materials with a relatively high resistivity which can be

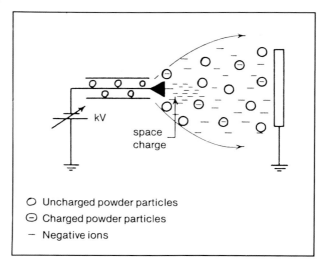

Figure 6.3 Corona charging of the powder particles [22] (Reproduced by permission of John Wiley & Sons, Inc.)

considered as a constraint to efficient particle charging. The influence of resistivity and powder particle size on charging efficiency is investigated by Corbett [24, 25]. Although there is no sound fundamental clarification of the charging mechanism, it is believed that ion trapping by the powder particles is purely a mechanical process as a result of direct collision between the flying powder particles and the ions when the latter are retained on the particle surface and 'locked into positions at the exact point of impaction' [22, p. 5].

It is obvious that the powder particle in the ionic cloud could accumulate ions until its potential equals that of the surrounding field. This means that a certain maximum surface charge density cannot be exceeded. The theory of charging predicts that the saturation charge, Q, obtained by a spherical particle in the size range of 1–200 μm by air ions in a uniform electric field is given by

$$Q = \frac{3kEa^2}{k+2} \tag{6.35}$$

where k is the dielectric constant of the particle, E is the charging field, which is in the case of corona charging proportional to the charging voltage, and a is the radius of the particle [26]. This maximum surface charge is known as the Pauthenier limit. For non-spherical particles, the charge is higher than that predicted from theory, but the difference is insignificant in practical applications [27].

The charge/mass ratio q/m is usually considered as a measurement for the charging efficiency. A somewhat modified version of equation (6.15) which

POWDER COATINGS APPLICATION TECHNIQUES

includes the permittivity of the free space can be used to calculate the maximum charge/mass ratio [28]:

$$\left(\frac{q}{m}\right)_{max} = \frac{3\varepsilon_0 E}{(r_0)a}\left(1 + 2\frac{\varepsilon_r - 1}{\varepsilon_r + 1}\right) \quad (6.36)$$

where

ε_0 = permittivity of free space
ε_r = relative permittivity of powder particles
a = particle radius
(r_0) = density of the particle
E = electric field to which the particles are subjected

Substituting corresponding values for the terms in equation (6.36) Hughes has obtained a maximum charge/mass ratio of 8.8×10^{-4} C/kg for an electrical field of 10^6 V/m [22, p. 7]. The same Pauthenier limit is valid for positive charging by corona guns, although the mechanism of ion generation is different. In this case the electrons have to be stripped from the neutral air molecules creating positive ions. On their way to the earthed object these ions collide with the powder particles, imparting a positive charge in a manner identical to that of the negative ions.

The basic principle of charging in corona guns is the same, independently of the producer and type of gun. The differences are in construction details such as the design and location of the electrode in the gun, differences in the air diffuser at the gun nozzle, improved handling convenience, etc. These minor differences and improvements are very important to the user because they greatly influence the deposition efficiency of the equipment.

The most common type of corona gun incorporates a single short pointed electrode at the gun nozzle located at the centre of the powder diffuser at the front of the spray gun. This is the so-called external charging gun (Figure 6.4) [13, p. 78].

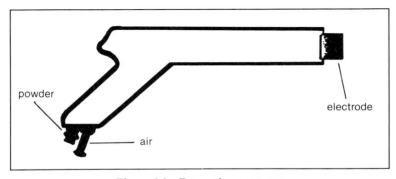

Figure 6.4 External corona gun

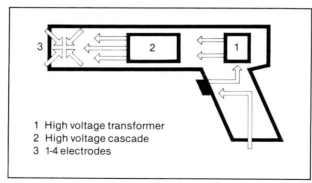

Figure 6.5 Corona gun with integrated high-voltage supply (Reproduced by permission of Ransburg-Gema AG)

The electrode is usually fixed; the only adjustment of the powder transfer efficiency that is possible is by changing the potential applied to the electrode. These guns generate high-voltage, low-amperage electrostatic fields of between 30 and 100 kV between the electrode and the workpiece. They provide reasonably effective particle charging, however, followed by an excessive population of free ions, which as will be discussed later creates considerable problems during the spray operation. The electrostatic charge on the electrode is usually of a negative polarity, although the contemporary guns offer the possibility of choosing different polarities. This is especially important when powders that can be easily charged by friction with a positive charge have to be applied with a corona. In such a case the negative corona charging gun is not the correct choice. Switching to positive corona charging considerably increases the charging efficiency.

Developments in the equipment for a high-voltage supply by switching from a conventional oil-filled high-voltage generator to an epoxy encapsulated cascade type contributed very much to its miniaturization. This resulted in an electrostatic corona gun with an integrated high-voltage generator (Figure 6.5) [29]. In the internal charging guns the electrode is located inside the tube behind the gun nozzle with an earthed counterelectrode (Figure 6.6). The charging is imparted by very thorough mixing in a turbulent flow in the region of the electrode. All powder emanating from the gun nozzle will have passed through the corona region, but few free ions are ejected from the nozzle. The result is a high specific charge with a fairly small voltage on the electrode. Under typical operating conditions with a charging voltage of 10 kV, internal barrel potentials greater than 55 kV have been measured with this type of gun [30]. Due to the fact that the electrical field is built up inside the gun, the strong field developed by the corona gun between the electrode and the workpiece does not exist.

Although remarkably good in terms of high-quality coatings, long-term tests with the low-voltage internal charging corona guns have indicated deterioration

POWDER COATINGS APPLICATION TECHNIQUES 331

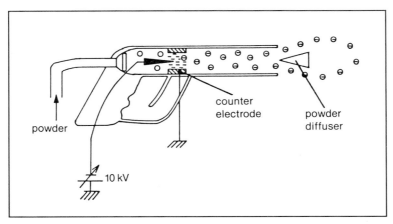

Figure 6.6 Internal charging corona gun [30] (Reproduced by permission of Elsevier Science Publishers BV)

in performance after long uninterrupted runs. This is associated with the growth of partially cured powder coating on the earthed counterelectrode inside the gun. However, back-ionization at the counterelectrode, or even nullification of the corona field, could account for this problem [30]. The only practical remedy for this problem is frequent cleaning of the gun at intervals of 1–2 hours. However, this is a time consuming operation and must be performed with the utmost care due to the complex internal design of the gun.

Another improvement in the corona guns is claimed by the design of the so-called flat-jet nozzle with an air cleaned central electrode that operates at a lower voltage than the standard gun and ensures optimum powder transport from the gun to the workpiece. This is achieved by the special nozzle geometry which produces a flat, well-directed spray jet which makes it easier to transport the powder into problematic zones such as recesses, internal edges and overlaps [29].

6.2.1.1.1 Problems Associated with Corona Spray Guns

The material leaving the gun nozzle consists of charged and neutral powder particles and ionized air or so-called free ions. While the trajectories of the uncharged particles will depend predominantly on the air flow pattern, those of the charged particles and the free ions will result from the interaction between the mechanical forces emanating from the movement of the turbulent air flow, and the electrostatic field forces between the gun nozzle and the workpiece. Since the free ions are much smaller than the charged particles and have a much higher mobility, their motion between the gun and the workpiece is very rapid. On the way to the workpiece they will collide with the neutral air molecules, inducing an additional motion of the air itself in the same direction as the ions. This additional

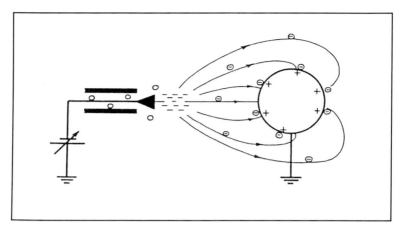

Figure 6.7 Typical electric line configuration [22] (Reproduced by permission of John Wiley & Sons, Inc.)

component of the air velocity to that created by the gun air supply is called the ion-wind [31]. In some cases the ion-wind air velocity can be compared with that of the gun air and can not be disregarded as a component participating in an all-over air dynamic pattern of the gun.

The trajectories of the charged particles, if not interrupted by the air movement, will tend to follow the lines of the electric field, as presented in Figure 6.7. Because of this, by the electrostatic spray technique it is possible to coat shadow areas that are not directly exposed to the air stream, leaving the gun as a result of wrap-around. Turbulent movement of the air in the environment of the workpiece is a prerequisite for achieving a wrap-around effect, but at the same time the linear air velocity near the object to be coated should not exceed the speed of the particle induced by the electrical attraction forces. Therefore, the air flow produced by the gun can be even more important than the electrostatic field forces created between the gun and the workpiece. Confirmation of this is given by the wrap-around effect, which is very effective in the case of tribo guns where the electrical field between the gun and the workpiece is essentially eliminated.

The air flow will play a predominant role for almost the whole distance between the gun nozzle and the workpiece, since the electrostatic attaction forces are shown to be significant only at distances shorter than 2 cm from the surface on which the particles deposit [32, 33].

The presence of the electric field between the gun nozzle and the workpiece creates enormous problems when articles with a convex geometry have to be coated by corona spraying guns. As a direct result of the classical electrostatic law, no field lines can exist or penetrate areas which are surrounded by the earthed metal boundary of the workpiece [22, p. 18]. If the air velocity is low (and

POWDER COATINGS APPLICATION TECHNIQUES

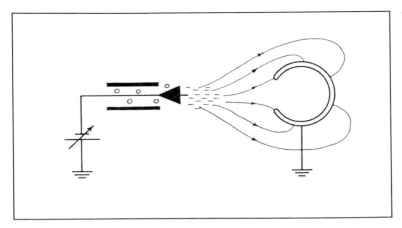

Figure 6.8 Typical flow pattern for corona charged guns [22] (Reproduced by permission of John Wiley & Sons, Inc.)

correspondingly the linear speed of the flying powder) the particles will follow the field line pattern that does not penetrate the inside surface of the workpiece (Figure 6.8). As a rule, the electrostatic forces will deposit the material down into an opening to a depth equal to the minimum internal dimension of that opening. This is known as the Faraday cage effect. The Faraday cage effect is less pronounced in the internal charging guns, as can be expected due to the lower voltage applied to the electrode, but this effect certainly cannot completely be eliminated.

A phenomenon typically associated with electrostatic spraying of powder coatings with corona guns is the so-called back-ionization which was the concern of many studies [34–40]. Figure 6.9 represents the development of a powder coating layer on the surface of the workpiece. As the powder layer grows, the potential across its thickness increases as a result of accumulation of charged particles and ions on the coated surface. At a certain moment the breakdown potential is exceeded and sparks will occur on the surface or within the layer. This breakdown potential is equivalent to that of the air, being 300 kV/m, since the coating layer is loosely packed material in uncured form. The result of the discharge is the creation of bipolar ions. While the negative ions will be retained by the substrate, the positive ions will drift away towards the gun, creating the phenomenon known as back-ionization.

It has been found that with most commerical corona-charged gun systems, the back-ionization commences immediately after the first monolayer of powder coating has been deposited on the earthed surface. In terms of time it is approximatively after one second of spraying time [22, p. 23]. The electrical discharge on the coated surface or within the coating layer is associated with a

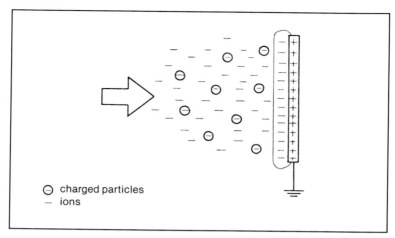

Figure 6.9 Back-ionization in the powder layer [22] (Reproduced by permission of John Wiley & Sons, Inc.)

release of energy, which could lead to disruption of the deposited layer. This will certainly have a negative effect on the quality of the applied coating in terms of uniformity, although an explanation of the orange peel effect by back-ionization [22, p. 24] is difficult to accept as the only contribution to this rather complex phenomenon.

Hughes and Ting [41] have found a considerable difference in the surface disruptions caused by the back-ionization effect between guns with different polarities. The spark associated with back-ionization in the case of the guns with negative charging polarity penetrates the depth of the deposited layer and creates surface defects such as pinholes and craters. When a gun with positive charging polarity is used, the spark during the discharge is a surface phenomenon. Although in this case there is no pinholing, the surface defects tend to be more severe. Huges explains these differences by the fact that the negative spark is more energetic and can penetrate the bulk of the layer. He recommends the choice of negative charging corona guns in the case of coatings whose aesthetic values are of primary importance and positive charging corona guns when the surface protection is the prime objective [22, p. 27].

The resistivity of the powder particles plays an important role in the system behaviour regardless of the type of application equipment. At the gun head or in the vicinity of the charging electrode the powder resistivity should be as low as possible in order to maximize charge acquisition. However, the resistivity should be as high as possible in order to ensure slow charge relaxation rates and good adhesion of the particles when they have reached the substrate. It has been stated that the lower acceptable limit of powder resistivity is about 10^{11}–$10^{12}\,\Omega\,m$. Below this value charging will be good but adhesion will be poor [42]. Almost all powder coatings today usually have a resistivity higher than $10^{14}\,\Omega\,m$ and display excellent adhesive properties prior to curing. Singh, O'Neill and Bright [33] found that a minimum charge/mass ratio of around $2 \times 10^{-4}\,C/kg$ was necessary for adequate adhesion of powder to the workpiece. However, the fact of having this minimum charge/mass ratio does not necessarily mean that a good coating efficiency should be expected. Hughes [43] suggests that typically about 40% of the powder in a normal coating operation emerges from the gun uncharged, which can be considered as one of the factors that reduces the efficiency of the coating process.

6.2.1.2 Tribo Charging Guns

Tribo charging guns make use of the principle of frictional charging known for several thousand years as one of the earliest phenomena associated with the electrical properties of solid materials. It is well known that when two insulating materials are rubbed and subsequently separated they become charged by opposite charges. Many examples in everyday life give evidence of frictional charging, such as combing the hair, walking on nylon-made carpets or changing clothes made out of synthetic fibres.

The mechanism of tribo charging has been studied by Davies [44] and Henry [45] who explain the phenomenon by the contact potential difference model in which the insulator is treated as a metal. It is well known that when two metals are brought in contact then an electron transfer will take place due to the difference in the mean energy level between the electrons in the different materials. However, since the metals are conductive materials, the charge transfer will occur simultaneously with neutralization, so that the net result of charge reorganization at the interface will not be obvious. If two insulators are brought into contact, the charge transfer will create a longer term unipolar charge on the surfaces in contact which will remain after their separation. It is believed that this mechanism itself cannot be good enough to explain the relatively high degree of charging obtained by sliding two insulators. An additional question that naturally occurs is why it is necessary for sufficient charge transfer to rub the materials. The explanation that in this way the contact surface is increased is not very convincing. Therefore another mechanism of charge transfer has been suggested which involves ionic transfer or bulk material transfer. Hughes suggests that charging of the powder coatings in the tribo gun charging installation is a result of combining these two mechanisms [22, p. 10].

The construction of the tribo gun is very simple. It is in most cases a Teflon-made tube through which the powder coating is propelled by means of compressed air, exhibiting a turbulent flow inside the gun. The tube walls are earthed, so that the charge generated on the walls is permanently drained to the earth (Figure 6.10).

The high-voltage supply in the tribo gun compared to the corona is completely

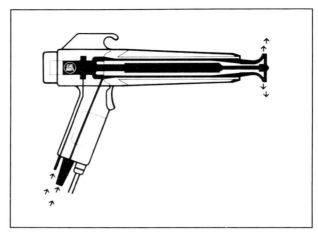

Figure 6.10 Tribo charging gun (Reproduced by permission of Ransburg-Gema AG)

eliminated. However, the cost savings because of this are not the most important gain. The Faraday cage effect is almost completely eliminated since there is only a very weak electric field created between the workpiece and the charged powder cloud. This makes it possible to coat objects having cavities and a convex shape since the powder particle trajectories are directed by the air stream rather than the weak electric field.

Long-term charging efficiency of the tribo guns is not always constant. In many cases, depending on the type of coating, there is a tendency for powder particles to adhere to the gun walls. This changes the gun performance with time and it is believed to be one of the reasons why results have not been consistent on long uninterrupted runs. Therefore it is suggested that for new unknown types of coatings, the tribo gun evaluation should be extended to at least five-hour continuous runs so that a realistic picture can be obtained about the suitability of this application technique.

Depending on the binder and crosslinker used in powder coating formulation, quite different results can be achieved with respect to the charging efficiency of the same gun. The charging efficiency depends very much on the dielectric constant of the binder/crosslinker system. The so-called dielectrical series tabulated in Table 6.3 in order of their relative dielectric constants can give an indication of the expected behaviour of different binders [21].

The charge achieved by rubbing two materials given in Table 6.3 will depend on the distance between these materials. Since all commercial tribo guns use Teflon as the material for the gun walls, it should be expected that epoxy powder coatings will be more suitable for tribo application than powder coatings based on polyester resins. However, powder coating has a rather complex composition, and the chargeability will also be affected by the other constituents of the coating.

Table 6.3 Dielectrical series of different materials [21] (Reproduced by permission of DSM Resins BV)

1. Teflon (∗)	13. Metals
2. Polyethylene	14. Cellulose
3. Polypropylene	15. Polyvinyl alcohol
4. PVC	16. Polyamide
5. Polyacrylonitrile	17. Wool
6. Nitrocellulose	18. Urea formaldehyde resins
7. Silicone	19. Glass
8. Anthracene	20. Polyvinyl acetate
9. Polystyrene	21. Epoxy resins (∗)
10. Polyester resins (∗)	22. Polymethyl methacrylate
11. Cellulose acetate	23. Polyurethanes
12. Rubber	24. Polyethylene oxide

Considerable work has been done to improve the charging efficiency in the case of polyester powder coatings. According to the patent literature [46–51] the problem of insufficient chargeability of the polyester powder coating has been successfully overcome.

Reports referring to the experience of using tribo guns in practice indicate that initial doubts about the potentials of this technique slowly disappear [52, 53]. Apart from elimination of the Faraday cage effect, the tribo application process offers additional advantages. The absence of high voltages increases the safety of the coating operation. When in use, all electrically conductive components of the gun are earthed, eliminating the possibility of sparks igniting. Since the tribo gun is not an electrical appliance it is not subject to a compulsory check before installation. More compact powder coating films are easily obtained as a result of the absence of free ions and a lower back spray tendency.

Tribo guns have usually a lower capacity than the corona charging guns. This can be considered as a serious disadvantage, especially in applications where large quantities of powder coatings have to be applied in a short time. Certain improvements in this respect have been achieved by the new construction of the tribo guns where the friction surface is interrupted by a large number of trucks. This allows the surface charge to flow easily to earth without having an adverse effect on the powder charge [21].

6.2.1.3 Alternative Guns

Several new types of electrostatic spray guns have been developed as an answer to the problems associated with the corona and tribo guns. The slow but steady decrease in the charging efficiency of the tribo guns is claimed to be overcome by adoption of internal corona augmentation [54].

The construction illustrated in Figure 6.11 has been tested at Southampton

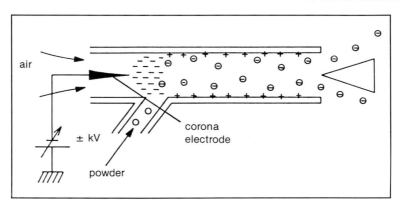

Figure 6.11 Corona augmented tribo gun [30] (Reproduced by permission of Elsevier Science Publishers BV)

University [30], leading to the conclusion that the gun gives uniform particle charging and high charging efficiency. High q/m values are obtained for epoxy/polyester powder coatings which are rather problematic for tribo application together with low free-ion emission, enabling good powder penetration in cavities.

Considerable efforts have been made to improve the long-term efficiency of the internal low-voltage corona gun. Two different approaches involving a piezoelectric ceramic ring electrode which undergoes an oscillatory deformation

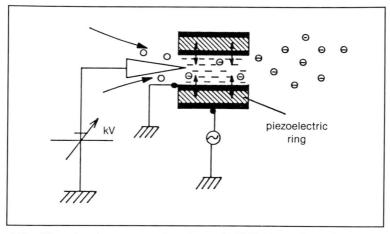

Figure 6.12 Piezoelectric attractor ring electrode [30] (Reproduced by permission of Elsevier Science Publishers BV)

POWDER COATINGS APPLICATION TECHNIQUES

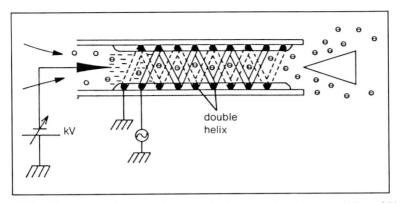

Figure 6.13 Electric curtain low-voltage gun [30] (Reproduced by permission of Elsevier Science Publishers BV)

(Figure 6.12) and a curtain electrode with a double helix configuration (Figure 6.13) have been considered as a solution to the problem [30, 55].

Despite the promising results, neither attempt resulted in a commercial gun system. It seems that the conventional porous metal ring electrode which is cleaned by air purge is the simplest and most efficient construction used in the previously described low-voltage internal corona gun.

6.2.1.4 Factors Affecting the Spraying Process

The coating process with electrostatically charged powder particles is controlled by both electrostatic and aerodynamic factors. In several excellent articles Hardy [56] and Golovoy [27, 57, 58] analyse the influence of these factors on the deposition efficiency of the coating process and the growth of the film thickness during the powder deposition.

Hardy has measured the potential at the surface of a layer of the powder after it has been deposited on a flat surface and compared the obtained results with those calculated from the following which he derived for such a case:

$$V_s = \frac{3E_{ps}m^2}{kBdD} \tag{6.37}$$

where

m = mass of the deposited powder per unit surface
B = bulk density of the powder layer
d = density of powder particles
D = powder particle diameter
k = dielectric constant of the powder layer
E_{ps} = electric field which the charge produces at the surface of the particle

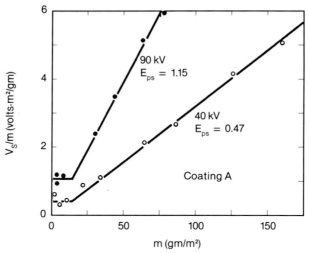

Figure 6.14 Plot of the surface potentials observed at various coating thicknesses, for two different gun voltages. The curves have been drawn according to equation (6.37) [56] (Reproduced by permission of the Federation of Societies for Coatings Technology)

Figure 6.14 confirms the excellent agreement between the experimental and calculated values.

Hardy has found that the E_{ps} values are independent of the nature of the powder coatings and the type of gun used for spraying. However, the E_{ps} values are dependent on the voltage applied to the gun as can be concluded from equation (6.35).

Equation (6.37) can be written in the following form:

$$\frac{V_s}{m} = \frac{3E_{ps}m}{kBdD} \quad (6.38)$$

The quantity V_s/m which is equivalent to the potential gradient across the insulating layer of deposited powder is obviously proportional to m, i.e. to the mass of deposited charged particles per unit coated surface. In other words, by increasing the coating thickness the potential gradient increases linearly as long as it reaches the dielectric strength of the powder layer. At this moment an electrical breakdown will occur and the excess of voltage will be drained to the earth. Powder particles arriving at the surface after this critical layer thickness has been reached will increase the surface potential again above the dielectric strength of the layer, causing a localized discharge, and will lose their charge. Since the charge of the particles is responsible for keeping the powder adhered to the substrate, they will be held weakly on the surface and will be blown away by the air flow or will detach from the surface due to the gravitational forces. This is the explanation offered by Hardy for the well-known self-limiting phenomenon, which can be useful for obtaining a uniform film build but creates problems when thick coating layers have to be applied.

Hughes [22] associates the self-limiting effect with the back-ionization phenomenon. According to him, the positive ions created by the electrical discharge when the dielectric strength of the powder layer has been exceeded drift away from the substrate towards the gun. On their way to the gun they collide with the oncoming negative ions or negatively charged particles. The resulting neutralization will decrease the amount of negatively charged particles reaching the workpiece. Depending on the thickness of the deposited layer and the voltage supplied to the electrode, the coating ceases at a certain moment due to this back-ionization. He uses the same arguments to explain the difference in the critical layer thicknesses between the tribo and corona gun application techniques. The back-ionization which commences almost immediately after trigging the corona appears in the case of tribo charging systems after 10–20 seconds. Since no ions contribute to the potential build-up in the deposited layer, it takes much longer for the breakdown potential to be reached. Therefore it takes 10–20 seconds before the onset of back-ionization when the tribo charging system is used compared to only one second in the case of corona guns. This makes it possible to apply much thicker films with tribo guns because the self-limiting effect does not play an important role. While this can be an advantage in the case of functional coatings when the protection efficiency is proportional to the coating thickness, in many other industrial applications it is not an acceptable situation. However, the absence of the self-limiting effect can be compensated by careful choice of the conveyor speed and the output of the gun.

Once adhered to the substrate, powder particles may lose their charge slowly by electric conduction through the layer to the earthed workpiece. Hardy has found that the powder layers applied by corona guns do not lose more than 10% of the initial charge during a period of 10 minutes after the application [56]. Even considerable delays in the transfer of coated objects to the curing oven will not affect the coating thickness very much. Moreover, some charge losses can be favourable if they cause the loosely held particles to drop off, decreasing in this way the roughness of the coated surface.

While the critical film thickness is dependent on the strength of the electrostatic field created between the gun and the workpiece, the deposition efficiency of the spraying process depends on a combination of electrostatic and aerodynamic factors and workpiece geometry.

The deposition efficiency of the electrostatic corona spray process increases with an increase in the charging voltage, although as already seen the critical layer thickness decreases. This is due to a higher average charge per particle, which can be achieved by a higher charging voltage, and a stronger electric field between the gun and the workpiece. Since the field is responsible for delivering the charged particles to the object, this will result in an increase in the deposition

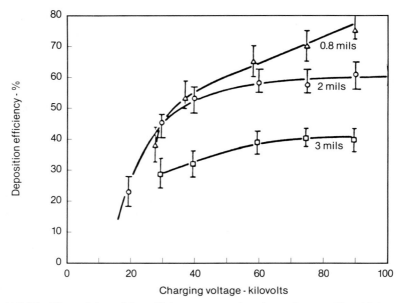

Figure 6.15 Plots of deposition efficiency versus charging voltage at film thicknesses of 0.8, 2 and 3 mils [27] (Reproduced by permission of the Federation of Societies for Coatings Technology)

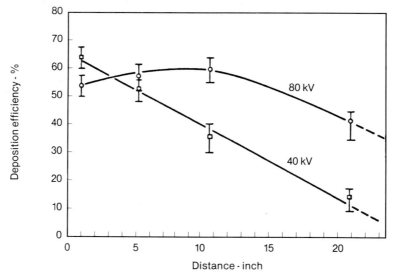

Figure 6.16 Plots of deposition efficiency versus spraying distance at charging voltages of 40 and 60 kV [27] (Reproduced by permission of the Federation of Societies for Coatings Technology)

efficiency. However, as can be seen in Figure 6.15, in the case of thicker films the percentage of powder particles adhering to the substrate decreases.

The electric field between the gun and the workpiece weakens when the distance between them is increased. At the same time the powder particles are exposed for a longer time to the air flow and self-repulsion effect on their way to the workpiece. This results in a decrease in the deposition efficiency by increasing the spray distance. This is much more pronounced when a lower voltage is supplied to the charging electrodes (Figure 6.16).

Once the powder particle is deposited on the surface it has to remain there until it enters the stoving zone. Particles with high electrical surface resistivity retain their original charge for a longer time. If the surface resistivity is below 10^8 ohm-metres the adhesion of the powder particles to the earthed substrate will be very poor, decreasing the overall deposition efficiency of the coating process [59]. On the other hand, too high a surface resistivity produces an adverse effect by the deposited particle field on powder particle transportation and deposition efficiency. For optimum adhesion and deposition efficiency Golovoy recommends a region of particle surface resistivity between 10^8 and 10^{14} ohm-metres [27].

It is normal to expect that the booth air velocity should have an adverse effect on deposition efficiency. This effect is, however, negligible when the booth air flow is parallel to the spraying direction. It is very pronounced when the booth air flow is normal to the spraying direction since it increases the velocity component of the particles parallel to the surface. The adverse effect of the booth air velocity can be drastically minimized by decreasing the spraying distance. The experiments performed by Golovoy showed that at a spraying distance of 5 inches and booth air velocity of 2 ft/s parallel to the spraying direction the deposition efficiency is ca. 58% compared to 35% in the case of the same air velocity normal to the spraying direction. The decrease of the spraying distance to 2 inches compensates almost completely for the negative effect of the latter, bringing the deposition efficiency to the same level as that of parallel booth air flow.

The booth air flow and the flow of compressed air used for transportation of the powder particles through the gun are responsible for the low deposition efficiency of powder coatings with a fine particle size. This is to be expected since the motion of the small particles is greatly affected by the air flow. The maximum deposition efficiency is observed by Golovoy at a particle size of 35 μm, after which the deposition efficiency decreases again. This is because above 35 μm fluidization and transportation of the particles from the powder reservoir to the gun becomes difficult, requiring an increase of the velocity of the carrying air and thus reducing the deposition efficiency.

Finally, the size of the substrate affects the deposition of the powder coating. This is especially emphasized for objects with small dimensions. Golovoy has found that for long objects with width W, the ratio W/D where D is the spray diameter at the gun exit plays an important role with respect to the deposition efficiency independently of the operating conditions. When $W/D < 3$, the deposition efficiency rapidly decreases. For values above 3, the efficiency of the coating process increases slowly, levelling off at a limiting value that depends on the other operating conditions [27].

6.2.2 FLUIDIZED BED PROCESS

Powder coatings which appeared in the 1950s were first applied by the fluidized bed technique. This rather simple application process is still in use although the electrostatic spraying plays by far the most important role among the different application techniques for powder coatings.

In this process the preheated workpiece enters the space where by means of the air stream the powder particles are kept in a fluidized bed. Powder particles coming in contact with the preheated surface melt and adhere to the substrate. Depending on the temperature, the heat capacity of the workpiece and its residence time in the fluidized bed, coating layers with various thicknesses can be applied. Postheating is usually not necessary in the case of thermoplastic powder coatings, especially when the workpiece is a massive object with a large heat capacity. When thermosetting powder coatings are used, additional postheating follows the application step in order to complete the curing of the coating.

The basic construction of the fluidized bed chamber is presented in Figure 6.17. In its simplest embodiment it comprises a container within which the powder coating is confined. The upper compartment of the container is open for dipping the part to be coated and for allowing the fluidizing air to escape. This upper chamber where the coating process takes place is separated from the bottom compartment by a porous membrane. Compressed air is blown into the bottom compartment to be dispersed uniformly through the porous membrane before entering the fluidized chamber. Fluidization of the powder in the upper compartment is effected by the lifting effect of the air stream. The amount of air required per cubic metre of the bed depends on the powder characteristics affecting the air dynamics of the fluidized bed process such as the density of the powder, the particle size and the particle size distribution. It is understandable that the even distribution of the air through the membrane is a prerequisite for proper operation of the whole installation. Complete aeration of the bed from the bottom to the top is necessary in order to obtain a uniform layer thickness all over the workpiece. It is very important to avoid the formation of a dead compact layer at the bottom of the fluidized chamber immediately above the membrane which leads to progressive stratification affecting the coating results in the dipping section [13, p. 62].

Although not essential it is preferable to have a continuous feed with fresh powder to compensate for the consumed powder, instead of discontinuous

POWDER COATINGS APPLICATION TECHNIQUES

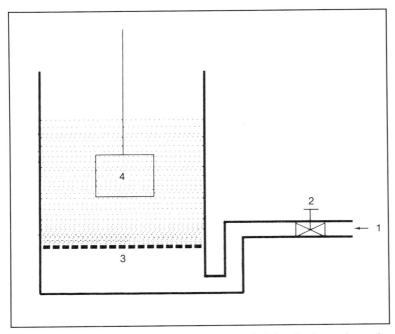

Figure 6.17 Scheme of fluidized bed installation: 1. air inlet; 2. air regulator; 3. porous membrane; 4. workpiece

addition. However, very consistent results can be obtained by supplying more frequent discontinuous additions of new powder to the dipping section.

The rate of coating deposition depends very much on the particle size. The fine particles are consumed in principle faster than the coarse particles. Therefore, the composition of the bed with respect to the particle size distribution changes continuously with time. However, this is a problem that exists only for a short period after the start of the coating operation. Very quickly, by subsequent addition of fresh powder, the fluidized bed reaches steady-state operation conditions. By the proper maintenance of a constant temperature of the preheated object, careful control of the air supply, frequent feed with fresh powder, constant residence time of the workpiece and constant time sequence between two dipping operations, good high-quality coatings can be obtained.

The fluidized bed process is a very attractive application technique in many respects. The powder is confined to an enclosure and therefore is much easier to handle without the need to recycle and reblend the overspray. Due to the absence of recycling and reblending equipment, fluidized bed installations are in principle very compact compared to the spray gun/booth combinations.

6.2.3 ELECTROSTATIC FLUIDIZED BED

A combination of fluidized bed and electrostatic charge of the powder particles is effected in the electrostatic fluidized bed process. The design of the electrostatic fluidized bed installation is almost the same as that of the normal fluidized bed. By a combination of an air flow very often accompanied by mechanical vibration and charging electrodes in the upper chamber or in the air plenum under the porous membrane (Figure 6.18), the powder is fluidized and at the same time electrically charged. The depth of the bed is only 5–10 cm above the porous membrane through which the air is blown. The charging electrodes are usually located, because of safety reasons, under the membrane so as to charge the air used to fluidize the bed. The electrodes may be sharp points or more usually small diameter wires. By applying high voltage to the electrodes, a high charge density ion cloud is created in the same way as in the case of corona charging guns. The ionized air charges the powder particles as it moves upwards through the powder bed. The electrostatically charged powder particles with an identical charge repel each other and move upwards. In this way a cloud of charged particles is formed above the upper surface of the powder bed. The earthed workpiece entering the cloud is covered by the powder particles due to the difference in the electrical

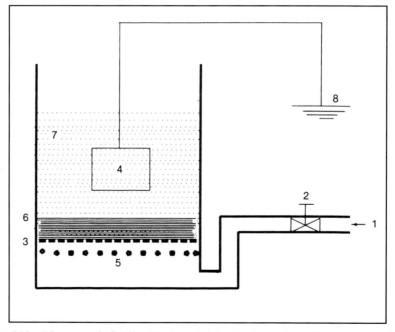

Figure 6.18 Electrostatic fluidized bed: 1. air inlet; 2. air regulator; 3. porous membrane; 4. workpiece; 5. electrodes; 6. bed level; 7. electrostatic cloud; 8. earthing

potential, creating an electrical field of attraction. In fact only this field is responsible for the deposition of the powder on the workpiece. Therefore, preheating the parts to be coated is not necessary. In contrast to the conventional fluidized bed process little air is used to control the volume of the powder cloud and to prevent stratification of the bed.

The cloud density depends more on the voltage applied rather than on the air flow rate. The higher the voltage the more dense the cloud. The deposition rate becomes less as the workpiece is moved away from the upper surface of the charged bed. Therefore this technique is especially suitable for coating small objects. Due to the Faraday cage effect the deposition of the powder particles into holes or other convex surfaces is rather poor. This is typical for any electrostatic spray coating technique characterized by the existence of a strong electric field between the charging electrodes and the workpiece. However, the electrostatic fluidized bed process maintains all the advantages of the fluidized bed with respect to the absence of recycling and reblending equipment. In fact the overspray is continuously recycled and recharged within the confines of the bed chamber.

The voltage employed with this method is in the range of 30–100 kV. This requires special safety precautions. The possibility of electrical sparking between the earthed workpiece and the high-voltage charging electrodes may lead to ignition of the powder. This can happen especially if the workpiece comes close to the electrodes. Because of this, contemporary electrostatic fluidized bed installations have a unit design that incorporates a so-called inherently safe charging system with the electrodes located below the porous membrane. In this way sparking between the workpiece and the electrodes is practically eliminated. Moreover, the particle trajectories are influenced much more by the air flow pattern than by the electrostatic field which is not the case when the charging electrodes are located in the upper compartment of the container. It is believed that superior powder penetration can be expected with such a design of the application unit, due to the less pronounced Faraday cage effect.

6.2.4 FLAME-SPRAY TECHNIQUE

The flame-spray coating technique has been developed recently for application of thermoplastic powder coatings. Polyethylene, copolymers of ethylene and vinyl acetate, nylon and polyester powder coatings have been successfully applied by flame spraying. This technique permits powder coatings to be applied to practically any substrate, since the coated article does not undergo additional heating to ensure film formation. In this way substrates such as metal, wood, rubber and masonry can be successfully coated with powder paints if the coating itself has a proper adhesion to the substrate [60]. The technique itself is relatively simple. Powder coating is fluidized by compressed air and fed into the flame gun. The powder is then injected at high velocity through a flame of propane. The residence time of the powder in the flame and its vicinity is short, but just enough to allow complete melting of the powder particles. The molten particles in the form of high viscous droplets deposit on the substrate forming high-build film upon solidification.

An example of a flame-spray gun disclosed in a patent of Oxacetylene Equi [61] is sketched out in Figure 6.19. The gun has a body (1) with air (7), combustion gas (9) and powder material (5) supply channels. The outlet of the powder channel is axially positioned at the gun mouthpiece (3) with the channels for the combustion gas outlet situated at equal distances on the circumference concentric to the axial powder channel. The efficiency is increased by preventing the powder from burning in the flame since the concentric circumference diameter is 2.85–4 times the powder outlet channel diameter. The coating quality is increased when using liquefied gas since the combustion gas outlet channel axis is at 6–9° to the powder channel axis, forming a diverging flame. The amounts of air and combustion gas are regulated by valves (8 and 10). The air passes through ejectors (11) creating a refraction in the channel (9). The air and liquefied gas mix in chambers (12) forming a combustible mixture which flows to the mouthpiece

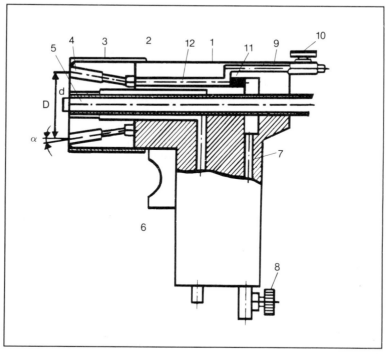

Figure 6.19 Flame-spray powder coating gun [61]

nozzles (4). The powder particles entering the flame are heated and in a molten form are supplied onto the surface being coated.

Since the flame spray process does not involve oven heating it is very suitable for field application on workpieces which are large or permanently fixed and thus not able to fit inside an oven. It has been reported that objects such as bridges, pipelines, storage tanks and railcars are suitable to be coated by this technique [20].

6.2.5 COMPARISON BETWEEN DIFFERENT APPLICATION TECHNIQUES

Electrostatic spray, fluidized bed and electrostatic fluidized bed processes are commercially well established, while the flame-spray process is relatively new and not widely used. The performances of the first three processes have been compared by Buren and Martin [20]. Despite the lack of relevant information about the practical experience of using the flame spray, some characteristics emanating from the nature of the technique can be compared with the other three processes (see Table 6.4).

6.3 DESIGN OF THE SPRAYING BOOTHS

One of the greatest advantages of the powder coating process is almost 100% utilization of the material achieved by recirculation of the oversprayed powder that does not adhere to the workpieces. In order to combine this advantage with a high quality of coating process, the design of the spraying booth system has to satisfy the following two requirements: recapturing the oversprayed powder as completly as possible and ensuring that the powder is continuously dressed to a consistent quality and with nearly particle size distribution [62]. There are three different principles used in the construction of the spraying booths to meet these requirements. A scheme of the so-called compact recovery system which makes use of air filters for recycling the powder is presented in Figure 6.20. The powder spray booth is provided with a replaceable filter which is part of the recovery unit. To obtain flawless surface quality, powder deposited on the filters must be continuously removed and added to the powder circuit as uniformly as possible. This is done by intermittent cleaning of the filters with a short blast of air. Two types of filters are used in this operation: sturdy cartridge filters with rotary wings and a relatively long service life and plate filters with exceptional separation efficiency for microfine powders and a virtually unlimited service life. The powder particles that are blown off the filters fall on a vibrating sieve or rotary sieve for automatic elimination of dirt particles. Reblending of the powder passing through the sieve is done in the fluidized bed powder hopper where fresh powder is supplied manually or automatically from the fresh powder hopper depending

Table 6.4 Characteristics of different powder coating application techniques (Reproduced by permission of Decision Resources, Inc.)

Characteristic of workpiece	Electrostatic spray	Fluidized bed and electrostatic fluidized bed	Flame spray
Size	Larger	Smaller	Not limited
Material	Metallic, must be conductive	Any, not necessarily conductive	Any, not necessarily conductive
Temperature resistance	Relatively high	High	Not relevant
Aesthetic value	High	Low, not suitable for decorative purposes	Low, not suitable for decorative purposes
Coating thickness	Thinner films	Thick high-build films with excellent uniformity	Thick high-build films; uniformity dependent on the operator
Type of coatings	Thermoplasts and thermosets	Thermoplastic and thermosets	Thermoplasts only
Colour change	Difficult	Relatively difficult	Easy
Capital investments	Moderate to high	Low	Very low
Labour	Low since highly automated	Moderate depending on the automatization	Relatively high
Energy consumption	Only postheating	Preheating and often postheating	Low, no preheating and postheating
Coating waste	Very little	Very little	Dependent on the workpiece geometry

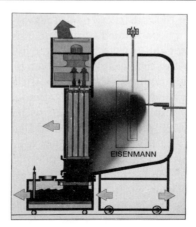

Figure 6.20 Compact recovery system (Reproduced by permission of Eisenmann Maschinenbau KG, Böblingen, West Germany)

POWDER COATINGS APPLICATION TECHNIQUES

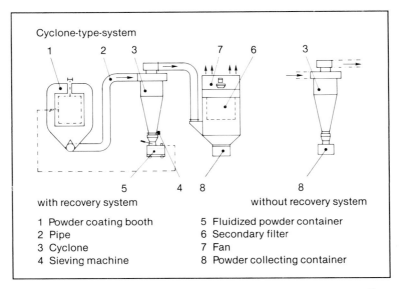

Figure 6.21 Cyclone-type-system (Reproduced by permission of Ransburg-Gema AG)

on the size of the coating operation. The continual addition of the virgin powder is essential if high-quality coating and efficient operation is to be maintained. The booth exhaust air is drawn through an exhaust fan, filtered and returned to the work area. The design of the powder booth provides a horizontal path for the powder to the workpiece, resulting in wrap-around coverage and a uniform coating. Although in principle any material can be used to make the booth, stainless steel booths are especially suitable for providing reduced cleaning expenditures, simple maintenance and extended service life.

The cyclone type powder recovery system is another option that is widely used in the coating process and is depicted in Figure 6.21. Any powder which does not adhere to the workpiece passes through an exhaust device usually located in the lower part of the booth and via a pipe enters the cyclone where it is deposited. Recovered powder is then passed through the sieving machine and collected in the fluidized bed container for recycling. Any powder particles not extracted in the cyclone are captured by the secondary filter and collected in the container beneath the filter.

The powder coating spraying booth is equipped with a filter belt system that is specially suitable for multicolour operation. Filter belt systems have two separate air circuits (Figure 6.22). The booth exhaust air is handled by a medium pressure fan while for powder recovery a side channel fan is used. The oversprayed powder is deposited on a continuously running filter belt and carried to the suction head located outside the booth. The suction head cleans the filter bed across its entire

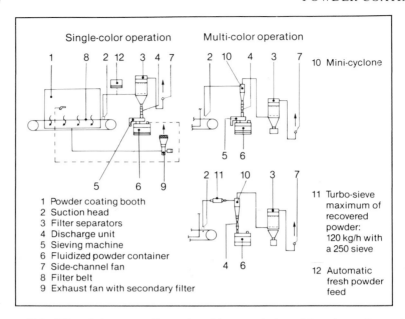

Figure 6.22 Filter belt system (Reproduced by permission of Ransburg-Gema AG)

width, conveying the particles either to the filter separators for single colour operation or to the mini cyclone for multicolour operation. The powder is collected in the fluidized bed container for recirculation.

In order to maintain the powder in the fluidized bed container at a constant level, the powder feed from the fresh powder hopper is controlled by a level sensor in the fluidized powder container.

Another option in multicolour operation with recovery rates up to 120 kg/h is replacement of the sieving machine by a turbo sieve in the powder air circuit presented in Figure 6.21. The delivery air filtered by the filter separators is returned to the working area via the side channel fan.

The exhaust air from the booth, after passing through the filter belt and an exhaust fan provided with a secondary filter, is returned back to the work area.

6.3.1 COLOUR CHANGE IN THE POWDER COATING PROCESS

Colour change in the powder coating process is a rather problematic operation. Contrary to the liquid coating systems, where there is no recirculation of the oversprayed material and the colour change is a matter of seconds, in the powder coating operation process transfer to a different colour can only be performed after being sure that the whole system is free of the powder particles from the previous colour. Many different solutions have been suggested and accepted by

POWDER COATINGS APPLICATION TECHNIQUES 353

the end users for reducing the costs of this time consuming operation that affects the capacity of the coating plant very much. Floyd [63] lists some of the possibilities such as mobile powder booths, detachable cartridge recovery modules, spray to waste and booths with moving parts with a minimum fixed structure.

The first option with movable powder booths is quite suitable for fast colour changes. However, it suffers from several obvious disadvantages. There is an enormous increase in the capital investments due to duplication of the system. The floor space is twice as large compared to a single booth system. The savings are made with respect to the production time only. In the case of applying multiple colours, the expenses for manhours necessary to clean down the off-line system are not reduced at all.

Separate recovery units are economical in cases of extended intervals between colour changes and low total output. It is still necessary, however, to clean down the interior surfaces of the booth, which is by far the most time consuming part of the colour change.

Spray to waste is obviously the fastest way to switch from one colour to another, with the complete loss of one of the main advantages of the powder coating process—the most economical one with respect to the material utilization.

Floyd claims that the time necessary to complete a colour change with full powder recovery in the case of spraying booths with movable parts is between 5 and 15 minutes. Since the booth parts are made from polyethylene they can easily be cleaned from the powder particles because the charged powder does not adhere to non-conductive surfaces.

For several colours and not very short intervals between the colour change, the booths with compact recovery systems making use of filter elements are a very economical solution. Spare filter units and powder containers are necessary for frequent and rapid colour changes. While powder deposition continues in the booth with a second recovery unit, the filters are automatically processed in a separate cleaning system in preparation for the next colour change. Economies of cost and space are the main advantages compared to the mobile powder recovery unit system.

Filter belt systems are mainly used for average to long product runs and are intended for high-quality manual and automatic work. Their flexibility allows more than one recovery system and a quick colour change. Due to the small size of the recovery unit, space requirements are small. Compared with both the cyclone type and compact recovery systems, longer intervals for colour changes are economical because of the better separation efficiency and powder recovery.

Cyclone type systems are the traditional solution for small batches with frequently changing colours. They are simple to operate, make few demands on the operators and are relatively inexpensive. Some limitations, however, exist due to the volume of powder to be handled and/or the volume of powder to be

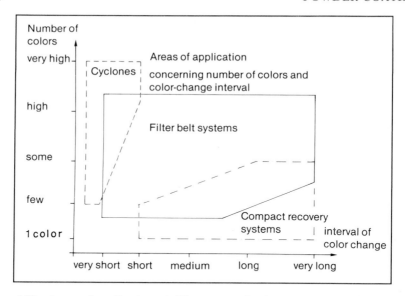

Figure 6.23 Areas of application of different spraying booth systems (Reproduced by permission of Ransburg-Gema AG)

recovered. Therefore this system is suitable for plants with small capacities with respect to the applied powder coating per unit time characterized by very frequent colour changes.

The diagram presented in Figure 6.23 can be used as a guide to choosing the correct system dependent on the number of different colours and the time interval between the colour change.

In addition to these criteria, the colour range with respect to similarities of the shades, the plant utilization, the quality standards, the available space and the price should also be considered when choosing a spraying booth system.

6.4 TROUBLESHOOTING

The liquid nature of solvent or water based coatings permits the coating composition to be corrected by postaddition of components which can improve certain properties or remedy certain deficiencies. This is very often done by adding flow, anticrater or defoaming agents in order to improve film formation properties, catalysts in order to increase the curing rate, crosslinker or binder to correct mechanical properties of the cured film, solvents to adjust the rheological properties, pigment pastes to match colour, etc.

Problems of this nature, which can occur during the production or application

POWDER COATINGS APPLICATION TECHNIQUES

Table 6.5 Troubleshooting guidelines for production and application of powder coatings [13, 64, 65] (Reprinted courtesy of the Society of Manufacturing Engineers. Copyright 1985, edited from *Users Guide to Powder Coatings*)

Possible causes	Possible solutions
A. Problems at production	
A1. Bad dispersion of the pigments	
1. Inhomogeneous preblend composition	1. Increase the residence time in the blender
2. Viscosity of the compounding mass too low	2. Decrease the extrusion temperature
3. Too high pigment volume concentration	3. Decrease the pigment and filler content
4. Too short residence time in the extruder	4. Decrease the extruder capacity
5. Pigment and filler particles too coarse	5. No help; contact the producer
A2. Sticking of the powder particles during grinding	
1. Too high temperature in the grinding mill	1. Increase the cooling efficiency.
2. Too low T_g of the coating composition	2a. Check the T_g of the binder and crosslinker
	2b. Reformulate the coating
3. Too high molecular mass of the binder (especially in the case of thermoplastic powders)	3. Use cryogenic grinding conditions
A3. Poor physical stability during storage	
1. Too low T_g of the coating composition	1a. Check the T_g of the binder and crosslinker
	1b. Reformulate the coating
A4. Poor chemical stability during storage	
1. Too reactive system	1a. Reduce the amount of catalyst
	1b. Reformulate the coating, adding resin with lower functionality
A5. Poor impact resistance of the cured film	
1. Undercured film	1. Increase the catalyst concentration
2. Inhomogeneous powder coating	2a. Increase the premixing time
	2b. Increase the dwell time in the extruder

Table 6.5 (*continued*)

Possible causes	Possible solutions
3. Non-stoichiometric binder/crosslinker ratio	3. Determine the acid, hydroxyl, isocyanate and epoxy equivalent values; rework the powder with corrected ratio

A6. Poor flow with orange peel effect

1. Too reactive system	1a. Reduce the amount of catalyst
	1b. Add resin with lower functionality

B. Fluidized bed operation
B1. Powder blowing out of hopper

1. Air pressure too high	1. Adjust the pressure
2. Too fine powder particles	2a. Too much reclaim added to the virgin powder
	2b. Too fine originally manufactured powder

B2. No air percolation through the powder

1. Insufficient air pressure	1. Check air supply line
2. Plugged membrane	2a. Check the membrane
	2b. Check the air purity
3. Compacted powder	3. Loosen the powder manually
4. Powder level too high	4. Remove part of the powder
5. Powder particles too fine	5a. Too much reclaim added to the virgin powder
	5b. Too fine original powder

B3. Air blowing large holes through the powder

1. Powder level too low	1. Add powder; keep the hopper two-thirds full
2. Packed or moist powder	2a. Loosen the powder manually
	2b. Check the humidity of the compressed air
3. Blocked powder	3. Check T_g of powder coating; if too low contact the paint manufacturer
4. Membrane obstructed	4. Check bottom of the bed
5. Plugged or broken membrane	5. Check the membrane

C. Electrostatic coating operation
C1. Fluctuation in the coating thickness

1. Irregular powder feed	1a. Clean powder feed hoses, pump venturis and guns
	1b. Check humidity of the air supply which can cause powder compaction

Table 6.5 (*continued*)

Possible causes	Possible solutions
	1c. Check powder free flowing properties
2. Inadequate gun distance to the workpiece	2. Readjust the distance between the gun and the workpiece
3. Inadequate conveyer speed in relation to the high voltage of the gun	3. Readjust the conveyer speed
4. High-voltage fluctuations	4. Systematically check electrical continuity from the voltage source to the electrode
5. Powder blown off from the workpiece	5a. Powder delivery air too high; turn down air setting
	5b. Move gun position further away from the workpiece
6. Pressure fluctuations in the air supply system	6. Check the air supply system

C2. Poor powder build-up and wrap on the workpiece

Possible causes	Possible solutions
1. Poor earthing	1. Check the earth connections; all contact areas must be free of powder build, grease or other insulating material
2. Powder output too high	2. Turn down powder feed until adequate charging is obtained
3. Electrode insulated by powder layer	3. Clean the electrode
4. Particle size distribution not ideal	4. Too much reclaim added to the virgin powder
5. Powder coating composition contains resins that are not adequate for the application equipment used	5. Contact the supplier of powder coatings

C3. Poor penetration into recesses and gaps

Possible causes	Possible solutions
1. Gun voltage too high	1. Turn voltage setting down
2. Powder delivery too low	2. Turn up powder delivery air setting
3. Poor earthing	3. Check earthing connections and contact areas
4. Inadequate gun position	4. Adjust the gun position so that powder cloud has a direct path to the recess area
5. The spray pattern is too wide	5. Wrong deflector plate
6. Delivery air speed in the gun too high	6. Change the air velocity

Table 6.5 (*continued*)

Possible causes	Possible solutions
C4. Back-ionization—powder repelled from the workpiece in spots	
1. Voltage too high	1. Turn down voltage setting
2. Gun too close to the workpiece	2. Relocate the gun
3. Too fine powder particle size	3a. Too much reclaim added to the virgin powder
	3b. Originally delivered powder is too fine
D. Properties of the cured film	
D1. Poor impact resistance	
1. Under cured film	1. Increase oven temperature or increase oven dwell time
2. Too high film thickness	2. Readjust the application equipment
3. Poor cleaning or pretreatment	3. Check pretreatment equipment and chemicals
4. Powder coating formulation problem	4. Contact powder manufacturer
D2. Poor adhesion	
1. Poor cleaning or pretreatment	1. Check pretreatment equipment and chemicals
2. Under cured film	2. Increase oven temperature or increase oven dwell time
3. Powder coating formulation problem	3. Contact powder manufacturer
D3. Poor corrosion resistance	
1. Poor cleaning or pretreatment	1. Check pretreatment equipment and chemicals
2. Under cured film	2. Increase oven temperature or increase oven dwell time
3. Powder coating formulation problem	3. Contact powder manufacturer
D4. Poor chemical resistance	
1. Under cured film	1. Increase oven temperature or increase oven dwell time
2. Powder coating formulation problem	2. Contact powder manufacturer
D5. Low film hardness	
1. Under cured film	1. Increase oven temperature or increase oven dwell time

POWDER COATINGS APPLICATION TECHNIQUES

Table 6.5 (*continued*)

Possible causes	Possible solutions
2. Powder coating formulation problem	2. Contact powder manufacturer

D6. Poor flow—orange peel effect

1. Too low film thickness	1. Increase the film thickness
2. Too slow heating rate	2. Increase the heating rate
3. Too fast curing system	3. Contact powder manufacturer
4. Too coarse powder particles	4. Contact powder manufacturer

D7. Craters in the film

1. Oil/grease residues on the metal surface	1. Check the pretreatment
2. Oil in the compressed air	2. Check the air supply system
3. Contamination with silicone oils from the chain lubrication	3. Clean the spraying area
4. Contamination with incompatible powder	4. Clean the application installation

D8. Pinholes in the film

1. Moisture content of the powder too high	1a. Check the powder storage conditions
	1b. Check humidity of the air supply
2. Air inclusions	2. Preheat the workpiece to 160°C before application

D9. Bubbles in the film

1. Water on the workpiece	1. Check the drier and the workpiece hangers
2. Powder film too thick (especially in the case of polyurethanes)	2. Decrease film thickness

of powder coatings, cannot be solved in the same way. Very often every large deviation from the required specification needs a complete reworking, starting from the very first production step. However, many of the problems can be overcome by simple interventions during the production, application or curing of the powder coatings. Certain guidelines for the causes of the problems with suggestions of how to overcome them collected from the published and presented data [13, 64, 65] are summarized in Table 6.5.

REFERENCES

1. Treacher, S., *Pretreatment Prior to Powder Coating, Powder Coatings 87 Conference*, Manchester, 1987.
2. Lund, P. J., *Pretreatment for Powder Coatings, Powder Coatings 87 Conference*, Manchester, 1987.
3. Short, N. R., Dennis, J. K., and Agbonlahor, S. O., *R/P Trans. Inst. Metal Finish.*, **66** (3), 107, 1988.
4. Bridger, R., and Burden, P., *Product Finishing*, p. 6, April 1988.
5. Marsh, C., *Product Finishing*, April 1988, p. 14; Brooks, M. E., *Product Finishing*, April 1988, p. 8.
6. Kent, G. D., and Petschel, M., *Product Finishing*, September 1988, p. 56.
7. Johnson, J. H., *Product Finishing*, September 1988, p. 48.
8. Gorecki, G., *Metal Finishing*, December 1988, p. 15.
9. Agbonlahor, S. O., Short, N. R. and Dennis, J. K., *R/P Trans. Inst. Metal Finish.*, **64**, 115, 1986.
10. Linck, W., *Oberflache + JOT*, **3**, 46, 1988.
11. *Pretreatments for Powder Coatings, Paint Research Association*, Notes to Industry No. 7, April 1989.
12. Gruss, B. B., *Metal Finishing*, May 1988, p. 29
13. Miller, E., *Users Guide to Powder Coatings*, Society of Manufacturing Engineers, Dearborn, Michigan, 1985, p. 35.
14. Adapted from Galjaard, D., and Leeuwen, A. van, *Chemische Oppervlaktebehandelingen van Metalen*, Spruyt, van Mantgem & de Does N. V., Leiden, 1969, p. 53.
15. Biestek, T., and Weber, J., in *Electrolytic and Chemical Conversion Coatings*, Portcullis press Ltd, Redhill, Surrey, 1976, p. 128.
16. Levitina, E. J., *Zh. Prikl. Khim.*, **34**, 984, 1960.
17. Ropper, M. E., *Metal Finishing*, **14**, 320, 1968.
18. Farr, J. P., and Kulkarni, S. V., *Trans. Inst. Metal Finish.*, **44**, 21, 1966.
19. Burns, R. M., and Bradley, W. W., *Protective Coatings for Metals*, Reinhold Publishing Corp., 3rd ed., New York, 1967, p. 553.
20. Buren, M. F. van, and Martin, J. F., *Spectrum, Advanced Materials and Chemical Specialities—Products and Technologies*, March 1988.
21. Lehmann, R. P., 'Optimized powder application of critical objects', in *DSM Resins—Powder Coating Symposium*, Nordwijk, May 1988.
22. Hughes, J. F., in *Electrostatic Powder Coating*, Research Studies Press Ltd, Letchworth, Hertfordshire, 1984, p. 1.
23. Moyle, B. D., and Hughes, J. F., *Institute of Physics Conference Series No. 66, Session VI, Electrostatics*, Oxford, 1983, p. 155.
24. Corbett, R. P., and Makin, B., *Trans. Inst. Metal. Finish.*, **51**, 160, 1973.
25. Corbett, R. P., *Dechema-Monographien*, **72**, 1, 1974.
26. Pauthenier, M. M., and Moreau-Hanot, M., *J. Phys. Radium*, **7** (3), 590, 1932.
27. Golovoy, A., *J. Paint. Technol.*, **45** (580), 42, 1973.
28. Hughes, J. F., *J. Oil Col. Chem. Assoc.*, **4**, 145, 1989.
29. Lehman, R., 'Electrostatic powder coatings, developments and trends', in *Procedures of Powder Coating Seminar*, Lucerne, October 1988.
30. Moyle, B. D., and Hughes, J. F., *J. Electrostatics*, **16**, 277, 1985.
31. Athwall, C. S., Conventry, P. F., and Hughes, J. F., *Institute of Physics Conference Series No. 66, Electrostatics*, Oxford, 1983, p. 167.
32. Singh, S., and Bright, A. W., *3rd International Congress on Static Electricity, Conference Proceedings 29G*, Grenoble, 1977, p. 729.

33. Singh, S., O'Neill, B. C., and Bright, A. W., *J. Electrostatics*, **4**, 325, 1978.
34. Cross, J., and Bassett, J. D., *Trans. Inst. Metal Finish.*, **52**, 112, 1974.
35. Masuda, S., Mitzuno, A., and Akutsu, K., *3rd International Congress on Static Electricity*, European Federation of Chemical Engineering, Grenoble, 1977.
36. Hughes, J. F., and Ting, Y. C., *IEEE/IAS Conference Proceedings 35G*, 1977, p. 906.
37. Sing, S., and Bright, A. W., *IEEE/IAS Conference Proceedings 29G*, 1977, p. 729.
38. Sing, S., and Bright, A. W., *IEEE/IAS Conference Proceedings 3F*, 1978, p. 105.
39. Sing, S., Hughes, J. F., and Bright, A. W., *Institute of Physics Conference Series No. 48*, 1979, p. 17.
40. O. Neill, B. C., *3rd International Congress on Static Electricity*, European Federation of Chemical Engineers, Grenoble, 1977.
41. Hughes, J. F., and Ting, Y. C., *3rd International Congress on Static Electricity*, European Federation of Chemical Engineers, Grenoble, 1977.
42. Hughes, J. F., *J. Oil Col. Chem. Assoc.*, **4**, 145, 1989.
43. Hughes, J. F., *Finishing*, **7**, 6, 1982.
44. Davies, D. K., *Institute of Physics Conference Series No. 4*, 1967, p. 29.
45. Henry, P. S. H., *Institute of Physics Conference Series No. 11, Static Electrification*, 1971, p. 195.
46. Misev, A. T., Binda, P. H. G., and Hardeman, G., to Stamicarbon BV, NL Pat. 8 800 840, 1988.
47. Misev, A. T., and Binda, P. H. G., to Stamicarbon BV, NL Pat. 8 802 913, 1988.
48. Binda, P. H. G., and Misev, A. T., to Stamicarbon BV, NL Pat. 8 802 748, 1988.
49. Hoechst AG, *Eur. Pat* 315 082, 1987.
50. Hoechst AG, *Eur. Pat.* 315 083, 1987.
51. Hoechst Ag, *Eur. Pat.* 315 084, 1987.
52. Stahlschmidt, K. H., *British–German Powder Coating Conference*, Stratford-Upon-Avon, 1–2 December 1986.
53. Kleber, W., *52nd Conference of the Coatings and Pigment Section*, Soc. Germ. Chem., 1985.
54. Bauch, H., Auerbach, D., and Kleber, W., *Elektrostatosches Beschichten Vom 24 Bis 27*, Hochschule fur Verkehrwren Friedrich List, Dresden, 1978, p. 109.
55. Masuda, S., *IEEE/IAS Conference Proceedings 35D*, 1977, p. 887.
56. Hardy, G. F., *J. Paint Technol.*, **46** (499), 73, 1974.
57. Golovoy, A., *J. Paint Technol.*, **45** (585), 68, 1973.
58. Golovoy, A., *J. Paint Technol.*, **45** (585), 74, 1973.
59. Corbet, R. P., *Conference On Electrical Methods*, IEEE, London, March 1970.
60. Chandler, B. W., *Proceedings SSPC Symposium on Technology for Long Term Protection of Steel Structures*, Atlanta, February 1986, pp. 205–7.
61. Oxacetylene Equi, SU Pat. 1423 176, 1985.
62. Kakossy, V. Von, *Polym. Paint Col. J.*, **179** (4236), 266, 1989.
63. Floyd, R., *Polym. Paint Col. J.*, **179** (4236), 270, 1989.
64. Misev, T., *Course on Polymer Coatings: Alternatives to Solvent Based Coatings*, CEI/Elsevier, Davos, September 1988.
65. Kakossy, V. von, *The Powder Circuit and Its Problem in Practical Application, Powder Coating Seminar*, Lucerne, October 1988.

7

Future Developments

7.1	The Market	362
7.2	Binders	364
7.3	Crosslinkers	365
7.4	Production and Application	366
7.5	Powder Coatings	367
7.6	Conclusion	368

7.1 THE MARKET

It is not an easy task to predict future developments in powder coatings with any degree of certainty. As is usually done in such cases, some of the trends of today will be extrapolated to the future in order to forecast the possible changes. This will be combined with an inventory of the wishes that may possibly catalyse some new developments. The major driving forces in the development of industrial coatings come from the market requirements combined with environmental compliances. This will be taken as a guideline in this short look to the possible future of powder coatings.

It could be assumed that the annual growth rates, which are at this moment 15% for Europe and about 20% for the United States, will slow down as powder coatings move slightly from the growth stage of the product life cycle into the mature stage. In the next decade America will probably remain the market with the highest growth by volume. However, because of their smaller base, growth rates are likely to be very impressive in the Far East, where at this moment there are remarkable investment activities either of domestic producers or powder paint companies of the West.

General industrial and domestic appliances, functional coatings, metal furniture and architectural applications are the major fields that consume almost 90% of the powder coatings production. No major changes are expected in these market segments in the short term. The replacement of the solvents has been one of the main driving forces for increasing powder coating consumption in many of these application areas and this has already been completed. The use of precoated metal should also be added to the previous element as a contraproductive factor that will also contribute to bringing the consumption of powder coatings to a consistent level. The technique of precoating metal (PCM) sheets with powder coatings in a continuous way, which was developed in the Far East, will certainly spread around the world. However, it is very difficult to predict to what degree the PCM powder coatings will penetrate the market since they have a very serious competitor in coil coatings. Trials are being carried out to use powder coatings for coil coating applications, although there is little evidence so far for success on a commercial scale.

The use of powder coatings in architectural applications has been well established in the last fifteen years. Major changes cannot be expected concerning the volume of powder coatings used for decorative purposes which will probably follow the average growth pattern. However, some qualitative shifts towards PVDF powders with an extremely long service life can be predicted.

The premature corrosion of reinforcing steel initiated a wide acceptance of the so-called FBE (fusion bonded epoxy) powder coatings for reinforcement bars in the United States. Excellent experience from North America for the last fifteen years will certainly support the tendency in Europe and the Far East to install high-speed fully automated rebar coating plants.

The biggest mistakes in forecasting the future come mostly from breakthroughs which are rather unpredictable. However, some potential breakthroughs can be named. One market segment is certainly the automotive industry which has not really started to apply powder coatings on car bodies. Several problems have to be solved before one can seriously count on success in this huge market. Relatively poor flow and an emphasized orange peel effect, poor mechanics of the acrylics and a high curing temperature not adequate for most existing automotive coating lines are typical drawbacks which can be overcome by mutual effort of the resin and paint producers. New electrostatic guns have to be developed to allow a more uniform thickness, better coverage of difficult shapes and improved penetration into recessed areas during spraying. Additional improvements in the application equipment will have to be made in order to also solve the problem of a relatively slow colour change, which is one of the biggest obstacles for getting into the automotive top finishes. In the short term it is very likely that the automotive clear top coat will be a top priority item to be tackled by the powder paint producers.

Breakthroughs can be expected in development powder coatings for wood, plastic and paper. These market segments are largely untapped areas regarding the use of powder coatings. Thermoplastic powder coatings are already in use as toners for copy machines. There are indications for the use of powder coatings on plastic surfaces such as bakelite, where the glass transition temperature of the substrate is rather high. For real success, however, low curing powder coatings have to be developed, followed by adequate solution of the problem of producing electrostatic coatings for non-conductive surfaces. Moreover, the temperature sensitivity of the substrate would require improvement of the infrared curing techniques in the direction of mid-wave or short-wave infrared curing.

7.2 BINDERS

Polyester resins are a typical example of binders with an excellent price/performance ratio. Moreover, this is combined with relatively easy chemistry, a big choice of raw materials, an easy-to-control production process and a very well-defined structure concerning the types and positions of the functional groups in the molecular chain. This is believed to be the main factor contributing to the excellent performance of the polyester resins.

It is very difficult to imagine any big change in the very stable market position of these binders for powder coatings. At the moment they fulfil most of the requirements in the application areas where they dominate.

New developments in polyester resin chemistry are expected in the direction of increasing their hydrolytic resistance, which is an inborn weak point of any polymer containing ester bonds in the backbone. Several types of new glycols that have appeared recently on the market as experimental products, such as 2,2,4-trimethyl-1,3-pentadiol; 2-ethyl-1,3-hexanediol; 2,2-diethyl-1,3-propanediol and 2-n-butyl-2-ethyl-1,3-propanediol, impart better boiling-water resistance and alkaline resistance to the polyester resins. This is due to the emphasized steric hindrance effect of the bulky alkyl groups in the alpha position next to the hydroxyl groups in the glycols. Substantial effects on outdoor durability can be expected due to the higher hydrolytic stability of the polyester backbone.

Outdoor durability of most of the existing polyester resins in terms of gloss retention of the cured coating during the QUV accelerated weathering tests is inferior compared to some acrylics. However, recent patents [Stamicarbon, NL 8901413 and NL 8900718, 1989] claim substantial improvements, reaching gloss retention levels almost comparable to the acrylic resins.

Considerable theoretical work has been done in the last fifty years in the field of non-linear polymerization and polymer network formation. Very often the mathematical expressions resulting from the theoretical considerations are not suitable for practical use. However, the penetration by computers into the resin laboratories together with the numerous programs and models based on already existing theories have made it possible to simulate the molecular weight build-up during synthesis of the polyester resin and the crosslinking process. These models allow very rapid computer analysis which greatly helps in the process of designing new resins and crosslinkers. Systems with improved flow properties and mechanical properties can be expected as a normal consequence of the use of these contemporary tools in the development work.

The introduction of on-line instruments for control of the relevant parameters during resin preparation will certainly contribute to an improved consistency of the resin quality and will narrow the typical polyester resin specifications for acid value, hydroxyl value, molecular weight, functionality and viscosity.

The use of bisphenol A epoxy resins in functional powder coatings in the form of pure epoxies or epoxy/polyester hybrids will follow a normal growth pattern without spectacular changes. Latest developments show trends towards modified bisphenol A resins that combine a high enough softening point to ensure good storage stability and a low melt viscosity in order to improve flow properties of the coating during film formation. Epoxy novolac resins could be expected to find broader application in functional coatings since they offer increased hardness, chemical resistance and toughness to powder coatings. Epoxy novolacs with a higher glass transition temperature could be a potential solution in formulating powders for electrical applications where resistance of the coating towards a high service temperature is an important consideration.

Acrylic resins have not yet found a proper place in the powder coating market. There are indications that a decrease in market shares of acrylic resins will occur, even in Japan, which is traditionally an acrylic powder market. High costs of the raw materials, an expensive production process of acrylic resins through the spray drying technique and relatively poor mechanical performance of the cured films are major drawbacks of the system. Two different streams can be predicted in the future: development of a sole acrylic resin/crosslinker system having good mechanical properties and production of acrylic/polyester hybrids. Although considerable improvements in mechanical properties have been achieved with hydroxyl functional acrylics crosslinked with polyisocyanates, the results are still far below the requirements.

The hybrid approach seems to be very attractive since it can offer an excellent price performance combination. However, present solutions offered to the market are certainly not a breakthrough in this area. The problem of incompatibility between acrylic and polyester resins causes considerable restrictions with respect to the choice of raw materials. Serious research work in order to elucidate the compatibility phenomena is a prerequisite for expecting successful acrylic/polyester hybrids.

Self-stratifying acrylic/polyester powder coatings which will form an acrylic top layer and a polyester bottom layer upon curing produce a system that will probably manifest the weatherability of the acrylics and the mechanics of polyester powders. Although it seems at this moment too nice a dream to be realized in the near future research work in this direction is expected.

7.3 CROSSLINKERS

Triglycidyl isocyanurate has been the only epoxy crosslinker for outdoor powder coatings for many years. TGIC fulfils most of the requirements defined as perfect for its use. However, the trend towards powder coatings with a lower curing temperature will certainly promote research work for developing a new generation of epoxy crosslinkers. Polyfunctional epoxy compounds of aliphatic or cycloaliphatic type having good u.v. resistance are expected to be possible solutions to the problem. Epoxy functional polymers with higher molecular weights and a lower diffusion rate through the human cell membrane can provide the way to decrease possible health hazards emerging from the presence of reactive epoxy groups.

Developments in the isocyanate crosslinkers would certainly be directed towards the internally blocked materials and blocked isocyanates with lower deblocking temperatures. It could also be expected that isocyanates with low toxicity hazards such as tetramethyl xylene diisocyanate will find wider acceptance by the end users.

The almost forgotten amino resins as cheap and non-toxic crosslinkers could be reincarnated in the next decade. Surface defects is the only reason why these materials currently used in wet coatings have not gained an adequate place in the powder world. With respect to the price/performance ratio they will certainly score higher than many of the contemporary crosslinkers for powder coatings. Surface defects in the form of blisters is a major concern and requires additional research and development efforts. The return on this investment could, however, be relatively high if an adequate solution can be found.

The number of curing reactions with potential use in powder coatings is comparatively high. It is difficult to predict what this rich chemistry will bring on a commercial scale. The real chances for commercialization should be counted only in cases of lower curing temperatures or shorter curing cycles. Curing reactions having no by-products as potential environmental pollutants will certainly have an advantage.

7.4 PRODUCTION AND APPLICATION

Previous predictions that new production techniques would appear next to the current hot melt extrusion process failed. The last decade was spent improving the basic production steps of the existing production concept.

The preblending step has been considerably improved by the use of high-speed mixers. Better material homogeneity obtained in this way is beneficial for the overall properties of the final product. At the same time the batch reproducibility has been improved and the number of off-spec batches reduced.

Basic changes in the extruders have not been made except for the introduction of computer programming leading to better control of the parameters of the extrusion process.

The in-line classifiers as a part of the grinding equipment have been widely introduced in the last decade. In this way not only is the production process rationalized but the particle size distribution has also been narrowed.

FUTURE DEVELOPMENTS

Contemporary in-line classifiers remove almost all particles below 10 μm. In this way, the impact fusion during spraying, the safety and health considerations have been reduced and the powder paint stability during storage has been improved.

The shape of the powder particle coming from the grinding and classifying equipment is irregular with many edges, corners and plane areas. The spherical shape which is considered to be beneficial for the flow and film forming properties can be obtained either by spray drying or by precipitation production techniques. However, these methods which were announced in the 1970s did not find commercial acceptance.

It is not likely that basic changes in the production process are expected in the near future. The same tendency of gradual improvement in existing equipment design will probably continue in the short term.

The use of the tribo application technique will certainly grow in the next decade. Problems connected with the inconsistency of the method still have to be solved. The efforts made simultaneously by the gun manufacturers, resin and paint producers will certainly contribute in establishing a solid position for this relatively simple application technique.

It could be expected that flame spraying will enjoy considerable growth. This new application method can contribute to penetration of powder coatings into market areas that have been dominated by air drying coatings. This is certainly a challenge that will initiate new developments.

Colour change is one of the oldest application problems associated with powder coatings. The present solutions include roll-on roll-off booths, detachable cartridge recovery modules and as the simplest solution the spray to waste method. Recent developments by the manufacturers of application equipment led to systems with entirely movable booth parts, offering a considerable time saving when changing the colours. This trend will certainly continue while keeping in mind the increasing demands from an aesthetic point of view. An additional impetus to this could be the possible acceptance of powder coatings as automotive top finishes.

7.5 POWDER COATINGS

Thermoplastic powder coatings were introduced on the market at the beginning of the 1950s followed by thermosetting systems in the early 1960s. Since then, development efforts have been concentrated on designing new binders or adjustments of the existing polymers for application in powder coatings. The result was the introduction of varieties of thermoplastic and thermosetting powder coatings which have been described in the previous chapters. Thermosetting types dominate worldwide with market shares almost equally distributed among the epoxy, polyester and polyurethane powder coatings.

New market areas such as automotive top finishes, corrosion protective coatings for application on cold substrates by flame spraying, finishes for wood, plastic and paper will certainly initiate further developments of new powder coatings. Even the fact that powder coatings are slowly reaching an asymptotic growth pattern in many of the present market segments does not mean that from a technical point of view new developments cannot be expected.

New developments in the fluoropolymers resulting in thermosetting low molecular weight resins can be considered as additional stimulants leading to greater use of these materials in the production of powder coatings. Next to the excellent outdoor durability they offer good decorative properties and ease of production and application compared to PVDF. The high price of the fluoropolymers, however, will remain a major obstacle for faster growth in this direction.

The general tendency for higher production rates and lower energy expenses will initiate the appearance of low curing powder coatings. This will be supported by the challenges of the wood, plastic and paper markets. However, because of the rather rigid relations between the flow behaviour, glass transition temperature and curing pattern of the system, powder coatings with a low T_g can be expected. This will affect the stability of the finished products during storage and recycling. Considerable changes could therefore be expected in the storage facilities and application equipment ensuring safe handling of the powders at low temperatures.

It could be expected that acrylic powder coatings or acrylic/polyester hybrids will gain wide acceptance in the exterior or automotive market. However, the present technical problems are still big enough to prevent a real breakthrough in the short term.

There are indications for development of functional coatings for electrical application having high glass transition temperatures after curing. These protective coatings could be used on places where the electrical equipment is permanently exposed to high temperatures.

7.6 CONCLUSION

Although entering their adult stage, powder coatings are still growing children. The early development work mainly supporting the efforts to break down the prejudices of the industry towards these new systems turned to quite exciting areas of new chemistries, new production and application techniques. Most of the problems of the adolescent period at this moment are almost over. Powder coatings have been widely accepted by the market and recognized as one of the few systems that enjoy exciting growth in the steady-state market pattern of the coating industry. The developments in the last twenty years erased the question *to*

FUTURE DEVELOPMENTS

be or not to be. The question now is—*what next*? This last chapter has tried to indicate some possible answers. No matter how accurate these predictions are it is obvious that powder coatings have a bright future. Complicated technical problems which have to be solved before being able to meet the sophisticated requirements of the new market segments should be regarded rather as a challenge than an obstacle.

Subject Index

Accelerated weathering
  of PVDF powder coatings 20
Acid anhydrides 48, 123
  in epoxy powder coatings 141
  resinous 124
Acid cleaners for
  aluminium 318
  galvanized steel 322
  steel 308
Acid value 152, 155
  determination 284
  functionality/molecular weight relations 180
ACM Classifying Mill
  layout of complete installation 257
  operational principle of 254
  rotor of 255, 256
  scheme of 254
Acrylic powder coatings 162
  carboxyl functional 166
  comparison with polyesters and polyurethanes 162
  glycidyl functional 165
  hybrids with polyesters 166
  hydroxyl functional 166
  selfcrosslinking 166
Acrylic resins
  functionality distribution 164
  future developments 365
  glass transition temperature of 164
  synthesis of 163
Acrylic/polyester hybrids 365
Acylurethanes 67
Adhesion
  determination of 299
  of nylon powder coatings 32
  of polyolefinic powder coatings 25
  promoters in polypropylene powder coatings 25
Air classifiers 269
  air velocity profiles of 271
  choice of 274
  classification limit 273
  critical particle diameter 272
  cut size of 272
  efficiency of 274
  operating principle 270, 274
  principle construction 274
  size distribution curves of 272
Air ionization in corona gun 327
Air velocity
  and deposition efficiency 343
  in corona guns 332
Aliphatic epoxies 366
Aliphatic polyamines 118
Alkaline cleaners
  for aluminium 316
  for galvanized steel 322
  for steel 307
  influence on grain size 307
  mode of application 308, 317
Allophanate formation 67
Aluminium surface
  acid cleaners for 318
  alkaline cleaners for 316
  cleaning and pretreatment 316
  electrocleaning 317
  properties of 316
  weight loss during cleaning 317
Amines
  as crosslinkers for epoxies 139
Amino resins
  as crosslinkers 124, 366
  curing reactions of 68-77
  HMMM 126
  synthesis 125
  TMMGI 128
  volatile products during curing 71
Application of techniques
  of powder coatings 324
    compared to conventional 325
    comparison between 349, 350
Aqueous cleaners 306
Architectural applications 363
Argand diagram 102
Aromatic polyamines 118

SUBJECT INDEX

Automotive clear top coat 6
AVP Baker Perkins extruder 240

Back-ionization 333, 334
  surface defects related to 334
Benzoguanimine
  crosslinkers of 125, 128
Bingham liquids 211
Bisphenol A 132
  condensation with epoxy resins 135
Blocking agents
  for isocyanate crosslinkers 63–67
Boron trifluoride 137
Bucholz hardness 297
Buss Ko-Kneader 240
  advantages/disadvantages 244
  capacity of 243, 244
  different types of 246
  in continuously operating plant 242
  metering unit of 241, 242
  residence time diagram of 245
  self cleaning action of 241
  temperature profile 241, 243

Capillary method
  for melting point determination 286
Caprolactam
  as blocking agent 64
Catalysts
  for amino resins 75
  for polyurethane powder coatings 59, 160
Catalyst level
  kinetic considerations 201
Catalytic oxidation
  of unsaturated compounds 131
Caustic soda
  in alkaline cleaners 316
Centrifugal sifters
  product loading 269
  benefits/limitations 270
  self cleaning 269
Charge losses
  after spraying 341
Charge/mass (q/m) ratio
  as charging efficiency 328
  determination 293
  related to adhesion prior to curing 335
Charged particles
  polarity of 330
Charging voltage
  and deposition efficiency 343

Chlorates
  in iron phosphate coatings 308, 310
Chromate conversion coatings
  for aluminium 321
Chromium phosphate conversion coatings
  for aluminium 319
  for galvanized steel 323
  properties 320
Classifier efficiency 274
Coating thickness
  measurement of 295
Cold-rolled steel
  cleaning and pretreatment of 306
Colour change 352, 367
  choice of the spraying booths 354
Compatibility
  determination of 290
Continous compounding 240
Conversion coatings
  comparison of 309
  for aluminium 318
  for galvanized steel 322
  for steel 308
  iron phosphate type 308
  reasons for use 305
Copper salts
  in zinc phosphate conversion coatings 312, 322
Corona charging gun 326
  back-ionization effect 333, 334
  choice of polarity 330, 334
  efficiency of 335
  electric field potential of 327
  electrostatic field strength 330
  external type 329
  Faraday cage effect 333
  field lines between workpiece 327
  flat-jet nozzle type 331
  functions of 327
  integrated high-voltage supply 330
  internal type 330
  problems associated with 331
Cratering 206
Critical layer thickness 341
Crosscut adhesion test 299
Crosslinkers for powder coatings 107
  acid anhydrides 123
  amino resins 124
  polyamines 117
  polyisocyanates 110
  polyphenols 120
  triglycidyl isocyanurate 108

Cryogenic grinding   251
Crystallinity
   of polyethylene   22
   grain size regulators   315
Curing reactions   44
   acid anhydride/epoxy   48
   acid/epoxy   45
   with amino resins   68
   epoxy/amino   51
   isocyanate/hydroxyl   56
      blocking agents   60
      catalysts of   58
      deblocking temperatures of   65
   polyetherification   54
   polyphenols/epoxy   53
Cut size
   of air classifiers   272
Cyclone
   collecing efficiency of   280
   combination with filters   279
   comparison with filter   281

Deblocking of blocked isocyanates
   mechanism of   60–63
   deblocking temperature of   61
Decane dicarboxylic acid
   in acrylic powder coatings   165
Deflection balance   94
Dehydrohalogenation   131
Deposition efficiency   339
   related to
      air velocity   343
      charging voltage   341
      particle electric resistivity   343
      spraying distance   343
      workpiece size   343
Dicyandiamide (DICY)   117
   in epoxy powder coatings   136–139
   substituted   118
Dielectric constant
   of powder particle   328
   related to charging efficiency   337
Dielectric series   337
Differential scanning calorimetry (d.s.c.)
   heat-flow type instrument   83
   kinetic analysis with   86
   melting temperature   92
   possibilities of   86
   power-compensational type   85
   temperature onset cure   92
   temperature onset flow   91
   T_g determination   90

Diglycidyl terephthalate   158
Dihydrazides   143
Domestic appliances   5, 362
Dust removal
   cyclon   263
   filters   263, 279
      Mikro Pulsaire type   281
   filter/cyclon comparison   281
Dynamic mechanical analysis (d.m.a.)   98
   applied on powder coatings   104
   determination of
      equilibrium modulus   105
      crosslinking density   105
      T_g 103
      viscosity   105
Dynamic modulus   102

Eddy current method
   for thickness measurements   296
Elastic body   98
Electric field
   line configuration of   332
   strength of corona gun   327
Electrical charging
   frictional   335
Electrical permittivity
   measurement of   292
Electrical properties
   measurement   291
Electrical resistivity
   measurement   291
Electrical resistivity
   of powder particles in corona gun   334
   relating to adhesion prior to curing   335
   related to deposition efficiency   343
Electromagnetic induction
   in film thickness measurements   296
Electrostatic field
   in corona gun   328, 329
Electrostatic fluidized bed   346
   Faraday cage effect of   347
   safety considerations   347
Electrostatic spraying   325
   principles of   326
Emila rheometer   287
Epoxidation by
   hydro/dehydrohalogenation   131
   by oxidation   131
Epoxy equivalent weight (e.e.w.)   135, 152
   determination   285
Epoxy molar mass (e.m.m.)   135
Epoxy novolacs   135, 365

SUBJECT INDEX

Epoxy powder coatings 131
  amine cured 139
    formulation of 139
  cured with acid anhydrides 141
  cured with acid dihydrazides 142
  DICY types 136
    types matt finishes 138
    use of 139
    formulation of 137
    properties of 138
  general properties 144
  major uses 143
Epoxy resins 131, 365
  aliphatic types 136
  functionality 136
  novolac types 136
  synthesis 131, 135
Epoxy/novolac powder coatings 141
Epoxy/phenolic powder coatings 140
  formulation 140
  properties and application 141
Exterior powder coatings
  comparison between 162
Extruder 240

Faraday cage effect 333
  in electrostatic fluidized bed 347
  in tribo guns 336
Faraday cup 293
Film leveling 195, 208
Fine grinding 251
  ACM Classifying Mill 253
  hammer mill 252
  opposed jet mill 258
  principles of 252
  scheme of 252
  ZSP Circoplex Classifier Mill 257
Flame-spray gun
  scheme of 348
Flame-spray technique 347, 367
Flexibility
  determination 298
Flow
  related to yield value 211
  related to particle size 215
Flow promoters 210
Fluidized bed 344
  electrostatic 346
  installation, scheme of 345
  operating principle 344
  in polyamide powder coatings 32

Fluorborates 319
Fluoride/chromate ratio
  in chromium phosphate coatings 320
Free ions
  in corona 327
  in q/m measurements 294
Free volume 182
Frenkel's equation 194
Functional groups
  in amino resins 69
Functionality 177
  acid value/mol. weight relation 180
  and crosslinking density 180
  and dynamic modulus 179
  and molecular weight buildup 177, 180
  viscosity changes during curing 179
Future developments 362

Galvanized steel
  alkaline and acid cleaners 322
  cleaning and pretreatment 321
  conversion coatings for 322
Gardner impact tester 298
Gel time
  determination 288
Glass transition temperature (T_g) 181
  and chemical structure 187
  and free volume 182
  and internal stress 185
  and melt viscosity 183
  and molecular weight 186
  and powder stability 183
  definition of 181
  determination of 285
  influence of the pigments 189
  of acrylic resins 164
  of branched polymers 187
  of polymer blends 189
Gloss
  determination 299
Gloss retention
  of PVDF powder coatings 19
Glycidol
  by product in epoxidation 134
Glycoluril 125, 128
Gradient oven test 295
  temperature profile simulation 296
Gun electrode
  curtain type 339
  piezoelectric ring type 338
  porous metal air purged type 339

SUBJECT INDEX

Hammer mill 252
Hardness
  measurement 297
Heating rate at
  melting point determination 286
  T_g determination 286
  softening point determination 287
High density polyethylene
  properties 23
High voltage generator
  cascade type 330
HMMM 126
Hookean solids 101
Hooke's law 98
Hot melt compounding 238
Hot-rolled steel
  cleaning and pretreatment of 306
Hydantoin epoxies
  in polyester powder coatings 158
Hydrohalogenation 131
Hydroxyl number
  determination 284

ICI cone and plate viscosimeter 287
Ideal liquid 99
Identation hardness 297
Imaginary modulus 102
Imidazolines 137
Impact resistance
  determination 298
In-mould powder coatings 167
  formulation 170
Infrared curing
  mid-wave/short-wave 364
Infrared ovens
  suitability for nylon powders 33
Integrated fluidity 195
Internal corona gun
  scheme of 331
Internal stress in powder coatings 185
Ion-wind 332
Iron hydrogen phosphate 312
Iron phosphate conversion coatings
  accelerators of 310
  for aluminium 318
  for steel 308
  mode of application 310
  pH control of 310
  properties of 310
Isophoron diisocyanate
  caprolactam blocked 111
  in polyurethane powder coatings 160

uretidione of 113
Isotactic polypropylene 24

Knoop hardness number 297
Knoop indenter 297
Konig pendulum 298

Laser diffraction analyser 291
Levelling additives 209
Loss tangent 103
Low density polyethylene
  properties 23

Maleic acid
  isomerization to fumaric 168
Maleic anhydride
  as crosslinker 123
Mandrel bend test 298
Maragony effect 206
Master batching 239
Maximum q/m ratio
  calculation of 329
Mechanical spectrometer
  measuring principle 100
Medium density polyethylene
  properties 23
Melamine 125
Melting point
  determination of 286, 295
  of polyester resins 36
Melting range
  determination 286, 295
Metal furniture 5, 362
Metal salts
  in zinc phosphate conversion coatings 314
Methylethyl ketoxime
  as blocking agent 64
Mikro Pulsaire filter
  bag assembly of 282
  different designs of 283
  operating principle of 282
Mill scale
  of hot rolled steel 307
Molecular weight
  acid value/funct. relations 180
  and crosslinking density 180
  and curing stoichiometry 176
  and dynamic modulus 179
  and T_g 186
  melt viscosity relations 178, 191
  number average 175

SUBJECT INDEX

of the binder 175
related to mechanical properties 43, 176
weight average 176
Monitoring the curing process 82
differential scanning calorimetry 83
thermal mechanical analysis 97
thermogravimetric analysis 92
Monosodium phosphate
in iron phosphate coatings 310

Network chains
elastically active 103
Newtonian liquid 99
Nitrites/nitrates
in iron phosphate coatings 308, 310
in zinc phosphate coatings 313, 322
mode of action 313
Non-linear polymerization 44
Novolacs 122
Null-type balance 94
Nylon based powder coatings 28
application 33
formulation 32
properties 31
Nylon-11 30
Nylon-12 31
Nylon-6 29
Nylon-66 28

Opposed jet mill 258
different types 262
granule size distribution 261
operational principle of 260
scheme of 260
Orchard equation 195
Outdoor durability
of polyester powder coatings 157
Overspray
liquid versus powder coatings 324
Oxidizing agents
in zinc phosphate conversion coatings 312, 322
mode of action 312

Particle charging
in corona guns 327
influence of resistivity 328
maximum surface charge density 328
Particle size
analysis 290
coalescence speed 217
critical diameter 272

classification 261
by sieving 262
distribution
changes during spraying 216
Pauthenier limit 328, 329
Pencil hardness 297
Pendulum hardness 297
Perfect elastic body 99
Permittivity
of free space 329
of powder particles 329
Peroxides
as catalysts for in-mould powder coatings 169
Persoz pendulum 298
Pfund
hardness number 297
indenter 297
Phase angle
in d.m.a. measurements 101
Phenol
as blocking agent 63
Phenolic resins 120
as crosslinkers for epoxies 120, 140
Phosphoric acid
in acid cleaners for aluminium 318
in iron phosphate coatings 309
in zinc phosphate coatings 312
Phthalic anhydride
as crosslinker 123
Pigment deagglomeration
theory of 239
Pigment volume concentration
in PVC powder coatings 15
Pigment wetting 205, 212
Pin disc mills
ACM Classifying Mill 253
ZSP Circoplex Classifier Mill 257
Pipeline coatings
polyolefinic 26
tests review 300
Plastic flow 211
Plate flow test 294
Policing 263
Polyamides
nomenclature 29
properties 29, 31
Polyamines 117, 119
Polyester powder coatings 34, 144
application 38
comparison with acrylic and polyurethane powder powder coatings 162

Polyester powder coatings (cont.)
  exterior use 154
  initial developments 150
  interior use 152
Polyester resins 144
  carboxyl functional synthesis 148
  comparison: TGIC/Hybrids 148
  crystallization 37
  epoxy functional 149
  high molecular weight production 34
  hydroxyl functional synthesis 146
  melting points 36
  new developments 364
Polyester/acid anhydrides
  powder coatings of 158
    formulation of coatings 159
    outdoor durability 159
Polyester/epoxy hybrids 152
  additives master batching 153
  catalysts 153
  formulations 153
  polyester/epoxy ratio 152, 153
  use 154
Polyester/TGIC powder coatings
  architectural applications 155
  curing temperature 156
  formulations 156
  initial developments 154
  outdoor durability 156
  use 157
Polyethylene
  production 22
  properties 22
  types 22
Polyethylene powder coatings
  for pipelines 1, 26
Polyisocyanates
  isomerization 112
  isophoron diisocyanate (IPDI) 111
  toluene diisocyanate 114
  TMXDI 115
Polyolefinic powder coatings 21
  properties 27
  use 27
Polyphenols 120
Polypropylene
  production 23
  properties 24
  types 24
Polypropylene powder coatings
  adhesion promoters 25
  production 25

Polyurethane powder coatings
  blocked IPDI types 160
  catalysts of 160
  comparison with acrylic and polyester powder coatings 162
  formulations of 161
  hydroxyl/carboxyl ratio 161
  levelling of 160
  matt finishes 161
Potential gradient
  across, powder layer 334
Powder coatings production
  collection and dedusting 279
Powder coating properties
  and melt viscosity 191
  and molecular weight 175
  and pigment volume concentration 211
  and resin/crosslinker ratio 196
  and T_g 181
  catalyst level 199
  cratering 206
  influence of functionality 177
  levelling 208, 212
  parameters influencing 175
  related to particle size 215
  resin/crosslinker ratio 196
  stoving temperature profile 217
  surface tension dependence 204
  wetting 205
Powder coatings
  application on wood, plastics, paper 363, 368
  application segments 7
  application techniques 304, 324
  automative applications 5, 363, 368
  economics of 7
  epoxy types 131
  growth figures 6, 362
  history 1
  market 3
  market shares 4
  polyester types 144
  polyolefinic 20
  production technology 224
  thermoplastic 9
  thermosetting 131
  vinyl 9
  world production 4
Powder collection
  closed circuit 279
  partially closed circuit 278
  plant design 279

SUBJECT INDEX

Powder density
   determination 289
Powder diagnostic kit 294
Powder stability
   determination 289
Precoated metal 363
Premixing 226
   conical mixers 237
   double cone blenders 229
   drum hoop mixers 228
   high-speed blender 233
   horizontal ribbon mixer 230
   tumbler mixers 228
   turbulent rapid mixer 230
Primer surfacer 6
Production technology
   schematic illustration 226
Pull-off adhesion test 299
PVC
   modulus-temperature behaviour 15
   production 10
   properties 12
PVC powder coatings 10
   application 17
   formulation 16
   lubricants 17
   pigment volume concentration 15
   plasticizers 14
   properties 17
   stabilizers 16
PVDF
   production and properties 18
PVDF powder coatings 18
   adhesion 20
   production 19
   properties 18
   use 21
   weathering resistance 19
Pyrazole
   as blocking agent 65
Pyromellitic dianhydride 123

q/m value
   determination of 293
Quality control 284
Quaternary ammonium salts 137

Rebar coatings 363
Reduction agents
   in zinc phosphate conversion coatings 312

Resin colour
   determination of 288
Resin/crosslinker ratio
   calculation of stoichiometric ratio 197, 198
Resistivity test cell 292
Resoles 121
Ring and ball softening point 286
Rubber elasticity
   theory of 103

Salt spray resistance 300
Scalping 262
Screen
   as classifying equipment 262
   air-jet 266
Screening machines
   centrifugal sifters 268
   pneumatic tumbler 266
      scheme of 267
   tumbler 263
      scheme of 265
   vibratory 264
      scheme of 266
Self-limiting effect
   and back ionization 341
   and potential gradient 341
Self-stratifying coatings 365
Shear deformation
   of elastic body 99
Shear modulus 99
   in-phase/out-of-phase components 101
Shear rate 99
Shear stress 99
Sieves
   air-jet 263, 266
Silicofluorides
   in zinc phosphate conversion coatings 319
Sodium carbonate
   in alkaline cleaners 307, 316
Sodium chlorate
   in zinc phosphate conversion coatings 314
Sodium chromate
   in alkaline cleaners 316
Sodium fluoride
   in zinc phosphate conversion coatings 323
Sodium fluorosilicate
   in zinc phosphate conversion coatings 323

Sodium nitrite
  stability of 313
Sodium phosphate
  in alkaline cleaners 307, 316
Sodium pyrophosphate
  in alkaline cleaners 316
Sodium silicate
  in alkaline cleaners 307, 316
Softening point
  determination of 286
Solvent cure test 296
Spraying distance
  and deposition efficiency 343
Specular gloss 299
Spray-drying technique 239
Spraying booths
  compact recovery type 349, 350
  cyclone types 351
  design of 349
  filter belt types 351
Steel surface
  alkaline and acid cleaners of 307
Storage modulus 102
Sulphuric acid
  in acid cleaners 308
Surface cleaners
  inhibited 307
Surface cleaning
  airblasting 305
  airless blasting 305
  by solvents 306
  chemical 306
  mechanical methods 305
  sandblasting 305
  sanding 305
  scratch-brushing 305
  vapour degreasing 306
Surface potential
  related to deposited material 339
Surface preparation 305
  by chemical pretreatment 306
Surface tension
  related to cratering 206
  related to film leveling 208
  related to wetting properties 205
Surface treatment
  comparison between different methods 325
Sward rocker 298
Syndiotactic polypropylene 24

Tensile stress and strain 98

Tertiary amines 137
Tetramethylmethoxy glycoluryl 128
Thermal mechanical analysis 97
Thermogravimetric analysis 92
  kinetic analysis
    derivative method 95
    integral method 96
Thermosetting fluoropolymers 368
Thermosetting powder coatings 42
Titanium salts
  as grain size regulators 314
  in surface cleaners 307, 322
TMMGU 128
TMXDI 115
Toluene diisocyanate (TDI) 114
Toners for copy machines 363
Transesterification
  in production of polyesters 35
Tribo charging
  mechanism of 335
Tribo charging gun 335, 367
  advantages/disadvantages 337
  corona augmented type 337
  efficiency of 336
  scheme of 336
Trichloroethylene
  in surface cleaning 306
Triglycidyl isocyanurate (TGIC) 108
  in exterior powder coatings 155
  isomers 109
    reactivity differences 109
    melting points 109
  polyester/TGIC ratio 155
  synthesis 108
Trimellitic anhydride 123
Troubleshooting 354
Tukon identation tester 297
Tumbler screening
  material movement 264
Turboplex classifiers 274
  different types 277
  multiwheel 276
  single wheel 275

Unsaturated polyester powder coatings 167
  catalysts of 169
  formulation of 170
Unsaturated polyester resins
  synthesis 168
Uretidiones 67

SUBJECT INDEX

Viscoelastic liquids 99
Viscoelastic properties 99
Viscoelastic solids 100
Viscosity
 and film forming properties 193
 and processing performances 192
 and T_g 183
 determination of 287
 temperature dependence 184
Viscosity in melt
 molecular weight relations 191

Water irrigated grid
 in q/m measurements 294
Weight per epoxy (w.p.e.) 135
Wetting promoters 210
WLF equation 184
Workpiece size
 and deposition efficiency 343

Yield value 211
Young's modulus 98

Z-blade mixers 239
Ziegler–Natta catalysts 21
Zinc dihydrogen phosphate
 in zinc phosphate conversion coatings 312
Zinc hydrogen phosphate
 in zinc phosphate conversion coatings 312
 precipitation of 313
Zinc phosphate conversion coatings 310
 accelerators of 312
 for aluminium 319
 for galvanized steel 322
 mode of application 315
 properties 315
ZSK Twin Screw Extruder 244
 advantages/disadvantages 250
 barrel types of 245
 building block principle 249
 cross-cut operational scheme 249
 different types of 251
 residence time 249
 screws of 247, 248
 self cleaning 247
 temperature control 249
ZSK Werner & Pfleiderer 240
ZSP Circoplex Classifier Mill 257
 classifying principle of 257
 different types 259
 scheme of 258